POPULARIZING SCIENCE AND TECHNOLOGY IN THE EUROPEAN PERIPHERY, 1800–2000

Science, Technology and Culture, 1700–1945

Series Editors

David M. Knight
University of Durham

and

Trevor Levere
University of Toronto

Science, Technology and Culture, 1700–1945 focuses on the social, cultural, industrial and economic contexts of science and technology from the 'scientific revolution' up to the Second World War. It explores the agricultural and industrial revolutions of the eighteenth century, the coffee-house culture of the Enlightenment, the spread of museums, botanic gardens and expositions in the nineteenth century, to the Franco-Prussian war of 1870, seen as a victory for German science. It also addresses the dependence of society on science and technology in the twentieth century.

Science, Technology and Culture, 1700–1945 addresses issues of the interaction of science, technology and culture in the period from 1700 to 1945, at the same time as including new research within the field of the history of science.

Also in the series

Writing the History of the Mind
Philosophy and Science in France, 1900 to 1960s
Cristina Chimisso

Science and Spectacle in the European Enlightenment
Edited by Bernadette Bensaude-Vincent and Christine Blondel

William Crookes (1832–1919) and the Commercialization of Science
William H. Brock

Popularizing Science and Technology in the European Periphery, 1800–2000

Edited by

FAIDRA PAPANELOPOULOU
University of Athens, Greece

AGUSTÍ NIETO-GALAN
Universitat Autònoma de Barcelona, Spain

ENRIQUE PERDIGUERO
Miguel Hernández University, Spain

ASHGATE

Published by
Ashgate Publishing Limited
Wey Court East
Union Road
Farnham
Surrey, GU9 7PT
England

Ashgate Publishing Company
Suite 420
101 Cherry Street
Burlington
VT 05401-4405
USA

www.ashgate.com

British Library Cataloguing in Publication Data
Popularizing science and technology in the European periphery, 1800–2000. – (Science, technology and culture, 1700–1945)
1. Science news – Europe – History – 19th century
2. Science news – Europe – History – 20th century
3. Communication of technical information – Europe – History – 19th century
4. Communication of technical information – Europe – History – 20th century
I. Papanelopoulou, Faidra, 1978– II. Nieto-Galan, Agustí
III. Perdiguero, Enrique
509.4

Library of Congress Cataloging-in-Publication Data
Popularizing science and technology in the European periphery, 1800–2000 / [edited by] Faidra Papanelopoulou, Agustí Nieto-Galan, and Enrique Perdiguero.
p. cm. – (Science, technology, and culture, 1700–1945)
Includes bibliographical references and index.
ISBN 978-0-7546-6269-3 (hard cover : alk. paper) 1. Science news–Europe, Western–History. 2. Science–Social aspects–Europe, Western–History. 3. Popular culture–Europe, Western–History. I. Papanelopoulou, Faidra, 1978– II. Nieto-Galan, Agustí. III. Perdiguero, Enrique.

Q225.2.E85P67 2008
306.4'50940903–dc22

2008023647

ISBN: 978-0-7546-6269-3

Mixed Sources
Product group from well-managed forests and other controlled sources
www.fsc.org Cert no. SA-COC-1565
© 1996 Forest Stewardship Council
FSC

Printed and bound in Great Britain by
MPG Books Ltd, Bodmin, Cornwall.

Contents

List of Figures

Notes on Contributors

Palmira Fontes da Costa is assistant professor of History of Science at the New University of Lisbon, Portugal. She received her PhD in 2000 from the University of Cambridge. Her research interests include the history of natural history and medicine in Portugal and England from the sixteenth to the eighteenth centuries, and also the history of the book and of visual representations. She has edited *O Corpo Insólito: Dissertações sobre Monstros no Portugal do século XVIII* (Porto, 2005) and *Ciência e Bioarte: Encruzilhadas e desafios Éticos* [*Science and Bioarte: Crossroads and Ethical Challenges*] (Lisbon, 2007). She is a contributor to Willem de Blécourt and Cornelie Usborne (eds), *Cultural Approaches to the History of Medicine: Mediating Medicine in Early Modern and Modern Europe* (Basingstoke, 2003) and to Daniela Bleichmar, Paula De Vos and Kristin Huffine (eds), *Science, Power and the Order of Nature in the Spanish and Portuguese Empires* (Stanford, 2008). She is also author of various articles published in international and Portuguese periodicals.

Matiana González-Silva graduated in Philosophy and holds a masters degree in Science Communication. After having worked as a journalist for eight years, she received her PhD in History of Science from the Universitat Autònoma de Barcelona in 2008. Her research interests cover the history of science popularization, scientific journalism, science policies and, in general, the intersection between science, medicine and the public sphere. She currently works at the Barcelona Centre for International Health Research within the Malaria Eradication Research Agenda (MalERA) initiative.

Paola Govoni is in the Faculty of Education, University of Bologna, and is attached to the International Centre for the History of Universities and Science (CIS). Her main research interests lie in the interconnection between science, education and development in nineteenth- and twentieth-century Italy, considered from a comparative perspective. She is the author of *Un pubblico per la scienza. La divulgazione scientifica nell'Italia in formazione* (Rome, 2002), on the popularization of science in Italy. On the historiography of science and the communication of the history of science to a broad public, she wrote: *Che cos'è la storia della scienza* (Rome, 2004) and 'Historians of Science and the "Sobel Effect"' (*Journal of Science Communication*, 4, 2005: 1–17). She is writing a book on Italian women and the culture of science from the belle époque to the Cold War.

Johan Kärnfelt is assistant professor at the Department of Literature, History of Ideas and Religion at the University of Goteborg, Sweden. His research is mainly on the history of Swedish popular science. He has published monographs on the development of the Swedish genre of popular science, and on the aims and motives behind Swedish astronomers' engagement in popularization. He is presently working on a monograph on the public appropriation of popular astronomy in early twentieth-century Sweden.

Àlvar Martínez-Vidal is senior lecturer in History of Science in the Universitat Autònoma de Barcelona. He has studied some medical issues concerning the so-called 'Spanish novator movement' in his books *Neurociencias y revolución científica en España* (Madrid, 1989) and *El nuevo sol de la medicina en la Ciudad de los Reyes* (Zaragoza, 1992). Since 1995, and together with José Pardo-Tomás of the Spanish National Research Council (CSIC, Barcelona), he has published a number of articles in this field, and also on the relation between medicine and nationalism in early twentieth-century Catalonia. He is currently editing a book on the history of the *Acadèmia de Ciències Mèdiques de Catalunya (1872–1939)*.

Agustí Nieto-Galan is senior lecturer in History of Science at the Universitat Autònoma de Barcelona. He has written widely on the history of chemistry and natural dyestuffs in the eighteenth and nineteenth centuries. Among his most important publications are *Natural Dyestuffs and Industrial Culture in Europe, 1750–1880*, edited in conjunction with Robert Fox (Canton, MA, 1999), *Colouring Textiles* (Dordrecht, 2001), *La seducción de la máquina* (Madrid, 2001), *Cultura industrial, historia y medio ambiente* (Barcelona, 2004) and *Chemistry, Medicine, and Crime: Mateu J.B. Orfila (1787–1853) and His Times* (Sagamore Beach MA., 2006), edited in conjunction with José Ramón Bertomeu-Sánchez. His research now focuses on the history of science popularization in the nineteenth and twentieth centuries.

Gábor Palló is a senior research fellow at the Institute for Philosophical Research of the Hungarian Academy of Sciences. His fields of research include history of chemistry and physics, twentieth-century history of natural sciences in Hungary, philosophy of science, history of migration of scientists and the relationship between science, politics and philosophy. His last book (in Hungarian) was *Zsenialitás és korszellem* [*Genius and Zeitgeist*] (Budapest, 2004).

Faidra Papanelopoulou is lecturer in History of Science at the University of Athens, Greece. She received her DPhil in 2004 from the Modern History Faculty of the University of Oxford, and held a post-doctoral position at the Centre A. Koyré (EHESS), and a Marie Curie post-doctoral fellowship at the Fondation Maison des Sciences de l'Homme in Paris. She also teaches History of Science at the Hellenic Open University. She is currently working on the history of artificial cold and low-temperature research in the late-nineteenth and twentieth centuries, as well as on the public image of science in early-twentieth-century Greece. Among her publications are 'Gustave-Adolphe Hirn (1815–90): engineering thermodynamics in mid-nineteenth France', *British Journal for the History of Science*, 39 (2006): 231–54 and 'The public image(s) of science and technology in the Greek daily press, 1908–1910', *Centaurus*, 52 (2009) [forthcoming].

José Pardo-Tomás is a research fellow at the Department of History of Science of the 'Milà i Fontanals' Institute (CSIC, Barcelona), and since 2002 has been vice-president of the Catalan Society of the History of Science and Technology. He has published on the social and cultural history of medicine, natural history, and scientific books in the early modern period. He is the author of *Ciencia y censura* (Madrid, 1991); *Las primeras noticias sobre plantas americanas* (Valencia, 1993, with Maria Luz López Terrada); *La influencia de Francisco Hernández* (Valencia, 1996, with José Maria López Piñero); *El tesoro natural de América* (Madrid, 2002); *El médico en la palestra* (Valladolid, 2004) and *Un lugar para la ciencia* (Tenerife, 2006).

Enrique Perdiguero is senior lecturer in History of Science at Miguel Hernández University, Alicante, Spain. He presented his PhD in the subject at the Universitat d'Alacant in 1989, and held a post-doctoral position at the Wellcome Institute for the History of Medicine (now The Wellcome Trust Centre for the History of Medicine at UCL) in London. His main interests are the popularization of medicine, the history of public health and the history of medical pluralism. He has written widely on these topics. Among his most important publications are 'The Popularization of Medicine during the Spanish Enlightenment' in Roy Porter (ed.), *The Popularization of Medicine, 1650-1850* (London, 1992); 'Healing Alternatives in Alicante, Spain, in the Late Nineteenth and Late Twentieth Centuries', in Gijswijt-Höfstrat M., Marland H. and de Waardt H., (eds), *Illness and Healing Alternatives in Western Europe* (London, 1997) and 'Magical Healing in Spain (1875–1936): Medical Pluralism and the Search for Hegemony', in de Blécourt W. and Davies O., *Witchcraft Continued: Popular Magic in Modern Europe* (Manchester, 2004).

Stefan Pohl-Valero graduated in Mechanical Engineering at the Universidad de los Andes (Bogotá, Colombia), and received his PhD in History of Science from the Universitat Autònoma de Barcelona in 2007. He has spent two terms as visiting student in the Department of History and Philosophy of Science at the University of Cambridge (2004–05). He is now assistant professor at the Department of History at the Universidad Javeriana (Bogotá). His research interests focus on the history of physics in the nineteenth century, the interaction of science and religion and the popularization of science in nineteenth-century Spain and Latin America.

Rikke Schmidt Kjærgaard is a post doctoral fellow at the MRC Mitochondrial Biology Unit, University of Cambridge and affiliated to the Centre for Membrane Pumps in Cells and Disease (Pumpkin), Aarhus University. Her research focuses on visual communication and the public understanding of structural biology. Her current project analyses how visualization is used in the creative process of scientific research and communication and, in the end, what makes a good scientific image within micro- and nano-scale science. Since 2005 she has directed various exhibitions on contemporary research and written a number of books and articles on science communication. She holds a PhD in science communication and has a background in mathematics, art history and history of science.

Josep Simon is a third-year PhD student at the University of Leeds, UK. His research has an international and comparative scope in the use of sources and historiographical perspectives, and one of its major aims is producing fruitful interactions between disciplines such as history of science, history of education, book history and museum studies. He has recently contributed to a book on the international family of scientific and medical booksellers Baillière, and to a volume edited by Robert Fox on Franco-British interactions. He is co-editor of *Beyond Borders: Fresh Perspectives in History of Science* (Cambridge Scholars Publishing, 2008).

Jonathan R. Topham is a senior lecturer in History of Science at the University of Leeds. He held the Munby Fellowship in Bibliography in the University of Cambridge in 1992–93, and worked for several years as an editor of the *Correspondence of Charles Darwin* (Cambridge, 1985–). More recently he was an AHRB institutional research fellow on the Science in the Nineteenth-Century Periodical (SciPer) Project at the Universities of Leeds and Sheffield. He is co-author of *Science in the Nineteenth-Century Periodical: Reading the Magazine of Nature* (Cambridge, 2004) and *Science in the Nineteenth-Century Periodical: An Electronic Index* (HRI Online, 2005), and co-editor of *Culture and Science in the Nineteenth-Century Media* (Aldershot, 2004). In 2002 he was awarded an AHRB innovation award for his book-length project on scientific publishing and the readership for science in early nineteenth-century Britain.

Geert Vanpaemel is professor of History of Science and Science Communication at the Katholieke Universiteit Leuven (Belgium). His research concerns the history of science in Belgium, in particular with respect to scientific institutions, the formation of a national science movement, popular support and the social legitimation of science. He is co-editor of a two-volume book on the history of science in Belgium 1815–2000 (Brussels, 2001). His current work focuses on the circulation of knowledge, on the role of popular science, and on the rise of laboratory culture and its wider social and epistemological implications.

Brigitte Van Tiggelen specialized in History of Chemistry after graduating both in physics and history, and is presently attached to the Centre de Recherche en Histoire des Sciences of the Catholic University of Louvain, Belgium. Her interests also focus on the history of science in Belgium, which very much remains to be investigated, and on the promotion of the field to the general public and teachers, through the foundation of Mémosciences. She has been a Neville Fellow at the Chemical Heritage Foundation (Philadelphia) with a project that examined the work of the physician J. Mongin and his mechanical philosophy of chemistry. She has also published extensively on History of Chemistry.

Series Editor's Preface

Political history used to be full of Kings and Queens, Lord Chancellors, Dukes and Archbishops; of Presidents, Prime Ministers, and Generals; of kingdoms, principalities and powers. This 'high politics' no longer seems the whole, or even the most interesting, part of the story. It has in the age of the common man and woman been supplemented by studies of popular politics: of Justices of the Peace, city bosses, ordinary voters, and single-issue campaigners; of town halls, Masonic lodges, and workhouses. The history of science similarly used to be focussed upon elite researchers, the people who gave their name to laws, units or chemical reactions, who transformed the way the world was seen; and of the Academies, Observatories, Research Universities, Museums and Laboratories where they worked. For the nineteenth and early twentieth centuries, the period covered by most of the essays in this book, that would be people like Karl Gauss, Michael Faraday, Justus von Liebig, Charles Darwin, Louis Pasteur, Rudolph Clausius, Marie Curie and Ernest Rutherford, and the institutions with which they were associated. Because critical mass is important in science, such scientists lived in, or moved to, great centres, such as Paris, Berlin, or London; and relations between these great centres and 'peripheral' places in the provinces or in other countries became a topic to be studied, with diffusion as a model.

Because in Britain especially, but also in France and Germany, a gulf opened by the end of the nineteenth century between eminent practitioners and everyone else as 'two cultures', scientific and humanistic, developed; and as mass literacy and cheap printing opened vast markets for accessible writings. Popularizers, usually despised by practitioners, wrote, lectured and arranged exhibitions and displays concerned with science, sometimes earning good incomes thereby. The model used to be that they were simply diluting the science, adding sugar to make it more palatable, and spooning it out: but this clearly does not fit the much-more-complex reality. Even in Britain, some eminent scientists were extremely good at getting their work across to audiences of working men, of leisured ladies and gentlemen, or of colleagues in different branches of increasingly-specialized sciences. Meanwhile popularizers were often disseminating science that professors might have groaned at: phrenology, the 'quinary system' of zoological taxonomy, anti-vivisection and anti-vaccination propaganda, or psychical research. And in the provinces, there were able and important people, in high local repute, running field clubs, mechanics' institutes, local museums, schools and colleges.

The essays in this book are written in the belief that 'high science' is only part of the story, and few great names of international repute appear in it. But if science is local and particular as well as international and cross-cultural, then its lower end is as worth studying as any other part, and because fresher to most of us, more interesting. That is where a scientific community and tradition is built up, and where science is used in technology or general culture. Popular science is a part of science. In this book, that is argued for, and then taken for granted as we look in one case at the fate of a French textbook translated into English; and in all the others, at science developing in Italy, in Belgium, in Denmark, and in Sweden, in all of which there was a long tradition of distinguished practitioners, but which experienced very much a new start in the nineteenth century. The authors explore different levels and characteristic vehicles of scientific understanding: book series; fairs and exhibitions; and adult education. For Hungary, an institution 'Urania' (taken up and adapted from one in Berlin) involved a theatre, a journal, meetings, dioramas, slide shows in a society where science was an important aspect of modernization: all of which collapsed in 1919 in the wake of defeat in the Great War. Four essays are concerned with Iberia. We see in Portugal an eighteenth-century patrician lady writing (but not publishing) on botany, in a man's world where the Inquisition was active; then in nineteenth-century Spain, how thermodynamics was understood and communicated in a country deeply divided between Catholics and liberals, and materialism was greatly dreaded. Those tensions were still there in Catalonia in the 1920s and '30s, as we find from the examination of a medical journal (intended also for a general readership). With the coming of the Republic, the editor became Mayor of Barcelona – but then Franco's victory in the Civil War put an end to it, and many Catalan doctors went into exile. Political history is an important aspect of these stories: and the last one on Spain, looking critically at science as reported in the post-Franco newspaper *El País*, takes us into happier times in a story not very different from what might be encountered in other countries of the European Union to which these countries all belong.

There is much to learn here, and approaches that could be readily extended into other countries as pure and popular science, and centre and periphery, are no longer thought of as polar opposites but rather as lying on continua, with different questions leading to different placings. It will be a truly useful book.

David Knight

Preface

Science and Technology in the European Periphery (STEP) gathers together more than 100 historians of science and technology whose research activities involve revisiting the historiography of knowledge transmission from centres to peripheries on a European level. This volume stems from the 5th STEP meeting devoted to the popularization of science and technology, which took place from 1 to 3 June 2006 in Mahon, on the island of Minorca (Spain). Over 35 speakers presented papers covering various aspects of the history of science popularization from the eighteenth century until the present day.

Arranged chronologically, the chapters presented in the current volume attempt to strike a balance between case studies from different countries dealing with a wide spectrum of practices of popularization. The selection of case studies was not grounded only on geography. We have attempted to include different practices of popularization that range from printed media (books, periodicals, encyclopedias, magazines, newspapers) to institutional spaces (learned societies and educational institutions), purposely designed events (exhibitions, public lectures, scientific theatres) and visits to sites (museums, botanical and zoological gardens). Because of the specificities of the countries of the European periphery – such as the lack of solid institutional structures, the blurred distinctions between professionalism and amateurism, and the merged identities of scientists, science teachers and science popularizers – the terms 'science popularization' and 'popular science' acquire a distinct character that we attempt to decipher here.

The volume begins with two historiographical chapters, which set the theoretical framework. In tune with recent developments in the field, Jonathan Topham questions the appropriateness of the use of the terms 'science popularization' and 'popular science' both as methodological categories and as subfields of the history of science. His preoccupation with the use of actors' categories and the need to situate them in their proper historical context has been a guiding thread for all contributors. The different meanings attributed to the term 'popular' over a relatively short stretch of time reveal the various appropriations of the term by different actors and for different purposes. In the same vein, Paola Govoni discusses historiographical issues concerning the popularization of science and technology, drawn from her own work on science popularization in late-nineteenth and early twentieth-century Italy.

Our case studies begin with Palmira Fontes da Costa's examination of the Portuguese Marquise of Alorna's *Botanical Recreations*. Her study not only offers us a deep insight into the making of a book belonging to the genre of didactic poetry, but also into the circulation of her writings through a variety of social networks. Geert Vanpaemel and Brigitte Van Tiggelen's chapter provides us with a fine discussion of two different strategies of science popularization in Belgium, which unveil the intricate relationship between the status of science and nation building.

Josep Simon's chapter focuses on the circulation and appropriation of two French physics textbooks in Great Britain, and associates the history of science popularization with the history of the book, the history of science education and studies of international scientific communication. Stefan Pohl-Valero's contribution examines the activities of a group of scientists regarding the introduction of thermodynamics to late nineteenth-century Spain. Pohl-Valero's chapter exposes the complex interplay between the public image of thermodynamics and the social and political agendas that shaped it.

Moving away from the printed media, Rikke Schmidt Kjærgaard takes us on a journey through various sites (museums, zoological and botanical gardens) where science and technology were exposed in nineteenth-century Denmark. Gábor Palló's contribution to the volume brings to the fore an alternative vehicle for science popularization: the Urania Scientific Theatre and the multifarious activities of the Urania scientific society in Budapest.

Johan Kärnfelt's contribution refers to the aims and motives of professional astronomers in engaging in the popularization of astronomy in early twentieth-century Sweden. Enrique Perdiguero, José Pardo-Tomás and Àlvar Martinez-Vidal's chapter tells us how the *Monografies Mèdiques*, a periodical series in the local language, served primarily as a vehicle for the creation and consolidation of a distinct Catalan scientific culture. Finally, for Matiana González-Silva, a current newspaper becomes a privileged, albeit understudied, arena for the presentation of science in the public sphere in late twentieth-century Spain.

The editors' concluding remarks discuss the double peripheral position that those issues occupy in the historiography of science.

Popularizing Science and Technology in the European Periphery, 1800–2000 is the result of our intention to draw new historiographical lessons from new case studies that deal not only with a wide variety of popularizing practices, but also with mainstream historiographical problems that are currently being discussed by the international community of historians of science.

Faidra Papanelopoulou, Agustí Nieto-Galan and Enrique Perdiguero
Paris, Barcelona and Alicante, December 2007

Acknowledgements

The success of the 5th STEP meeting and the publication of this volume is owed, in the local context, to the excellent intellectual environment provided on the island of Minorca by a good number of institutions: the Museu de Menorca, the Institut Menorquí d'Estudis (IME), the Conselleria d'Economia, Hisenda i Innovació del Govern de les Illes Balears and the Ajuntament de Maó. Our special thanks go to Josep Miquel Vidal, the research director of the IME, for his unfailing encouragement and support. We also thank the Universitat Autònoma de Barcelona and the Spanish Ministerio de Educación y Ciencia for their economic support. In particular the Ministerio contributed substantially through the following projects: HUM2005-25478-E; HUM2005-25426-E; HUM2006-7206-03.

The editors are especially grateful to all the STEP members who participated in the discussions in Minorca. We would like to give particular thanks to Jonathan Topham and Paola Govoni, whose historiographical interventions sparked off fertile discussions and opened the ground for further reflection. We would also like to thank the scientific committee, who encouraged and supported the decision to publish this volume. Special thanks are also due to the contributors to the volume, who worked in close collaboration with us and within the deadlines imposed upon them. David Owen and his team at the Language Service of the Universitat Autònoma de Barcelona have been of great help in polishing the English in all the chapters that were originally written by non-native speakers. Finally, we would like to thank the editors of the Ashgate Series 'Science, Technology and Culture, 1700–1945', David Knight and Trevor Levere, for their support to our project.

Chapter 1
Rethinking the History of Science Popularization/Popular Science

Jonathan R. Topham[1]

The history of science popularization at the European periphery seems to be a project that is doubly subject to marginalization within the history of science. In both its aspects, the main event seems to be taking place elsewhere. While the 'European periphery' as conceived in this book is primarily geographical, it nevertheless has connotations of Edward Shils's socio-cultural coinage, in which the periphery is defined not primarily by geography, but by distance from a central zone in which authority is invested.[2] Similarly, the history of science popularization has often seemed to be secondary to the history of 'science proper'. Popularization, it is supposed, is what happens to scientific knowledge once it has been successfully made and passes out of the specialist realm.

In this chapter, I seek to demonstrate that this marginalization of science popularization is based on a misconception of the nature of science. I start by sketching out what I consider to be the contemporary crisis in the historiography of science popularization. My argument here is that an alternative focus on communication provides the best means of reconceptualizing this troubled field, and I join Jim Secord in suggesting that such an approach breaks down the artificial distinction between 'popular science' and 'science proper'. Having thus advocated dispensing with 'science popularization' as a neutral analytical category, however, I then urge the need to understand this and related categories

---

[1] Earlier versions of this chapter were presented at the 5th STEP Meeting, 'Popularization of Science and Technology in the European Periphery' (June 2006) and at the informal History and Philosophy of Science seminar in the University of Leeds (March 2006). I am grateful to the discussants on those occasions for their comments and criticisms. I would also like to thank the organizers of the STEP meeting for their invitation to speak, and the editors of this volume for their helpful comments on an earlier draft. I am particularly grateful to Greg Radick, Anne Secord, Josep Simon, Leucha Veneer and Adrian Wilson for their helpful advice and assistance in the writing of this chapter.

[2] Edward Shils, *Center and Periphery: Essays in Macrosociology* (Chicago and London, 1975), pp. 3–16. On the difference between geographical and socio-cultural peripheries in science see also Svante Lindqvist, 'Introductory Essay: Harry Martinson and the Periphery of the Atom', in Svante Lindqvist (ed.), *Center on the Periphery: Historical Aspects of 20th-Century Swedish Physics* (Canton, MA, 1993), pp. xxiv–xxvi.

from the perspective of historical actors in different national and linguistic contexts, pointing out that such categories have relatively short, but shifting and varied histories. In the third section, I argue that as historians we need to pay more attention to the different uses to which the concepts and language of 'science popularization' have been put by historical actors, stressing particularly that the exclusionary usage with which we are most familiar is by no means the only one. In the final section, I emphasize that the notion of 'popular science' provides scientists with important means of communicating with each other, across disciplinary and national boundaries, and even within disciplines, especially when exploring the boundaries of orthodox knowledge. The important role which 'science popularization' thus plays in the practice of science re-emphasizes the historiographical conclusion that it must be incorporated as a core element in a communication-focused history of science.

Science Popularization/Popular Science in Crisis

While studies of science popularization and popular science have proliferated in recent years, the historiography of the field has become increasingly uncertain. It is now de rigueur for historians ritually to reject that stereotypical notion of popularization in which science is first made by scientists and then subsequently passes to passive lay publics by a process of diffusion involving simplification or even distortion. Yet we have struggled to replace that discredited notion with a conceptual framework that gives the history of science popularization coherence. One of the main reasons for this difficulty is the elision signified by the composite phrase in my title, 'science popularization/popular science'. Amongst historical actors, discussion of science popularization has habitually graduated into discussion of the scientific knowledge or skill of those on the receiving end of attempted acts of popularization. Typically, indeed, the object of science popularization is asserted (albeit sometimes insincerely) to be the generation of scientific competency among those at whom the process is aimed. Thus, the measure of success for self-conscious acts of *science popularization* has been the production of people knowledgeable about or even themselves engaged in science – what has sometimes been called *popular science* or *science in popular culture*. Science *for* the people and the science *of* the people (or even *by* the people) have been inextricably intertwined.

Not surprisingly, the same elision frequently occurs in the historical scholarship in this field. In the past, the elision of science *for* and science *of* the people has often been rather crude, with public attitudes to and knowledge of the sciences read off 'popularizations' in a manner which implies passive consumption. More recently, of course, the relationship between science *for* and science *of* the people has been analysed with increasing sophistication, as historians have sought to understand how consumers of popularizations have appropriated them to their own ends. Yet, as Roger Cooter and Steven Pumfrey pointed out a decade ago, approaching science popularization/popular science from the popularization end can give a distorted view. For this reason, they express reservations about the 'utility of "popularization" as an umbrella term' for an area of history including both 'the study of the myriad processes of "popularization"' and 'the study of "popular" science in all its conceivable operations over time'.[3] 'The very idea of studying "the popularization of science"', they claim, reflects the dominance of 'concepts, models and practices developed by and for the study of authorized science' and has 'stunted the investigation of science in popular culture as a result'.[4]

Are we, then, to resist the elision, and to insist on a separate discipline to investigate science in popular culture? Cooter and Pumfrey consider this possibility only to reject it. Early students of popular culture, they observe, made the mistake of representing popular culture as an authentic sphere autonomous from the dominant culture, but more recent work has emphasized its close relation to the dominant culture not only through domination, but also through resistance and appropriation. Nowhere is this more evident, they claim, than in regard to 'science in popular culture', where the 'elitism of scientific discourse immediately delegitimates popular experiences and epistemologies of "nature"'.[5] We cannot escape the slippage between science popularization and popular science by describing the science of the people as, say, 'ethno-natural knowledge', because, they argue, that would involve the false assumption that there is an epistemologically level playing field.[6] Instead, the science of the people is ineluctably enmeshed – through domination, resistance and appropriation – with the science of the dominant culture, often as encountered through science popularization.

We thus have good reason to embrace rather than repudiate the interconnectedness of science popularization and science in popular culture. Moreover, as work since Cooter and Pumfrey's paper has shown, there is a coherent framework within which such research can take place, namely the history of scientific communication. In particular, as I have argued elsewhere, work on print culture by such scholars as Robert Darnton and Roger Chartier has suggested an approach highly relevant for the histories of science popularization and popular science.[7] In this approach, printed works and other communicative media are seen as the products of a wide range of agents (whether publishers, taxidermists,

---

[3] Roger Cooter and Stephen Pumfrey, 'Separate Spheres and Public Places: Reflections on the History of Science Popularization and Science in Popular Culture', *History of Science*, 32 (1994): 237–67, p. 239.

[4] Ibid., p. 253.

[5] Ibid.

[6] Ibid., p. 254.

[7] See, for example, Jonathan R. Topham, 'Beyond the "Common Context": The Production and Reading of the *Bridgewater Treatises*', *Isis*, 89 (1998): 233–62; Roger Chartier, 'Texts, Printings, Readings', in Lynn Hunt (ed.), *The New Cultural History* (Berkeley, 1989), pp. 154–75; and Robert Darnton, 'What is the History of Books?', in *The Kiss of Lamourette: Reflections in Cultural History* (London and New York, 1990), pp. 107–35.

or TV producers), all of whom have a hand in giving them meaning. Moreover, the readers and users of such communication media are active in appropriating and resisting the intended meanings. Such an approach keeps in focus the crucial interconnectedness of science popularization and popular science, but provides a framework for understanding the dynamic of history, often involving vigorous cultural contestation.

As Jim Secord argued in his keynote address to the joint British–North American conference entitled 'Circulating Knowledge', this emphasis on communication can serve to reunite the history of science popularization/popular science with other aspects of the history of science from which it has become divorced. Secord urges that the focus on 'science in context' that has underpinned much work in the field for a generation has a tendency to degenerate into 'parochial antiquarianism'.[8] He consequently recommends a shift of focus to consider 'science as a form of communicative action – to recognize that questions of "what" is being said can be answered only through a simultaneous understanding of "how," "where," "when," and "for whom"'.[9] This approach, which Secord dubs 'knowledge in transit', reinscribes the formerly marginal field of science popularization/popular science as a core element of a reconceptualized history of science and has important consequences for historical practice. To begin with, by 'eradicating the distinction between the making and the communicating of knowledge', it focuses attention on how science popularization contributes to the project of science as a whole.[10] In addition, it breaks down boundaries between the history of science popularization and other important communicative histories, such as science education and international scientific communication (a point of obvious relevance in the context of this book). In so doing, however, Secord's approach raises the question of whether there is anything distinctive left that we can identify as the history of science popularization.

Despite, or perhaps because of all he does to integrate the history of science popularization as a core aspect of the history of science, Secord is wary of being accused of 'advocating the creation of a subfield within history of science devoted to the study of popular science'. This is a project he explicitly repudiates, recommending, indeed, that 'popular science' and its cognates should be abandoned as 'neutral descriptive term[s]'.[11] According to Secord, the price to pay for the history of science popularization/popular science ceasing to be marginal is that it ceases to be a separately conceived focus of study. Should we, then, abandon such books as this? Secord's concerns about the use of the language of the popular are certainly well founded. Identifying past phenomena in terms of 'science popularization' and 'popular science' unquestionably introduces diffusionist implications that are difficult to eliminate. Moreover, the terms do have 'exceptionally rich and

8 James A. Secord, 'Knowledge in Transit', *Isis*, 95 (2004): 654–72, p. 659.

9 Ibid., pp. 663–4.

10 Ibid., p. 661.

11 Ibid., p. 670.

multivocal' as well as unstable histories, which makes them problematic for use as categories of historical analysis.[12] This latter point, however, should give us pause. For, while it entails the loss of a fixed category of 'popular science' by which to categorize and analyse the past, it at the same time suggests that we should as historians be interested to recover and analyse the diverse historical uses of the concept and language of 'popular science'.

Increasingly in recent years, historians of science have sought to understand how the culture of science is shaped by its language. Examples include Lorraine Daston's exploration of the rich layers of meaning in the term 'objectivity', Simon Schaffer's analysis of the concept of 'genius' in Romantic natural philosophy, Timothy Alborn's work on the use of the language of political economy in organizing scientific work in early nineteenth-century Britain, Ruth Barton's account of the self-designation of mid-Victorian men of science, and Graeme Gooday's recent work on the development of the term 'expert'.[13] Thus, while I am keen to reconceptualize the history of science popularization within a wider history of 'knowledge in transit', I am also keen to advocate a closer attention to actors' own categories of the 'popular'.

What I have said thus far might suggest that I should focus in the remainder of this chapter on the wider questions of communication in science, of which science popularization is but one facet. However, these are topics on which I have already published, urging the value of focusing on what Darnton calls the 'communication circuit' in understanding the creation and maintenance of scientific knowledge.[14] Moreover, James Secord has provided a dazzling and instructive exemplar of how to understand works which have previously been considered under the rubric of 'popular science' in terms of a history of science as communication, and it will not easily be improved upon.[15] Instead, I propose to focus in the remainder of this chapter on the second aspect of the subject identified here, which to date

---

[12] Ibid.

[13] Lorraine Daston, 'Objectivity and the Escape from Perspective', *Social Studies of Science*, 22 (1992): 597–618; Daston, 'Scientific Objectivity with and without Words', in Peter Becker and William Clark (eds), *Little Tools of Knowledge: Historical Essays on Academic and Bureaucratic Practices* (Ann Arbor, MI, 2001), pp. 259–84; Simon Schaffer, 'Genius in Romantic Natural Philosophy', in Andrew Cunningham and Nicholas Jardine (eds), *Romanticism and the Sciences* (Cambridge, 1990), pp. 82–98; Timothy L. Alborn, 'The Business of Induction: Industry and Genius in the Language of British Scientific Reform, 1820–1840', *History of Science*, 34 (1996): 91–121; Ruth Barton, '"Men of Science": Language, Identity and Professionalization in the Mid-Victorian Scientific Community', *History of Science*, 41 (2003): 73–119; and Graeme Gooday, 'Liars, Experts and Authorities', *History of Science*, 46 (2008): 431–56.

[14] Topham, 'Beyond the "Common Context"'; Darnton, 'What is the History of Books?', pp. 110–13.

[15] James A. Secord, *Victorian Sensation: The Extraordinary Publication, Reception, and Secret Authorship of 'Vestiges of the Natural History of Creation'* (Chicago, 2000).

has received rather less attention, namely, the development and deployment of distinctively modern notions of 'popular science' and 'science popularization' among historical actors in different cultures and languages over the last two centuries. This approach is particularly relevant within the context of this book, since it focuses attention both on the distinctive histories of 'science popularization' in different national contexts and on what those histories share in common.

Locating 'Popular Science' in History

Once we accept that the language and concepts of 'science popularization' and 'popular science' have histories, one of the first things that strikes us is how short those histories are. By the middle of the nineteenth century, the notion of 'popular science' had become widely familiar in an Anglophone context, but it was generally acknowledged to be a relative novelty. In 1852, an anonymous reviewer of Dionysius Lardner's *Handbook of Natural Philosophy and Astronomy* (1851), writing under the title 'Popular Physics' in the *Dublin University Magazine*, observed that one of the most prominent characteristics of modern literature, especially in the Anglophone world, was that 'scientific men of the highest order' wrote works 'for the million', explaining the 'laws of the material world' in a 'language intelligible and familiar to all'. The reviewer continued: 'It is not more than five-and-twenty years since this movement commenced, and to be conscious of the rapidity of its progression we have only to glance at the catalogue of works on popular science which issue from the press from year to year.'[16] Three years later, another author observed that, while there was nothing new in the notion that science should form a constituent of universal education, it had 'of late risen into importance amongst us, and hence "Practical Science" and "Popular Science" have become fashionable phrases'.[17] Such a significant change clearly warrants further historical attention. If the concept and language of 'popular science' is so recent, why did it develop when it did? And how does this development in the Anglophone world relate to conceptual and linguistic developments elsewhere? In what follows, I will briefly

---

16 'Popular Physics', *Dublin University Magazine*, 39 (1852): 27–32, p. 27. On the development of popular literature more generally in this period, see 'Literature for the People', *The Times*, 9 February 1854, p. 10.

17 'Science for the People', *Eclectic Review*, n.s. 9 (1855): 293–309, p. 294. Some authors at mid-century sought to give 'popular science' a longer history, but they had nevertheless to concede both that it was generally considered to be a new phenomenon, and that there were few unambiguous instances of 'popular science' before the 1820s. See [Whitwell Elwin], 'Popular Science', *Quarterly Review*, 84 (1849): 307–44; and Jonathan R. Topham, 'Publishing "Popular Science" in Early Nineteenth-Century Britain', in Aileen Fyfe and Bernard Lightman (eds), *Science in the Marketplace: Nineteenth-Century Sites and Experiences* (Chicago, 2007), pp. 135–68, p. 135.

consider the first of these questions before merely glancing at the second, by way of suggesting the fruitfulness of such an approach.

Studying the development of a language of 'popular science' in English is more feasible now than ever before. A significant proportion of English-language books of the period before 1800 are now available as digital full text. As a result, we can ascertain with some certainty that the phrase 'popular science' was not used at all in English-language books prior to that date. Likewise, the word 'popular' was only used in the sense of 'intended for or suited to ordinary people' in the titles of six scientific books before 1800 (only two before 1780).[18] A similar pattern emerges from the new digital full-text editions of nineteenth-century periodicals. *The Times* newspaper did not use the phrase 'popular science' before the 1830s.[19] Moreover, while more than a third of the 130 periodicals in Chadwyck-Healey's *British Periodicals Collection 1* date from before 1830, only one use of the phrase 'popular science' is recorded before 1828, and then in an older sense (a reference to the current enthusiasm for the science of botany). Thereafter, however, use of the phrase 'popular science' became commonplace.[20] Thus, in an important sense, there was no 'popular science' in Britain before the early nineteenth century.

The words 'popular' and 'science' are, of course, both older than the early nineteenth century. However, the combinations of these words that interest us depend on definitions that were new in this period. We too readily forget that keywords like these have complex histories, acquiring new layers of meaning over time. As Lorraine Daston has shown for the word 'objectivity', the layers of meaning need to be teased apart in order to understand the unacknowledged conceptual tensions.[21] In English, the earliest (fifteenth-century) usage of the word 'popular' was in a legal context, describing an action 'Affecting, concerning, or open to all or any of the people', but very quickly the word came to be applied to that which was 'Of, pertaining to, or consisting of the common people, or the people as a whole'. By the seventeenth century, the word encompassed such additional meanings as 'Studious of, or designed to gain, the favour of the common people', 'Finding favour with or approved by the people', and 'Prevalent or current among, or accepted by, the people generally'. What was new in the late eighteenth and early nineteenth centuries was the sense of popular as 'Intended for or suited to ordinary people', in the sense of being adapted to their understanding, taste, or means.[22] For the first time, 'popular' described cultural products intended *for* the people.

---

[18] Thomson-Gale, *Eighteenth-Century Collections Online* <http://galenet.galegroup.com/servlet/ECCO> (April 2006).

[19] Thomson-Gale, *The Times Digital Archive, 1785–1985* <http://infotrac.galegroup.com/itw/infomark/1/1/1/purl=rc6_TTDA> (April 2006).

[20] Chadwyck-Healey, *British Periodicals Collection 1* <http://britishperiodicals.chadwyck.co.uk> (October 2007).

[21] Daston, 'Objectivity and the Escape from Perspective'; Daston, 'Scientific Objectivity with and without Words'.

[22] *Oxford English Dictionary Online* <http://www.oed.com> (April 2006).

This novel meaning of the adjective 'popular', and the simultaneous emergence of the verb 'to popularize', are clearly interlinked, as Morag Shiach has shown, with increasing literacy, developments in communications technologies resulting in the cheapening of print and changing patterns of distribution of printed materials.[23] In Britain, as I have argued elsewhere, the change is associated with the commercialization and industrialization of the book trade and cultural developments in the new industrial towns and cities in the half-century between 1775 and 1825.[24] As newly entrepreneurial publishers sought to expand their markets to include new groups of middle- and even working-class readers, the language of the 'popular' developed to signal the targeted audiences for certain products. Here were works accessible to all, in terms of their price, their mode of address and their educational requirements. Yet, as Morag Shiach also shows, the new usage represented 'a telescoping of several debates'. The concern with a popular mode of address, for instance, readily ran into a concern with reception, in the older sense of 'popular' as 'finding favour with [...] the people'.[25] The elision with which we started, between science *for* and the science *of* the people, was there at the outset.

The combination of the word 'popular' with 'science' presents further complexities. At the same time that the new sense of 'popular' emerged, the word 'science' in English came to apply increasingly to 'natural and physical science', rather than merely to any recognized department of learning.[26] The shift is clearly related to the increasingly specialized and disciplinary condition of natural enquiry in Britain at the end of the eighteenth and start of the nineteenth century, as tangibly represented by the burgeoning of specialist scientific societies and the rapid development of both society transactions and commercial periodicals. Linguistically and conceptually, it makes sense to think of these changes in an Anglophone context, not as a second scientific revolution so much as 'the invention of science'.[27] The new, disciplinary science was developed in part by the withdrawal of natural enquiry from a general learned culture, and by a

23 Morag Shiach, *Discourse on Popular Culture: Class, Gender and History in Cultural Analysis, 1730 to the Present* (Cambridge, 1989), pp. 27–8.

24 Jonathan R. Topham, 'Scientific Publishing and the Reading of Science in Nineteenth-Century Britain: A Historiographical Survey and Guide to Sources', *Studies in History and Philosophy of Science*, 31 (2000): 559–612; Topham, 'Publishing "Popular Science" in Early Nineteenth-Century Britain'.

25 Shiach, *Discourse on Popular Culture*, pp. 28–9.

26 *Oxford English Dictionary Online*.

27 Simon Schaffer, 'Scientific Discoveries and the End of Natural Philosophy', *Social Studies of Science*, 16 (1986): 387–420. See also Andrew Cunningham, 'Getting the Game Right: Some Plain Words on the Identity and Invention of Science', *Studies in History and Philosophy of Science*, 19 (1988): 365–89; and Cunningham and Perry Williams, 'De-Centring the "Big Picture": *The Origins of Modern Science* and the Modern Origins of Science', *British Journal for the History of Science*, 26 (1993): 407–32.

growing concern to address scientific findings differently to specialist and popular audiences. In this sense, the language of 'popular science' not only represented the social hierarchy typical of 'popular' productions arising from the constant slippage between the people as a whole and the common people, but it also involved a cultural hierarchy of knowledge, understanding or skill. The 'popular science' that first appeared in the years after the Napoleonic wars often represented the consolidation of a putatively legitimate community of scientific practice distinct from a wider audience.

While the development of a vocabulary of 'popular science' and the restriction of the community of those involved in the making of scientific knowledge were clearly interlinked in early nineteenth-century Britain, it is important to appreciate that neither at the outset, nor subsequently, has the language of 'popular science' been used with only one end in view. Writers in the 1820s used the phrase to maximize a market, to exclude non-specialists from commenting on specialist findings and, conversely, to engage new workers in the scientific enterprise. Moreover, the use of this vocabulary has changed significantly over time, though little work has been done to chart the changes. One of the few exceptions is Bernadette Bensaude-Vincent's genealogical study of the 'increasing gap between science and the public', which provides a valuable overview of the shifting conceptualization of 'popular science'.[28] Bensaude-Vincent begins in the eighteenth century, when 'enlightened amateurs' contributed to a science that was part of the public sphere, before turning to the development of science popularization in the nineteenth century. Here, her argument is that popularization was ambiguous, involving a commodity intended for consumption by a paying public, but also presenting the possibility of the public playing a role in the making of science. According to Bensaude-Vincent, it was not until the years after World War I that popular science came to rest on the assumption that there was an unbridgeable gap between scientists and the public, and that the latter had nothing to contribute to science. She relates this development particularly to the emergence of the new physics, and to associated epistemological changes, and correlates it with a linguistic change in English from 'popular science' first to 'science popularization' and then to 'science communication'.[29] While the somewhat schematic character of this account perhaps rather obscures the complex layers of meaning in the language of the popular, it nevertheless nicely highlights the need to be sensitive to the radical changes in meaning and use which have occurred over the last two hundred years.

The account I have given so far has related entirely to the Anglophone world, and much could clearly be learned by exploring the conceptual differences in different linguistic contexts. Bensaude-Vincent has contributed a further valuable element to this literature in her and Anne Rasmussen's helpful discussion of the

---

[28] Bernadette Bensaude-Vincent, 'A Genealogy of the Increasing Gap between Science and the Public', *Public Understanding of Science*, 10 (2001): 99–113.

[29] Ibid., p. 106.

development of the language of popular science in French.[30] It is striking, for instance, that, in contrast to the English case, the French adjective '*populaire*' was used to denote accessibility and comprehensibility as early as the end of the seventeenth century, and the related sense of the verb '*populariser*' dated from the early eighteenth century. Nevertheless, it is also significant that, as in English, the meaning of '*populaire*' was extended at the very end of the eighteenth century to include that 'which addresses the people; which is put within the reach of the people, adapted to the tastes of the people', while '*populariser*' was shortly afterwards extended to include the sense 'to carry to the knowledge of the general public'.[31] In a further twist, however, there is no English counterpart for the mid-nineteenth-century French coinage *vulgariser* (from the Latin root *vulgus*, the common people), which Bensaude-Vincent and Rasmussen see as increasingly smothering a more inclusive notion of the popular.[32]

According to Andreas Daum, a similar, but nevertheless distinctive pattern is to be found in Germany.[33] There, as in Britain, the concept of '*popularität*' (popularity) was early associated with politics, and it was only from the end of the eighteenth century that it began to be used in connection with a mode of address accessible to all. By 1813, the adjective '*populär*' began to be applied in connection with the natural sciences, starting with astronomical works. However, as this discourse developed, use of the adjective '*populär*' contended in shifting ways with '*volkstümlich*', a word which, when reasonably translated as 'folksy', is seen to have no proper English equivalent. A further linguistic complication is involved when we consider the introduction of the adjectival '*populärwissenschaftlich*' (popularly scientific)

---

30 Bernadette Bensaude-Vincent and Anne Rasmussen, 'Introduction', in Bensaude-Vincent and Rasmussen (eds), *La science populaire dans la presse et l'édition XIXe et XXe siècles* (Paris, 1997), pp. 13–30. See also Bensaude-Vincent, 'La science populaire: Ancêtre ou rivale de la vulgarisation?', *Protée: Théorie et pratiques sémiotiques*, 16 (1988): 85–91.

31 Louis Guilbert, René Lagane and Georges Niobey (eds), *Grand Larousse de la langue française*, 7 vols (Paris, 1978).

32 In other Latin languages, the vocabulary is that of divulging (also from *vulgus*). See Bensaude-Vincent and Rasmussen, *La science populaire dans la presse et l'édition*, p. 14. However, Paola Govoni suggests that *divulgazione* and *divulgatore* were supplemented by *scienza popolare* in nineteenth-century Italy, perhaps reflecting the fact that French and English science popularization provided models for the Italian case. Paola Govoni, *Un Pubblico per la Scienza: La Divulgazione Scientifica nell'Italia in Formazione* (Rome, 2002). I am grateful to Josep Simon for assistance in understanding the Italian text of Govoni's work.

33 Andreas W. Daum, *Wissenschaftspopularisierung im 19. Jahrhundert: Bürgerliche Kultur, naturwissenschaftliche Bildung und die deutsche Öffentlichkeit, 1848–1914*, 2nd edn (Munich, 2002), pp. 33–42. I am grateful to Leucha Veneer for assistance in understanding the German text of Daum's work. See also Nick Hopwood, '[Review of *Wissenschaftspopularisierung im 19. Jahrhundert*, by Andreas Daum]', *British Journal for the History of Science*, 37 (2004): 357–8.

and its cognates from around the middle of the nineteenth century, since, of course, '*Wissenschaft*' does not map neatly onto English usage.[34]

These brief glances at non-English conceptual histories of 'popular science', and the summary account given here of the British case, serve to indicate not only how recently such terminology has been introduced, but also how complex and revealing its history is. Of course, for some historians, the recency of the language and concepts of popular science will seem problematic.[35] Why should it be that, while what used to be conceived of as the history of science popularization/popular science is to be generally subsumed in a larger history of 'knowledge in transit', historians of nineteenth- and twentieth-century science alone have the added opportunity of focusing on what historical actors themselves meant by 'popular science'? The straightforward answer is that each period has its own linguistic and conceptual histories to explore, and once 'popular science' is historicized, new questions will arise. How, for example, did contemporaries conceptualize or speak about such a book as William Buchan's enormously widely circulated *Domestic Medicine* (1769), which pre-dated the notion of 'popular science' by half a century? In answering such questions, one would expect historians of eighteenth-century Britain to explore further the conceptual histories of 'public' and 'polite' science, as well as related concepts, such as 'useful knowledge' and 'diffusion', whose histories straddle the eighteenth and nineteenth centuries.

Using the Language of 'Popular Science'

As the brief foray we have made into linguistic and conceptual history has already made clear, using the terminology of the popular to describe certain ventures in, and engagements with, science has served a variety of purposes over the last two centuries. As historians, however, we have been slow to examine these purposes. Those who have begun the analysis have tended to focus on the use of 'popular science' talk to do important boundary work, building an insurmountable barrier

---

[34] In Swedish, as Johan Kärnfelt has shown, *populärvetenskap* involves a notion like the German *Wissenschaft* which is again far broader than the English 'science'. See Johan Kärnfelt, *Mellan Nytta Och Nöje: Ett Bidrag Till Populärvetenskapens Historia I Sverige* (Stockholm, 2000).

[35] Guido Giglioni, for instance, recently suggested that such 'obsessing about linguistic precision betrays residues of positivism disguised in philological garb', since it makes nineteenth-century science 'the reference point for historians of science who wish to establish what science is'. Such a criticism seems to me only to be fair if the definition of such terms is taken to be unchanging. As I have sought to indicate, however, historians need to be alive to the change and complication which such terminology undergoes. I am not claiming that the language of 'popular science' appeared in a definitive form in the early nineteenth century and has since remained the same. See Guido Giglioni, '[Review of *Irritating Experiments*, by Huber Steinke]', *Bulletin for the History of Medicine*, 81 (2007): 662–4, p. 663.

between knowledge producers and knowledge acquirers, and thus securing the authority of scientists against the criticism of outsiders. As Stephen Hilgartner has shown, a significant reason for the continued pervasiveness of the diffusionist notion of science popularization is that it 'shores up the epistemic hierarchy which ranks scientists above such actors as policy-makers, journalists, technical practitioners, historians and sociologists of science, and the public'.[36] Moreover, Hilgartner argues that the widespread notion that popularization involves simplification and distortion 'provides scientific experts with a repertoire of conceptual and rhetorical tools' that can be used to justify selective simplification of scientific findings in garnering public support while at the same time attributing public opposition to misunderstandings due to distortion in popular media.

Such uses of the notion of 'the popular' to exclude non-specialists from knowledge production and render them passive consumers have clearly been common for many years. However, they do not exhaust the range of possibilities. Nor should we assume that exclusionary uses predominated or even necessarily existed when the language of 'popular science' was first introduced in the pre-professional age. As the brief history I have given suggests, the origins of a discourse of the popular in the British context (and very probably in the French and German contexts), are intimately connected with commercial developments in print culture. For many of those who first used the phrase 'popular science' to describe their products in the 1820s and 1830s, the primary motivation was to increase market share by signalling affordability, accessibility, utility and attractiveness. One of the first to use the phrase was John Timbs, editor of Britain's first successful cheap weekly magazine, the *Mirror of Literature, Amusement, and Instruction*. Timbs began his 'Arcana of Science' column in 1827 to condense 'such new facts in science as are connected with the arts of social life, and which from being scattered through elaborate and expensive works, might thereby be lost to some portion of our readers'. The column was to detail '*popular* discoveries in science' avoiding the 'perplexing' details of abstract research and puerile scientific jargon.[37] Whereas, under its previous editor, the magazine had been generally open to scientific contributions from readers in the manner of gentlemanly miscellanies, Timbs now packaged accessible gobbets of science from more specialist publications for a mass audience. Both his practice and his rhetoric was designed to render the magazine commercially attractive, and his success was such that

[36] Stephen Hilgartner, 'The Dominant View of Popularization: Conceptual Problems, Political Uses', *Social Studies of Science*, 20 (1990): 519–39, pp. 533–4.

[37] 'Arcana of Science; or, Remarkable Facts and Discoveries in Natural History, Meteorology, Chemistry, Mineralogy, Geology, Botany, Zoology, Practical Mechanics, Statistics, and the Useful Arts', *Mirror of Literature, Amusement, and Instruction*, 10 (1827): 252–4, p. 252.

he was able to launch a separate annual entitled *Arcana of Science and Art* the following year, again trumpeting the rhetoric of 'popular science'.[38]

The use of the epithet 'popular' to make science sell, whether in printed form or in exhibitions, lectures and museums, is clearly distinct from the exclusionary usage identified by Hilgartner. Nevertheless, as we have seen, the commercial usage can contribute to the exclusionary when 'mere popularizers' are represented as having distorted science in order to sell it. Conversely, however, many of those engaged in commercial ventures of popularization in early nineteenth-century Britain had mixed motives – wishing to make a living, but also committed to an inclusive ideology of science. As Susan Sheets-Pyenson has shown, for instance, popular science periodicals were often produced with the intention of engaging readers in scientific practice. There were, she insists, forms of 'low science' in the period that were not merely 'a simplified reflection of "high" or "academic" science'. On the contrary, she argues, many popular science periodicals developed 'their own canons of scientific investigation, criticism, and explanation'.[39] This optimistic account of the inclusiveness of science in the pre-professional period might reasonably be criticized for attributing an idealized autonomy to 'low science', but subsequent work has provided a more considered basis for understanding the importance of the notion of 'popular science' in including individuals in the practice of science.

In this regard, Anne Secord's work is helpful. In an article on the role of pictures in promoting early nineteenth-century botany, Secord argues that many expert naturalists considered 'popular botany' to be 'the means by which private individuals could best be encouraged to extend their aesthetic appreciation and love of plants to an active and participatory pursuit of science'.[40] Popular science, far from being 'that which is excluded from "science proper"', was intended by some specialists to be an inducement to scientific practice.[41] As Secord observes, however, such popularization also involved a concern with standardization of practice – in the case of botany, of observational practice. In addition to pleasure, therefore, involving new participants in botany also entailed a notion of discipline. Moreover, while the fluid social basis of scientific disciplines in early nineteenth-century Britain makes it anachronistic to expect the same kind of divide between professional and popular science that we might encounter in the early twenty-first

[38] Jonathan R. Topham, '*The Mirror of Literature, Amusement and Instruction* and Cheap Miscellanies in Early Nineteenth-Century Britain', in Geoffrey Cantor et al., *Reading the Magazine of Nature: Science in the Nineteenth-Century Periodical* (Cambridge, 2004), pp. 37–66 (see pp. 63–6). See also Topham, 'Thomas Byerley, John Limbird, and the Production of Cheap Periodicals in Regency Britain', *Book History*, 8 (2005): 75–106.

[39] Susan Sheets-Pyenson, 'Popular Science Periodicals in Paris and London: The Emergence of a Low Scientific Culture, 1820–1875', *Annals of Science*, 42 (1985): 549–72, p. 551.

[40] Anne Secord, 'Botany on a Plate: Pleasure and the Power of Pictures in Promoting Early Nineteenth-Century Scientific Knowledge', *Isis*, 93 (2002): 28–57, p. 28.

[41] Ibid., p. 29.

century, Secord points to the work of Timothy Alborn to remind us that in Britain by the 1830s 'those aiming to reform scientific practice idealized a self-governing hierarchical division of labor in the sciences'.[42] Science was neither democratic nor entirely meritocratic, but practitioners did use the language of popularization in order to recruit workers into a divided body of practitioners.

Secord's focus on the complex social organization of science is pivotal in moving beyond the merely exclusionary uses of the concept of popular science to a consideration of the complexities of its other uses. In this context, Alborn's work on the organizing language of a scientific division of labour finds useful parallels in Martin Rudwick's study of the 'social topography' of geology in the 1830s, in which 'a series of graduated zones of "ascribed competence" [...] shade insensibly into one another'. Rudwick envisions concentric circles, moving out from an elite, through a circle of '"accredited" geologists', to an outer circle of 'amateur geologists', beyond whom are the public at large. He thus presents a graduated range of more or less visible boundaries within the body of scientific practitioners.[43] Work such as that by Secord causes us to reconsider the extent to which 'popular science' talk was used to do boundary work *among* practitioners, rather than to do boundary work between practitioners and a wider public.

Rudwick's use of the term 'amateur' in relation to the 1830s is, of course, an anachronism which to some extent undermines his own project of recognizing the complexity of the pre-professional scientific body. Yet the organization of that body changed rapidly in nineteenth-century Britain, and it was not long before new notions of 'professionalism' and 'amateurism' began to cut across the notion of the 'popular'. More generally, of course, we need to be aware that the rapidly changing terminology of the 'popular' was intimately connected with other changing terminology, such as 'expert', 'useful', 'practical', 'diffusion' and 'pseudoscience'. Even before the inception of the professional/amateur divide, as Alborn has shown, William Whewell's introduction of the term 'scientist' in the 1830s represented an attempt to supplant notions of a scientific division of labour with the individual hero of scientific discovery – an entrepreneur who had earned the right to apportion tasks for the rest of the community.[44] By the 1850s, professionalizers like Thomas Huxley were happy to use the notion of 'popular science' in exclusionary ways. Huxley famously savaged the anonymous evolutionary work *Vestiges of the Natural History of Creation* in the *Westminster Review* by deriding it as popular science drawn from *Chambers's Edinburgh Journal* and the *Penny Magazine*. It was, he claimed, a 'lumber-room of second-hand scientific furniture'.[45] Moreover, Susan Sheets-Pyenson and Ruth Barton

42 Ibid.

43 Martin Rudwick, 'Charles Darwin in London: The Integration of Public and Private Science,' *Isis*, 73 (1982): 186–206.

44 Alborn, 'The Business of Induction', p. 95.

45 [T.H. Huxley], 'The Vestiges of Creation', *British and Foreign Medico-Chirurgical Review*, 13 (1854): 425–39, p. 438. See also Secord, *Victorian Sensation*, pp. 498–504.

have argued that the new popular science periodicals of the 1860s 'sought not participation from amateurs, but support for professionals', and Ernst Homburg has described a similar situation in regard to the popularization of chemistry in the Netherlands.[46]

It is thus perhaps not surprising that, inasmuch as they have considered the changing use of the notion of 'popular science', historians have inclined to see it as a process of degeneration from an inclusive past to an exclusive present. However, the thrust of my argument here is that the layers of meaning which the vocabulary conveys, and the changing situation of the several sciences in different local and national contexts, means that 'popular' talk has continued to serve a variety of contingent purposes which it is our task as historians to uncover. To give a much more modern example, Stephen Jay Gould repeatedly deplored the 'equation of popular writing with pap and distortion', preferring instead to view his own practice as a popular journalist as '*vulgarisation*' within 'the highest traditions of humanism'.[47] It is perhaps no surprise, then, that in the 1970s Gould was also a member of the left-wing academic activist group 'Science for the People', which opposed the government and corporate interests that establishment science was perceived to serve.[48] Gould's inclusive rhetoric of 'popular science' has thus been claimed by some as 'serving a democratic socialist agenda', although it has also been critiqued as enmeshing 'the "facticities" of science with the master narratives of liberal bourgeois culture'.[49] Either way, however, as Gillian Fuller has argued, uses of 'the popular' like Gould's, which seem to defend the people from aggression, have great moral power.[50] The language of 'popular science' continues to serve multiple ends.

---

[46] Sheets-Pyenson, 'Popular Science Periodicals', p. 563; Ruth Barton, 'Just before *Nature*: The Purposes of Science and the Purposes of Popularization in Some English Popular Science Journals of the 1860s', *Annals of Science*, 55 (1998): 1–33, p. 3; Ernst Homburg, 'From Chemistry for the People to the Wonders of Technology: The Popularization of Chemistry in the Netherlands during the Nineteenth Century', *HYLE – International Journal for Philosophy of Chemistry*, 12 (2006): 163–91.

[47] Steven Jay Gould, *Bully for Brontosaurus: Reflections in Natural History* (New York, 1991), p. 11.

[48] See, for example, John Walsh, 'Science for the People: Comes the Evolution', *Science*, n.s. 191 (1976): 1033–5.

[49] Gillian Fuller, 'Cultivating Science: Negotiating Discourse in the Popular Texts of Stephen Jay Gould', in J.R. Martin and Robert Veel (eds), *Reading Science: Critical and Functional Perspectives on Discourses of Science* (London and New York, 1998), pp. 35–62, pp. 36–8.

[50] Ibid. Fuller's analysis draws on the perceptive critique of Pierre Bourdieu. See Bourdieu, 'Did You Say "Popular"?', in John B. Thompson (ed.), *Language and Symbolic Power*, trans. Gino Raymond and Matthew Adamson (Cambridge, MA, 1991), pp. 90–102; and Bourdieu, 'The Uses of the "People"', in Bourdieu, *In Other Words: Essays Towards a Reflexive Sociology*, trans. Matthew Adamson (Stanford, 1990), pp. 150–55.

'Popular Science' *as* Science

The diverse uses to which 'popular science' and cognate language has been put further undermine the binary opposition between a 'real' science generated by scientists and a 'popular science' divorced from the knowledge production and validation process. We have seen, indeed, that the concept of 'popular science' has often been used to organize important aspects of the scientific enterprise and the structuring of the body of scientific practitioners. In this last section, I want to consider in a little more detail how those cultural products designated as 'popular science' and those activities designated as 'science popularization' have constituted a core aspect of the scientific enterprise.

Martin Rudwick's account of the social topography of early Victorian science, with its concentric rings of ascribed competence, is strikingly reminiscent of Ludwik Fleck's portrayal of the esoteric and exoteric circles of what he calls the 'thought collective' of modern science, in which the esoteric circle also includes internal gradations from special to general experts.[51] Fleck's account of the dynamics of communication within the scientific thought collective highlights important aspects that are particularly pertinent here. He argues that the 'popular science' through which scientific 'facts' are presented to the general public depends on a process of communicative transformation, as key features of factuality are produced, namely certainty, simplicity and vividness. However, unlike in the diffusionist model, Fleck considers that this popular science 'furnishes the major portion of every person's knowledge', including that of the 'specialized expert'.[52] Thus, there is a feedback mechanism, by which what is conceived of as 'popular science' crucially shapes the thought style of research scientists.

It is perhaps easiest to see this when a scientist in one discipline draws on popularizations of other sciences. Thus, for instance, it is striking that James Clerk Maxwell adopted a statistical approach in his work on the kinetic theory of gases partly in consequence of having read the first volume of Henry Buckle's *History of Civilization in England* (1857). Buckle's use of social statistics to find underlying regularities in individual action within large populations provided Maxwell with a suitable model for molecular behaviour.[53] Moreover, this kind of inter-disciplinary role for works of popularization functions at a level of very great generality. As Fleck observes, when an economist 'speaks of the *organism* of the economy' he uses concepts derived from his 'fund of popular knowledge'.[54] Such conceptual appropriation has, of course, been explored by numerous scholars, perhaps most

[51] Ludwik Fleck, *Genesis and Development of a Scientific Fact*, eds Thaddeus J. Trenn and Robert K. Merton, trans. Fred Bradley and Thaddeus J. Trenn (Chicago, 1979).

[52] Ibid., p. 112.

[53] Theodore M. Porter, *The Rise of Statistical Thinking, 1820–1900* (Princeton, 1986), pp. 113–14.

[54] Fleck, *Genesis and Development of a Scientific Fact*, p. 112.

famously in the case of Darwin by Robert Young and Gillian Beer.[55] However, the tendency to consider the phenomenon in relation to the 'common cultural context' in which Victorian science is supposed to have developed tends to undermine the extent to which the reliance of scientists on popularizations of other disciplines is, if anything, an increasingly important aspect of modern disciplinary science. As the several sciences have become more specialized, popularizations have often been consciously addressed to scientists in other disciplines. Similarly, and of particular pertinence in this book, ventures in 'popular science' often serve important purposes in communicating between scientists working in different countries.

It is now some years since Richard Whitley pointed out the need to examine the manifold audiences for popularizations of science, including 'researchers from other fields and disciplines', but work in this area is still in its infancy.[56] Historical studies of science popularization still often fail to challenge the notion inherited from the diffusionist model, that this is a form of communication between a unitary body of scientists on the one hand and an undifferentiated lay public on the other. Of course, closer study reveals a range of intended audiences being addressed, and a range of often unintended usages being developed. A good example from my own work is provided by William Buckland's Bridgewater Treatise, *Geology and Mineralogy Considered with Reference to Natural Theology* (1836). Buckland sought in the work to give 'a popular general view of the whole subject avoiding technical detail as much as possible'.[57] Yet, while the book was welcomed as a species of popular science suitable for those with no knowledge of geology, it was also praised for its value to scientific practitioners. In his presidential address to the Geological Society, Charles Lyell applauded Buckland for skilfully combining 'several distinct objects' in the treatise, laying particular emphasis on the extent to which, by giving 'a general view of the principal facts brought to light by the study of organic remains', he had contributed 'towards the filling up one of the greatest blanks which existed in the literature of our science'.[58] Moreover, as Nicolaas Rupke has argued, Buckland's book gave 'form and definition' to the 'geological

---

[55] Robert M. Young, *Darwin's Metaphor: Nature's Place in Victorian Culture* (Cambridge, 1985) and Gillian Beer, *Darwin's Plots: Evolutionary Narrative in Darwin, George Eliot and Nineteenth-Century Fiction*, 2nd edn (Cambridge, 2000).

[56] Richard Whitley, 'Knowledge Producers and Knowledge Acquirers: Popularisation as a Relation Between Scientific Fields and their Publics', in Terry Shinn and Richard Whitley (eds), *Expository Science: Forms and Functions of Popularisation* (Dordrecht, 1985), pp. 3–28, p. 6.

[57] Bodleian Library, University of Oxford. Buckland to [Anon.], n.d. Ms.Eng.Lett.d.5, fol. 251.

[58] Charles Lyell, 'Address to the Geological Society, Delivered at the Anniversary, on the 17th of February, 1837', *Proceedings of the Geological Society of London*, 2 (1833–8): 479–523, p. 517.

progressivism' of the English school of geology.[59] Writing the book for a wide audience gave Buckland licence to provide an overview of the subject which was of considerable importance for practitioners in his own scientific field.

The notion that works conceived of by their authors as 'popular' might serve important functions within a scientific discipline applies equally to modern science. Richard Dawkins's million-selling book *The Selfish Gene*, for instance, published in 1976 within the genre of popular science, was, by his own admission, intended also to address his peers in evolutionary biology at a level of generality that was otherwise unattainable. When writing, Dawkins had three 'imaginary readers' in mind: the 'general reader, or layman', the 'expert' and the 'student, making the transition from layman to expert'. His 'greatest hope' was that the expert would 'find something new' in what he wrote: 'a new way of looking at familiar ideas perhaps; even stimulation of new ideas of his own'.[60] In the preface to the second edition (1989), Dawkins elaborated, observing that he preferred 'not to make a clear separation between science and its "popularization"'. Popular exposition of technical ideas required 'new twists of language and revealing metaphors', which, if pushed far enough, might result in 'a new way of seeing'.[61] Certainly, Dawkins's book, with its controversial metaphor, has been highly significant in the ongoing development of evolutionary theory. As one of Dawkins's former students has claimed, a 'generation of biologists learnt about natural selection from *The Selfish Gene*' and the 'clarity of vision' which it offered 'encouraged field biologists to apply the newfangled ideas they now understood, and theoreticians to take up and further develop the theories that now seemed so central to the concerns of biology'.[62]

More generally, addressing a wide audience has often afforded opportunity for exploring and expanding the limits both of science and of the scientific community. As the Science in the Nineteenth-Century Periodical (SciPer) project of the Universities of Leeds and Sheffield has shown, for instance, non-specialist periodicals in nineteenth-century Britain often provided forums for the development of heterodox, marginal or novel sciences. Sometimes, as with James Sully and George Henry Lewes's development of 'baby science' in the 1860s and 1870s, a new scientific field emerged from such a liminal state to be accepted as orthodox. However, the process was not predestined to succeed. Lewes and Sully were 'relative outsiders to the scientific community, struggling initially to make names for themselves in scientific circles whilst financing their work through

---

59 Nicolaas A. Rupke, *The Great Chain of History: William Buckland and the English School of Geology, 1814–1849* (Oxford, 1983), p. 183.

60 Richard Dawkins, *The Selfish Gene* (Oxford, 2006), pp. xxi–xxii.

61 Ibid., p. xvi.

62 Alan Grafen, 'The Intellectual Contribution of *The Selfish Gene* to Evolutionary Theory', in Alan Grafen and Mark Ridley (eds), *Richard Dawkins: How a Scientist Changed the Way We Think* (Oxford, 2006), pp. 66–74, pp. 71–2.

extensive popular journalism'.[63] Unlike Buckland or Dawkins, they faced the difficult task of establishing their own claims to scientific expertise as well as the claims of their new science. In such a situation, the price of failure was to have one's work described as 'mere' popular science.

A corollary of the perspective I have been developing here – that scientists rely on works of science popularization in order to advance research – is the point made by Fleck, that the transformation which scientific knowledge undergoes when presented to exoteric audiences feeds back into the practice of esoteric science. This, of course, returns us to the perspective outlined under my first heading, in which the history of science popularization becomes a core element of the history of science, rather than an optional sub-field. Christopher Hamlin recently expressed the concern that the kind of approach I am advocating here implies that 'science is made in the public sphere, by the people, not in the laboratory by the elite'.[64] I do not, however, mean to present an exclusive choice concerning where science is made, merely to suggest that the various activities conceived of as 'science popularization' at different periods and in different places cannot be detached from the rest of the practice of science as the diffusionist model implies that they can. Hamlin refers to the 'need to rehabilitate popularization, the category that dares not speak its name'.[65] My argument here is that once the diffusionist baggage has successfully been removed, so that the history of 'science popularization' is seen as part of a wider history of knowledge in transit, the historian is freed up to consider the multifarious ways in which 'popularization' and its cognates have been conceived and used over the course of the last two hundred years and across national and linguistic boundaries.

Conclusion

The first part of my chapter was perhaps a little subversive of the project of this book, inasmuch as it suggested reconceptualizing the basis of research on 'science popularization' in terms of a wider history of communication in science. My argument was that the category 'popular science' carries so much historical baggage as to be seriously compromised for use as a term of historical analysis. In particular, it does not define a topic of study in a way that is either intellectually rigorous or heuristically rewarding. In its place, I have suggested the need to develop an approach in which all scientific knowledge is understood as part

63 Sally Shuttleworth, 'Tickling Babies: Gender, Authority, and "Baby Science"', in Geoffrey Cantor et al., *Science in the Nineteenth Century Periodical: Reading the Magazine of Nature* (Cambridge, 2004), pp. 199–215, p. 199.

64 Christopher Hamlin, 'Games Editors Played or Knowledge Readers Made?', *Isis*, 96 (2005): 633–42, p. 642.

65 Ibid.

of a communicative process, involving appropriation, resistance and cultural contestation. This has the benefit of reintegrating 'science popularization' with other forms of communication in science, including education and international communication, and it also provides a range of new historiographical resources, drawn from, among other places, the cultural history of the book, translation studies and the history of education.

If the first part of the chapter was subversive, the second was emphatically not. For, I have argued, there is a rich conceptual and linguistic history of 'popular science' and 'science popularization' that needs to be explored as part of this larger history of communication. We need to know when these kinds of terms were introduced in different countries and why, and we need to know how they changed over time, and how they related to other keywords like 'amateur', 'expert' and 'diffusion'. Reviewing work done to date on the English case, it is clear that the language of the popular has had multiple layers of meaning at any given time. Importantly, however, it has often been used as a means of organizing people in relation to the scientific exercise – situating people within a hierarchy of competence and both excluding people from and including people in the scientific project. This view confirms the wider claim of my chapter, that 'popular science' needs to be considered as part of science, and I have sought to indicate towards the end the practical outworkings of such a view. By advocating in some sense abandoning the 'history of science popularization' as a field of study, I thus do not mean to suggest that we should all stop doing what we are doing. Rather, I have suggested that we can reinvigorate our researches by incorporating them into a wider history of communication in science, and by focusing on what is particular to the historical vocabulary of 'popular science'.

Chapter 2

The Historiography of Science Popularization: Reflections Inspired by the Italian Case

Paola Govoni

For a long time, the world of scientific research and that of its popularization were, to all appearances, parallel and autonomous universes. In order to 'shift' concepts and data from one universe to the other, it was thought necessary to 'translate' or 'adapt' them to the requirements and abilities of their different audiences, just as with their publishing outlets. That the reader might play a role in the dynamics of the production of knowledge – either of research or popularization – was naturally rejected. On the contrary, the 'strong impression' popular science would have on its public was taken for granted, even though opinions on how this actually occurred sometimes radically diverged. Remarkable examples of roughly contemporary yet conflicting declarations are to be found in the words of two Nobel prize-winners, Albert Einstein and Czeslav Milosz.

On the very first page of his famous 1949 autobiographical notes, Einstein wrote:

> Thus I came – though the child of entirely irreligious (Jewish) parents – to a deep religiousness, which, however, came to an abrupt end at the age of twelve. Through the reading of popular scientific books I soon reached the conviction that a good deal in the stories of the Bible could not be true. The consequence was a positively fanatical [orgy of] freethinking, coupled with the impression that youth was deliberately being deceived by the state through lies; it was a crushing impression.[1]

Only two years later, in 1951, Milosz wrote:

> In the nineteenth century, with the rise of literacy, brochures popularizing scientific theories made their appearance. … The simplified and vulgarised version of Darwin's theory of the origin of species and the struggle for existence [however] is not the same concept that it was for Darwin or for his scientific opponents. … The leaders of the twentieth century like Hitler for instance,

---

1 Albert Einstein, *Autobiographical Notes*, trans. and ed. Paul Arthur Schilpp (La Salle, IL, 1996), pp. 3, 5 [1st edn 1949].

> drew their knowledge from popular brochures, which explains the incredible confusion of their minds ... [2]

These two quotations sum up the clichés that for a long time relegated the success of nineteenth-century popular science in Europe to a marginal role in the history of science. On the one hand, we have popular science depicted as a tool capable of awakening a dormant genius in an apparently ordinary boy, Einstein, and on the other, the same popular science is portrayed as a harbinger of confusion that may lead to the worst of crimes – genocide.

When I began to work on this subject, and in particular on the case of Italy in the second half of the nineteenth century, I knew that there was much truth in that double image of the genre. The episode involving Einstein is little more than a footnote in the biography of the scientist, but what can be said about the formation of tens of millions of young Europeans who, like Einstein (and Hitler), read those popular science books? Those young Europeans would have seen – and in most cases had to live through – two world wars, the birth of Fascism and National Socialism, the enactment of racist laws and the Holocaust: events that were made possible also through the 'scientific' legitimacy of theories born in the age of positivism. Those young readers of 'popular science', for the most part belonging to families in the European middle classes, were formed in the decades after the 1880s, just as scientific popularization in Europe was enjoying its peak. The challenge inherent in my reconstruction of some aspects of popular science in Italy towards the turn of the nineteenth century, however, was to show that there were other interesting aspects of popular science that could reveal something about science and scientists of the period in general.

In the first part of this chapter, after dealing briefly with the meaning and use of the expressions 'popular science' and 'popularization of science' in the Italian context, I shall recall several basic features of the historical case I am most familiar with: the success of popular science literature in Italy in the 1870s and 1880s and its subsequent decline. The second part presents a survey of the historiography of science popularization, and highlights some of the interdisciplinary tools I have found most useful when examining that topic. [3] I will conclude by advocating a

---

2 Czeslav Milosz, *The Captive Mind* (New York, 1990), pp. 200–201 (1st edn, 1951). However, on the slight 'effects' of scientific popularization, see John C. Burnham, *How Superstition Won and Science Lost: Popularizing Science and Health in the United States* (New Brunswick, 1987).

3 In a paper delivered at the 5th STEP meeting I gave an overview of the literature which was basic to my reconstruction of the Italian case up until the most recent publications. In this published version, I shall be reducing references to the wealth of recent secondary literature to the minimum; for further references, see Chapter 1 of this volume by Jonathan R. Topham. See also Jonathan R. Topham, 'Scientific Publishing and the Reading of Science in Nineteenth-Century Britain: A Historiographical Survey and Guide to Sources', *Studies in History and Philosophy of Science*, 31 (2000): 559–612. For some French,

greater effort in comparative studies of the popularization of science, technology and medicine in the long period; comparative studies basic to the construction of big pictures – transcending national boundaries and disciplines – that at this stage in the development of research are both possible and to be hoped for, for reasons I shall try to make clear.

On Uneasiness about Labels

For readers of a book that compares cases in several different countries, a brief note on the use of the expressions 'popularization of science' and 'popular science' in different contexts may be of interest.

Since around the mid-1980s, scientists, historians and sociologists of science have begun asking themselves about the ways the worlds of experts and the worlds of people far removed from science communicate. Back in 1985 the scientists of the Royal Society launched a controversial programme to popularize science characterized by a new label, the 'Public Understanding of Science'.[4] Those were the years when the first studies on so-called popular science began to appear, studies in which historians and sociologists of science sometimes wondered if it were not time to substitute 'popular science' and 'popularization of science' with expressions more suitable to the sophisticated analyses being carried out.[5]

Spanish, German and Italian references, as well as plentiful references to the international literature before 2002, see Javier Ordóñez and Alberto Elena (eds), *La Ciencia y su público: perspectivas históricas* (Madrid, 1990); Bernadette Bensaude-Vincent and Anne Rasmussen (eds), *La science populaire dans la presse et l'édition XIXe et XXe siècles* (Paris, 1997); Andreas Daum, *Wissenschaftspopularisierung im 19 Jahrhundert: Bürgerliche Kultur, naturwissenschaftliche Bildung und die deutsche Öffentlichkeit, 1848–1914* (Munich, 1998); Bensaude-Vincent, *L'opinion publique et la science: à chacun son ignorance* (Paris, 2000); Paola Govoni, *Un pubblico per la scienza. La divulgazione scientifica nell'Italia in formazione* (Rome, 2002).

4 Walter F. Bodmer (ed.), *The Public Understanding of Science* (London, 1985).

5 Roger Cooter, *The Cultural Meaning of Popular Science: Phrenology and the Organization of Consent in Nineteenth-Century Britain* (Cambridge, 1984), pp. 1–11; Terry Shinn and Richard Whitley (eds), *Expository Science: Forms and Functions of Popularisation* (Dordrecht, 1985); Stephen Hilgartner, 'The Dominant View of Popularization: Conceptual Problems, Political Uses', *Social Studies of Science*, 20 (1990): 519–39; Roy Porter (ed.), *The Popularization of Medicine, 1650–1850* (London, 1992), pp. 1–16; Roger Cooter and Stephen Pumfrey, 'Separate Spheres and Public Places: Reflections on the History of Science Popularization and Science in Popular Culture', *History of Science*, 32/97 (1994): 237–67; Barbara T. Gates and Ann Shteir (eds), *Natural Eloquence: Women Reinscribe Science* (Madison, 1997), pp. 3–24.

Though I share the basic premises of those studies,[6] I nonetheless believe that historians will inevitably have to resign themselves to using the expressions of the past in the case of 'popular science', just as they do with 'phlogiston' when dealing with eighteenth-century chemistry, or of the human 'races' when discussing nineteenth-century anthropology. It is clear that if historians of the Victorian age make use of the expression 'popular science' it will not be because they adhere to the 'deficit model' condemned by recent historiography, but because they cannot discard the expressions used by the protagonists of their history, however aware they may be of the changeable meanings of 'popular science'. As a straightforward example, we may take *Popular Science*, the American monthly founded in 1872, still being published under the same title: it is hardly likely that writers, publisher and readers of the monthly today allot the same meaning to the title as those of the 1870s. Using the expression 'popular science' in English today on the one hand helps to define the context – and the level – of the magazine's scientific knowledge, and on the other alludes to a world of amateurs who have always been important in English-speaking scientific culture. Finally, perhaps, using the expression 'popular science' in English today means referring to the awareness of the existence of a longstanding tradition of relationships – not always easy, but at any rate uninterrupted in two and a half centuries – between science and the public.

In the Italian linguistic and cultural context, the situation is very different. No one today uses the expression *scienza popolare* (popular science), and I, too, prefer to restrict myself to using *divulgazione scientifica* (science popularization/popularization of science). In modern Italy, as a matter of fact, to use the adjective 'popular' in a scientific context risks generating a fair number of misunderstandings.

As in many other European countries, in the Italy in the second half of the nineteenth century that I shall be discussing shortly, the actors (writers, publishers and public) did use the expression 'popular science'. Yet the science presented to their readers by the popularizer-scientists I have examined, those writers with such extraordinary success in the 1870s and 1880s, had very little in common with popular or 'low culture' in its usual sense. In Italy the publishing market, or the market for culture, at that time did not rival the wealth and variety of those in Paris and London, where, as Susan Sheets-Pyenson has shown, 'On occasion, "low science" periodicals vigorously opposed the "high" scientific establishment'.[7]

---

6 For a first attempt of mine to 'popularize' social studies on the popularization of science in Italy, see Paola Govoni, 'La Divulgazione Scientifica: Un Genere Marginale?', *Intersezioni*, 3 (1991): 553–64.

7 Susan Sheets-Pyenson, *Low Scientific Culture in London and Paris, 1820–1875*. PhD thesis, University of Pennsylvania (Philadelphia, 1976): 15. Subsequently Sheets-Pyenson's thesis became the basis of the classic 'Popular science periodicals in Paris and London. The emergence of a low scientific culture, 1820–1875', *Annals of Science*, 42 (1985): 549–72. In France, research has been done on the presence of popular culture in the field of science, from 'alternative' medicine up until the compromise between academic and

In Italy in the second half of the nineteenth century, the adjective 'popular' (*popolare*) meant *for* the people (*per il popolo*), not produced *by* the people.

At present – and the history of science popularization in Italy is in its infancy – our knowledge of popular science is limited to its 'top–down' manifestations, that is, we are familiar with the literature produced by the professional scientists involved in popularizing in their battle to promote science; and we are aware of the role of scientists (and doctors) and the culture they embodied, producing a literature whose main aim was to replace what is usually meant by 'low culture' because it furthered superstition – a superstition that was believed to have been tolerated and even encouraged by the Catholic Church – which was felt to be the main brake on 'progress' and modernization. In Italy over the course of the nineteenth century, as back in previous centuries,[8] it is very likely that a scientific lowbrow culture already existed, as it did in other countries,[9] although there is a lack of systematic research on this. A good deal more needs to be done, studying the production of popular almanacs and the dissemination of popular lectures and popular cultural circles.

For the reasons briefly cited here, I prefer to use the expression *divulgazione scientifica* (science popularization/popularization of science),[10] an expression that I feel is acceptable to generically indicate a literature – at various levels and for manifold purposes – that brought science and lay audiences in Italy into communication over the last three centuries.[11]

---

popular science, carried out by a popularizer such as Raspail. See Jean-Pierre Poirier and Claude Langlois (eds), *Raspail et la vulgarisation médicale* (Paris, 1988).

[8] See one of the best-known cases, that of the cosmologies of the miller Menocchio that Carlo Ginzburg writes about in *The Cheese and the Worms: The Cosmos of a Sixteenth-century Miller* (Baltimore, 1980; 1st edn, Turin, 1977).

[9] The case studied in greatest detail is that of Robert Chambers. See James Secord, *Victorian Sensation: The Extraordinary Publication, Reception, and Secret Authorship of 'Vestiges of the Natural History of Creation'* (Chicago, 2000).

[10] The use in Italian of the terms '*divulgazione*' [popularization] and '*divulgatore*' [popularizer] goes back to the sixteenth century; see *Grande dizionario italiano dell'uso*, ed. Tullio De Mauro (6 vols, Turin, 1999), vol. 2, p. 711. In English, the use of 'popularization' in our sense dates back to between 1797 and 1801: see *The Oxford English Dictionary*, 2nd edn (20 vols, Oxford, 1989), vol. 12, p. 126. The appearance in French of the verb '*populariser*' seems to date back to 1789: see *Trésor de la langue française: Dictionnaire de la langue du XIXe et du XXe siècle (1789–1960)* (16 vols, Paris, 1988), vol. 13, pp. 778–9. The appearance of the French verb '*vulgariser*' goes back to 1827, whereas the term '*vulgarisation*' came about between the 1850s and the 1870s. See Bernadette Bensaude-Vincent, 'La Science Populaire, Ancêtre ou Rivale de la Vulgarisation?', *Protée: Théorie et Pratiques Sémiotiques*, 16 (1988): 3.

[11] When not otherwise indicated, for subjects, data and bibliography referring to the Italian case, see Govoni, *Un pubblico per la scienza*. The book reconstructs the history of 'popular science' in Italy in the period 1861–1900, with an introductory chapter on the premises of that success (1500–1800), focusing in particular on the age of Enlightenment.

The Success and Later Decline of the 'Science for All' Movement in Italy, 1861–1900

In the Europe of the second industrial revolution, colonial expansion and the birth of modern nations, as well of the research university, scientists understood the importance of consolidating their dialogue with the various audiences of science. Politicians, industrialists and the educated middle classes, as well as artisans, workers and women, became the targets of an amazing variety of writings and public events which featured science: a 'popular' science 'for all', produced in many European languages.

In Italy, too, for the many intellectuals and scientists whose ideals of liberty and modernization derived from the tradition of eighteenth-century reformism, commitment to popular education was a 'natural' choice as soon as the political unification of the country was achieved (1861). Often, political refugees in Paris and London during the *Risorgimento* were more familiar with the situation in central and northern Europe than that of central and southern Italy. And it was with those models of civilization in mind – and especially the cultural markets of cities like Paris and London – that some Italian scientists and publishers believed they could solve the problem of the backwardness of the country in no time: by spreading education and scientific culture at every level of the population. In fact, when in 1870 the Vatican militia was defeated and Rome was annexed to the new kingdom, many people felt a contagious optimism on the outlook for Italy – and also for science – in Europe. The British weekly *Nature* confidently declared:

> Italy has become a nation. It is no longer enslaved by the barbarous despotism of a single city, nor divided into mutual throat-cutting republics, nor diplomatically parcelled into heir-looms for royal families. It has at last become a country of its own people. The moral and intellectual laws of Natural Selection are now freely operating, and they will soon show what manner of people these Italians are ...[12]

In the 1870s, many scientists, museum curators and astronomers became well known as writers of successful popularizing works, and came to play a prominent role in public life. In Italy, too, the moment had finally arrived for the public to allow recognition for the scientist, who thus joined the prelate, the state functionary, the politician and the teacher (and to a lesser extent the industrial entrepreneur), as one of the main players of the public sphere. The public image of the scientist was formed south of the Alps in the second half of the nineteenth century thanks to the success of the popularization of science, a phenomenon known in Italy by the label *scienza per tutti* (science for all, *science pour tous*).

The history of the popularization of science in Italy after unification helps us to understand something about how the educational and cultural policies were set up and oriented at an important time for the formation of the scientific institutions

---

[12] William Mattieu Williams, 'Science in Italy', *Nature*, 4 (8 June 1871): 98.

of the modern state. For the first time, Italian middle-class public opinion, mainly in the name of the 'Darwinian revolution', was confident it could take its destiny in its own hands. Thanks also to the enormous success of the popularization of science, the conviction spread that it was finally possible for Italy to take over responsibility for its own affairs, sweeping away all interference – barring that of reason. This was the basic message of the popularizing scientists, a message rooted in Enlightenment culture, to which an important novelty for Italy was added: the targets were no longer an elite of 'philosophers' and frequenters of aristocratic *salons*, but the 'people', that is, the middle classes in their various stratifications. Because of the very high rates of illiteracy in Italy, the eager readers of nineteenth-century popularization of science were mainly from the middle classes.[13]

Italian scientists involved in the success of the popularization of science dealt with popularizing at several levels and for a variety of reasons. The upper echelon of the popularization of science included works for an educated public and for the specialists in related fields. One of its most visible products was the large number of contributions to the debate on evolutionism. Another was the burgeoning production of publications for technicians and engineers.[14] One example of the genre, created in the USA and well known in many European countries, was the *Biblioteca Scientifica Internazionale* (International Scientific Library), published in Italy by Dumolard of Turin. The series was created in 1875 as an Italian branch of the International Scientific Series, thought up by Edward L. Youmans. The series allowed a certain number of Italian works to reach beyond national frontiers, thus putting writers of different disciplines and countries in touch. In Italy it encouraged the circulation of excellent-quality works among both experts and the non-specialist educated public.[15]

---

13 In English, see the classic Carlo M. Cipolla, *Literacy and Development in the West* (Harmondsworth, 1969).

14 On technical education, see Carlo G. Lacaita, *Istruzione e sviluppo industriale in Italia, 1859–1914* (Florence, 1973); Carlo G. Lacaita, *L'intelligenza produttiva: Imprenditori, tecnici e operai nella Società d'Incoraggiamento d'Arti e Mestieri di Milano (1838–1988)* (Milan, 1990). On the reception of Darwinism, see Giuliano Pancaldi, *Charles Darwin: 'Storia' ed 'economia' della natura* (Florence, 1977), pp. 161–206; Giuliano Pancaldi, *Darwin in Italy: Science Across Cultural Frontiers* (Bloomington, IN, 1991); Giovanni Landucci, *Darwinismo a Firenze: Tra scienze e ideologia (1860–1900)* (Florence, 1977). On the age of positivism, see Antonio Santucci (ed.), *Scienza e filosofia nella cultura positivistica* (Milan, 1982) and Paolo Rossi (ed.), *L'età del positivismo* (Bologna, 1986).

15 Roy MacLeod, 'Evolution, Internationalism and Commercial Enterprise in Science: The International Scientific Series, 1871–1910', in Arthur Jack Meadows (ed.), *Development of Science Publishing in Europe* (Amsterdam, 1980), pp. 63–5; Leslie Howsam, 'An Experiment with Science for the Nineteenth-Century Book Trade: The International Scientific Series', *British Journal for the History of Science*, 33 (2000): 187–207. On the Italian branch, see Michele Nani, 'Editoria e culture scientifiche nell'Italia postunitaria: Appunti sulle edizioni Dumolard', *Ricerche Storiche*, 2 (1999): 257–97; Govoni, *Un pubblico per la scienza*, pp. 132–8.

In addition to this upper echelon of popularization of science and technology, often easily comparable to what was being produced north of the Alps, in Italy there was another type of popular science addressed to a wider range of readers (see Figures 2.1 and 2.2). It was a popular science of a lower standard, but one in which some clever publishers such as Emilio Treves became involved, together with professional scientists such as naturalist Michele Lessona, anthropologist Paolo Mantegazza, astronomer Giovanni Virginio Schiaparelli, geologist and abbot Antonio Stoppani, engineer Giuseppe Colombo and science professor Carolina Magistrelli, as well as (albeit seldom) professional popularizers such as Paolo Lioy.

Figure 2.1 A volume in the series Piccola Biblioteca del Popolo Italiano (Italian People's Little Library). The ideal reader, a worker, is surrounded by books, scientific and technological instruments, and the traditional '*putti*'.

As was happening almost everywhere else in Europe, to encourage the circulation of *scienza per tutti*, popularizers and publishers made use of every available medium and genre: periodicals and newspapers, encyclopedias and novels, dictionaries and 'how-to' manuals, almanacs and 'moral' works in the tradition of Samuel Smiles. And, of course, there were popular scientific conferences and evening schools for workers and artisans, with the participation of many Italian scientists.

Figure 2.2 Street vendor of almanacs and matches, 1887. In Italy, until the 1930s, colportage continued to be the most effective way to circulate popular books.

If all this was also happening in several European countries, other circumstances fostering the success of popular science in Italy were linked to the specific national context. The politics of unification, in the first place, with the exhortation to 'make Italians' – after having made Italy – and the spread of the positivism vulgate, created an ideological climate that drove many scientists to commit themselves to popularizing science: a science regarded as the driving force behind progress, modernity and hopefully the new nation. Following this, the elites in power were determined to foster an image of the close links between education and development in public opinion. That conviction dovetailed with the need to give the newly unified Italy a social and economic infrastructure comparable to that of the leading European countries. The elites' ideal of a widely diffused scientific culture coincided with the need to aid an economy that was in fact timidly progressing along the industrial road only in parts of the north of the country. The expansion of an industrial, eventually national, publishing sector took root thanks also to popular science literature. The *Fratelli Treves* (Treves Brothers), the main publishing house in Italy up until World War I, owed its strong foundations to the success that the popularization of science enjoyed between the 1870s and the 1890s. The founder, Emilio Treves, was a political refugee in Paris in the 1850s, and after his marriage to Suzette Thompson got to know the publishing market in London inside and out thanks to several sojourns in London and an everyday acquaintance with the English press. He apparently knew how to import some of the successful models of French and English popular science, both in the journal and book sectors, and adapt them to the Italian market.

If what I have briefly described so far contributed to a climate of general enthusiasm for science and 'progress', in the specific context of scientific popularization, nevertheless, several different approaches coexisted, depending on the prevailing ideas of culture and modernity.

There were militant Catholics such as geologist Antonio Stoppani, who was both a successful popularizer and the author of the book *Il dogma e le scienze positive*[16] (Dogma and the positive sciences), in which he vigorously attacked the popular science literature with a secular and evolutionist orientation. Yet there were also moderate Catholics such as Paolo Lioy. Personally involved in education and politics, Lioy was a representative of that current of Catholic opinion that wanted to find routes of mediation with secularized scientists in the name of a shared adherence to Darwinism[17] and a shared interest in a public education system based largely on the natural sciences. There were also secular figures such as zoologist Michele Lessona, who as a popularizer often worked in conjunction with his wife, Adele Masi, the ghost translator of some of Darwin's works into Italian. The kind of education the Lessonas had in mind for their readers reflects the view of an

[16] Antonio Stoppani, *Il dogma e le scienze positive, ossia la missione apologetica del clero nel moderno conflitto tra la ragione e la fede* (Milan, 1886).

[17] Stefano Bertani, *L'ascensione della modernità: Antonio Fogazzaro tra santità e evoluzionismo* (Soveria Mannelli, 2006).

educational programme in which science, the humanities and ethics were deeply interconnected. With regard to women popularizers, research has so far brought to light two: Evangelina Bottero and Carolina Magistrelli, the first two women science graduates in Italy from the University of Rome in 1881, who together wrote a manual on the recently introduced telephone.[18] Magistrelli later continued to publish science textbooks of great interest and high quality (see Figure 2.3).[19]

Figure 2.3 'Gorilla' from Carolina Magistrelli, *Elementi di zoologia* (Turin, 1888), p. 2.

[18] Evangelina Bottero and Carolina Magistrelli, *Il telefono* (Turin, 1883).

[19] Paola Govoni, 'Studiose e scrittrici di scienza tra età liberale e fascismo. Il caso Bottero e Magistrelli', in Teresa Bertilotti and Maria Pia Casalena (eds), *Esercizi di stile*, *Genesis*, 6 (2007): 65.

Finally, there were the supporters of the 'religion of science' genre, such as anticlerical anthropologist Paolo Mantegazza, who was a Member of Parliament and later of the Italian senate. His popular books on hygiene and medicine were of high quality, while others, like his anthropological works, were often phoney and dishonest and spread the seeds of sexist and racist ideas for generations of Italian readers to come. Tens of thousands of copies of some of Mantegazza's books on popular science were sold in translation and reissued for decades in several European countries and the United States. Indeed, in the peak years of the popular science movement in Italy, during the 1880s, writers like Mantegazza were the most successful.

The popularization of science, the so-called *scienza per tutti* movement, had its moment of glory in the 1870s and 1880s. In a scientifically marginal country like Italy, the popularization of science was then credited by many as the instrument capable of breaking the link between scientific backwardness and religious superstition,[20] providing a much-wanted freedom for research: the inevitable victory of science would at the very least erase the Church's insult to Galileo and to Italian science. Thus, it was in the wake of the political enthusiasm generated by unification that many Italian scientists engaged in popular science, and the educated middle classes, encouraged by publishers acquainted with international publishing markets, were the protagonists of the success – mainly in the north of the country – of popular science literature.

Yet if these were the causes of success, what then were the reasons for the decline of popular science literature, a trend that had already started by the 1890s? What stopped Italy from developing a stable tradition of popularization of science comparable to that of England, which is so often cited as a model?

To explain the decline, we should focus on two points: on the one hand, the scientists and their commitment to popularizing science, and on the other, the public and its capacity to consume popular science, keeping up a vigorous demand for it.

In Italy, the more successful authors of popular science were scientists who looked to authors such as Faraday, Lockyer, Huxley and Tyndall as models, scientists who were famous in their own right as well as popularizers emerging as prominent public figures: the 'leaders of science in London'.[21]

However, in Great Britain, unlike in Italy, for decades scientists had already realized the importance of having public opinion participate in scientific matters.

---

[20] However, the Vatican was naturally also capable of carrying out scientific projects which had a great impact on the public sphere: see Massimo Mazzotti, 'I significati della precisione. Per una storia socioculturale dell'astrofisica italiana', in Paola Govoni (ed.), *Storia, scienza e società. Ricerche sulla scienza in Italia nell'età moderna e contemporanea* (Bologna, 2006), pp. 143–73.

[21] See Ruth Barton, 'Just before *Nature*: The Purposes of Science and the Purposes of Popularization in some English Popular Science Journals of the 1860s', *Annals of Science*, 55 (1998): 1–33.

The involvement of the public in questions connected to science, education and development had been an important component in the process of turning the new figure of the scientist into a professional figure, increasingly close to positions of power. From the early Victorian age, British scientists' way of bringing common objectives before the public and the government alike had been a means for them to emerge as a new and powerful social group.[22] This did not happen in any comparable way in Italy.

These circumstances, which had led among other things to the birth of the journal *Nature* in 1869,[23] were not perceived clearly by Italian scientists, despite the fact that they frequently drew attention to the British case and example. Rather than identifying themselves as a professional group, with at least partly shared values and objectives, Italian scientists apparently preferred to side with some academic or political faction. This circumstance seems to have had a detrimental effect on research as well as on the quality of popular science publications. Italian scientists of course perceived the need to have a voice in society, but they tended to find expression more often through efforts to gain a seat in Parliament, where scientists, like Mantegazza, were comparatively numerous – seven out of one hundred in the 1880s. This did not in itself encourage the formation of either a national professional community or strong ties with public opinion. The label *scienza popolare* was in fact often used to help circulate political programmes – typically Catholic, or anti-Catholic – which did not *per se* foster the dissemination of scientific education. Another factor seems to have been even more decisive in hindering the consolidation of the popularization of science movement in nineteenth-century Italy: the pitfalls of the national general education system.[24]

Scientists' and publishers' commitment to promoting popular science literature could do little to provide the kind of education among all ranks of society that only schools, public libraries and associations for the dissemination of knowledge could guarantee. The poor results obtained by the new state on the education front, regardless of who was in power, reveal in the Italian elite, even in its intellectual

---

22 On this subject, the bibliography is especially extensive, starting from Susan F. Cannon, *Science in Culture: The Early Victorian Period* (New York, 1978); Jack Morrell and Arnold Thackray, *Gentlemen of Science. Early Years of the British Association for the Advancement of Science* (Oxford, 1981); Joseph Ben-David, *The Scientist's Role in Society: A Comparative Study* (Chicago and London, 1984); Bernard Lightman (ed.), *Victorian Science in Context* (Chicago and London, 1997); Arthur J. Meadows, *The Victorian Scientist: The Growth of a Profession* (London, 2004).

23 Roy MacLeod, 'The first issue', *Nature*, 224 (1969): 440; Roy MacLeod, 'Securing the foundations', *Nature*, 224 (1969): 443; Roy MacLeod, 'Lockyer: Editor, civil servant and man of science', *Nature*, 224 (1969): 453–6; Donald Roos, 'The "Aims and Intentions" of Nature', in James Paradis and Thomas Postlewait (eds), *Victorian Science and Victorian Values: Literary Perspectives* (New York, 1981), pp. 159–80.

24 Paola Govoni, 'The Rise and Fall of Science Communication', in Martin W. Bauer and Massimiano Bucchi (eds), *Journalism, Science and Society: Science Communication between News and Public Relations* (London, 2006), pp. 21–32.

and most progressive sectors, an inability – with few exceptions[25] – to see how crucial the education of all citizens was for a modern nation.

Thus, while by around 1900 illiteracy rates in France had dropped to 5 per cent and to 3 per cent in England, in 1901 only Italian regions such as Lombardy and Piedmont had illiteracy rates as low as 20 per cent, while Liguria, Veneto, Emilia, Tuscany and Lazio hovered at around 40 per cent, and illiteracy in other regions was as high as 70 per cent.[26] How could the diffusion of popular science beyond the elites be expected to prosper in a context like this? When the generation of enthusiastic popularizers who had taken part in the unification of the country reached the age of retirement, popular science in Italy went back to being a minority genre.

On the Usefulness of Many Tools

As already mentioned, not until the 1960s did historians of science find the popularization of science of much interest from either a historiographical or epistemological perspective. Things changed in the 1960s when in Europe and the United States reinterpretations of the relationships between science and society paved the way for new research agendas, pursuing directions seldom tried before. Important at this time was a now well-known book by Ludwik Fleck, containing statements such as the following:

> Cognition is the most socially-conditioned activity of man, and knowledge is the paramount social creation [*Gebilde*]. … Thoughts pass from one individual to another, each time a little transformed, for each individual can attach to them somewhat different associations. Strictly speaking, the receiver never understands the thoughts exactly in the way that the transmitter intended them to be understood.[27]

---

[25] These were scientists and intellectuals of great worth, but often isolated voices, for example Carlo Cattaneo and Gaetano Salvemini, or scientists such as Giuseppe Colombo and Francesco Brioschi, up to the geneticist Adriano Buzzati Traverso. In English, see Carlo G. Lacaita and Filippo Sabetti (eds), *Civilization and Democracy: The Salvemini Anthology of Cattaneo's Writings* (Toronto, 2007).

[26] Tullio De Mauro, *Storia Linguistica dell'Italia Unita* (Rome, 2001), p. 95.

[27] Ludwik Fleck, *The Genesis and Development of a Scientific Fact*, eds. T.J. Trenn and R.K. Merton, trans. F. Bradley and Trenn (Chicago and London, 1979), p. 42 [1st edn, Basle, 1935]. As is well known, the book was rediscovered thanks to its mention in Thomas Kuhn's *The Structure of Scientific Revolutions* (Chicago, 1962). Interesting for the history of the historiography of science in Italy is the anti-Kuhnian introduction by Paolo Rossi to the Italian translation of Fleck's book. Paolo Rossi, 'Ludwik Fleck e una rivoluzione immaginaria', in Ludwik Fleck, *Genesi e sviluppo di un fatto scientifico. Per una teoria dello stile e del collettivo di pensiero* (Bologna, 1983), pp. 9–42.

The reinterpretations of the processes of production and dissemination of scientific knowledge eventually resulted in a new sociology of scientific knowledge,[28] often combined with a new sensitivity towards popular culture and greater attention towards anthropological studies, all leading to the first important studies on the history of science popularization. In the 1970s, the path-breaking works of Susan Sheets-Pyenson, William Brock and Roy MacLeod, among others, shed new light on the relationships between scientists and society and raised new issues concerning the role played by science's audience.[29]

Sub-fields like the study of the publishing industry – such as by Elisabeth Eisenstein[30] – or the history of the book were important additions to the trend.[31] A contribution of special significance in this connection has been Robert Darnton's work highlighting a 'books' circular life', a useful concept for our understanding of the world of popular science. According to Darnton, the book's life:

> could be described as a communication circuit that runs from the author to the publisher (if the bookseller does not assume that role), the printer, the shipper, the bookseller, and the reader. The reader completes the circuit because he [sic] influences the author both before and after the act of composition. Authors are readers themselves.[32]

---

28 Jan Golinski, *Making Natural Knowledge: Constructivism and the History of Science* (Chicago, 2005; 1st edn, Cambridge, 1998).

29 The main contributions by Brock and MacLeod have been included in William H. Brock, *Science for All: Studies in the History of Victorian Science and Education* (Aldershot, 1996); Roy MacLeod, *Public Science and Public Policy in Victorian England* (Aldershot, 1996); Roy MacLeod, *The 'Creed of Science' in Victorian England* (Aldershot, 2000). See also: Steven Shapin, 'The Audience for Science in Eighteenth Century Edinburgh', *History of Science,* 12 (1974): 95–121; Gerald Holton and William A. Blanpied (eds), *Science and Its Public: The Changing Relationship* (Dordrecht, 1976); Philippe Roqueplo, *Le partage du savoir: Science, culture, vulgarisation* (Paris, 1974); David Knight, *The Age of Science: The Scientific World-View in the Nineteenth Century* (Oxford, 1986).

30 Elisabeth Eisenstein, *The Printing Press as an Agent of Change*, 2 vols (Cambridge, 1979). See also: Arthur J. Meadows (ed.), *Development of Science Publishing in Europe* (Amsterdam, 1980); Adrian Johns, *The Nature of the Book: Print and Knowledge in the Making* (Chicago, 1998); Marina Frasca Spada and Nick Jardine (eds), *Books and Sciences in History* (Cambridge, 2000).

31 Guglielmo Cavallo and Roger Chartier (eds), *A History of Reading in the West* (Amherst, 2003) (1st edn, Rome-Bari, 1995). See Ann Blair, Jonathan R. Topham and Lorraine Daston, 'Scientific reading', *Isis*, 95 (2004): 420–48.

32 Robert Darnton, 'What is the History of Books?', in Darnton, *The Kiss of Lamourette: Reflections in Cultural History* (New York, 1990), pp. 111 (originally in *Daedalus*, 1982). A basic text is Richard D. Altick, *The English Common Reader: A Social History of the Mass Reading Public, 1800–1900* (Chicago, 1957).

It was around 1985 that the collaboration between historians and sociologists of science – mainly writing in English or French – led for the first time to an in-depth exploration of the popularization of science. It was the beginning of a research tradition that would soon yield fertile results and render definitively obsolete what has since been known as the 'conventional view' of the popularization of science. Richard Whitley, co-editor with Terry Shinn of *Expository Science*, described it as follows:

> In the conventional view, scientific knowledge is disseminated to a lay audience after it has been discovered and this process is separated from research. Since the scientific community is autonomous and distinct from the general public, the latter's acquisition of translated knowledge cannot affect the production and validation of new knowledge. Thus feedback from popularisation to scientific research is non-existent.[33]

With the publication of books by Bruno Latour, especially *Science in Action* and *The Pasteurization of France*,[34] the 'traditional view' of the popularization of science was further superseded. Awareness has grown that the production, circulation and reception of knowledge are not independent of each other.

With the emergence of social studies of science tradition, based on the study of primary sources, the question of popular science has been fully addressed from a historical perspective. The works by Bernadette Bensaude-Vincent and Christine Blondel on nineteenth- and twentieth-century France, and by James Secord on Victorian England, were among the most significant in the field from the end of the 1980s through the 1990s. In the new research tradition, landmark publications were a special 1988 issue on science popularization in the French journal *Cahiers d'Histoire et de Philosophie des Sciences*, another 1989 special issue of *Romantisme*, followed in 1994 by a special issue of *History of Science*.[35] Books devoted to case studies, such as the impact of phrenology in nineteenth-century England (Roger Cooter), the diffusion of Newtonian physics in the eighteenth century (Larry Stewart) or chemistry around 1800 (Jan Golinski), [36]

---

[33] Richard Whitley, 'Knowledge Producers and Knowledge Acquirers: Popularisation as a Relation Between Scientific Fields and Their Publics', in Shinn and Whitley (eds), *Expository Science*, p. 8.

[34] Bruno Latour, *The Pasteurization of France* (Cambridge, MA, 1988); Bruno Latour, *Science in Action. How to Follow Scientists and Engineers Through Society* (Milton Keynes, 1987).

[35] Bernadette Bensaude-Vincent and Christine Blondel (eds) 'Vulgariser les sciences, 1919–1939: acteurs, projets, enjeux', *Cahiers d'Histoire et de Philosophie des Sciences*, 24 (1988): 33–44; Bensaude-Vincent (ed.), 'Sciences pour tous', *Romantisme*, 65 (1989): 93–104.

[36] Roger Cooter, *The Cultural Meaning of Popular Science*; Larry Stewart, *The Rise of Public Science: Rhetoric, Technology, and Natural Philosophy in Newtonian Britain,*

provided models that could be adopted when following scientists and science in other social contexts as well. A brilliant example of how the new awareness of the relationships between science and society could be used to revive biography was set by Adrian Desmond and James Moore with their *Darwin*.[37]

Another sector affected was the history of medicine, which Roy Porter studied so effectively. Medicine as seen from the point of view of the patients, the history of quacks and the cures offered as alternatives to those approved by academies, and the history of the doctor–patient relationship provided new perspectives on science and medicine, and a model for scholars working on the history of popular science.[38]

Finally, one other historiographical perspective favoured a greater understanding of the popularization of science and history *tout court*: studies on gender and science. I shall restrict myself to mentioning Evelyn Fox Keller with her *Reflections on Gender and Science* (1985):

> [The studies on gender and science came] into being with the meeting of two apparently independent developments in recent scholarship: feminist theory and the social studies of science. The second has changed our thinking about the relations between science and society – without, however, considering the role of gender – and the first has changed our thinking about the relation between gender and society but has been only peripherally concerned with science. As productive as each of these developments has been in its own terms, each leaves critical gaps in our understanding that the other can help to fill. Furthermore, their conjunction enables us to identify the critical role of gender ideology in mediating between science and social forms.[39]

1660–1750 (Cambridge, 1992); Jan Golinski, *Science as Public Culture: Chemistry and Enlightenment in Britain, 1760–1820* (Cambridge, 1992).

37 Adrian Desmond and James Moore, *Darwin* (New York, 1992).

38 Among the most interesting books from the 1970s and 1980s, see John Woodward and David Richards (eds), *Health Care and Popular Medicine in Nineteenth-Century England: Essays in the Social History of Medicine* (London, 1977); Roy Porter (ed.), *Patients and Practitioners: Lay Perceptions of Medicine in Pre-Industrial Society* (Cambridge, 1985); William F. Bynum and Roy Porter (eds), *Medical Fringe and Medical Orthodoxy, 1750–1850* (London, 1987); Roger Cooter, *Studies in the History of Alternative Medicine* (London, 1988); Matthew Ramsey, *Professional and Popular Medicine in France, 1770–1830: The Social World of Medical Practic*e (Cambridge, 1988); Roy Porter, *Health for Sale: Quackery in England, 1660–1850* (Manchester, 1989); Dorothy and Roy Porter, *Patient's Progress: Doctors and Doctoring in Eighteenth-Century England* (Stanford, CA, 1989); Roy Porter (ed.), *The Popularization of Medicine, 1650–1850.*

39 Evelyn Fox Keller, *Reflections on Gender and Science* (New Haven, 1985), p. 4.

In my own personal experience, Fox Keller's analysis (together with that of several other scholars, of course)[40] helped me to 'see' and recognize for the first time a number of certain neglected social actors, whether women or rarely considered audiences, authors or science publishers. It helped me to understand, incidentally, why in Italy until recently there were so few studies on gender and science, as well as on science and society: without a social studies of science approach, such subjects tend to remain invisible. The literature on gender and science helped historians of science to realize the importance of inquiry into the social construction of notions such as race[41] and gender. Race was a crucial category when popularizing topics like evolutionary and anthropological theories, which played a vital role all around Europe and the United States, as 1951 Milosz's insight reminds us, and as the Italian (and international) fame of Mantegazza in the nineteenth century demonstrated.

Despite the seminal works mentioned so far, it took many years before historians of science approached the history of popular science systematically rather than in piecemeal fashion, aiming to understand it as a major aspect of science rather than as a separate literary genre. The idea that the popularization of science was a by-product of science has continued to keep historians away from the genre until recently. The situation has undergone a substantial change only since around 2004 with the publications of several books, mainly on the English cultural context.[42]

---

[40] Here there is room only to cite a few of the classics, such as: Carolyn Merchant, *The Death of Nature: Women, Ecology and the Scientific Revolution* (San Francisco, 1980); Ruth Swartz Cowan, *More Work for Mother: The Ironies of Household Technology from the Open Hearth to the Microwave* (New York, 1983); Margaret Rossiter, *Women Scientists in America: Struggles and Strategies to 1940* (Baltimore, 1982); Margaret Rossiter, *Women Scientists in America. Before Affirmative Action, 1940–1972* (Baltimore, 1995); Pnina G. Abir-Am, Dorinda Outram (eds), *Uneasy Careers and Intimate Lives: Women in Science, 1789–1979* (New Brunswick, 1987); Londa Schiebinger, *The Mind Has No Sex? Women in the Origins of Modern Science* (Cambridge, MA, 1989); Ludmilla Jordanova, *Sexual Visions: Images of Gender in Science and Medicine between the Eighteenth and the Twentieth Centuries* (London, 1989); David F. Noble, *A World Without Women: The Christian Clerical Culture of Western Science* (New York, 1993).

[41] Cynthia Eagle Russett, *Sexual Science: The Victorian Construction of Womanhood* (Cambridge, 1989).

[42] See G. Cantor, G. Dawson, G. Gooday, R. Noakes, S. Shuttleworth and J.R. Topham (eds), *Science in the Nineteenth-Century Periodical: Reading the Magazine of Nature* (Cambridge, 2004); L. Henson, G. Cantor, G. Dawson, R. Noakes, S. Shuttleworth and J.R. Topham (eds.), *Culture and Science in the Nineteenth-Century Media*, (Aldershot, 2004); G. Cantor and S. Shuttleworth (eds), *Science Serialized: Representations of the Sciences in Nineteenth-Century Periodicals* (Cambridge and New York, 2004); Aileen Fyfe, *Science and Salvation: Evangelicals and Popular Science Publishing in Victorian Britain* (Chicago and London, 2004); Bernard Lightman, *Victorian Popularizers of Science: Designing Nature For New Audiences* (Chicago, 2007); A. Fyfe and B. Lightman (eds), *Science in the Marketplace: Nineteenth-Century Sites and Experiences* (Chicago, 2007).

Still lacking are works comparing the production of popular science at an international level and over a long period, this in fact being a limitation in the history of science in general.[43] We do not have sufficiently broad, comparative, long-term reconstructions enabling us to verify the migration of scientific ideas in the popular press of different countries, or the public's different reactions to the same scientific topics. Just to cite one example, the production of history of science literature on the Enlightenment has considerably increased in recent years, yet there is no study comparing the *international* circulation of authors such as the French Fontenelle, the Italian Algarotti and the Englishman Henry Baker. Another kind of gap concerns the production of popular science by women, which in the nineteenth century was an important phenomenon both in the USA and all over Europe. Margaret Rossiter once remarked that 'probably the most widespread (and easily documented) aid to women's informal education in science in the early to mid-nineteenth century were the popular books and textbooks designed for female readers'.[44] Those books were often written by women, but generally speaking women popularizers are considered as cases apart. Typically, we have on the one hand the history of science, which means overwhelmingly the history of *men* in science (or men in popular science), and on the other we have the history of *women* in science (or women in popular science).

A history of the popularization of science would indeed seem an ideal place to restore features of science that for a long time have been kept separate from their original unity. Let us take as an example the history of the English weekly journal *Nature*. (The lack of a major study on *Nature* is another gap that continues to amaze me). *Nature* was founded by scientists – indeed some of the most prominent ones in Victorian Britain – who were also famous popularizers, and the journal was simultaneously a frequent milieu for amateurs. A history of *Nature* should very likely rely on primary sources somehow combining lab protocols and scientific correspondence, popular articles and daily politics, as well as the trade correspondence between scientists, publishers, the editor, authors and readers. *Nature* was founded at a time when the first colleges for women opened in England: years in which the debate on women's education – just like the education of the masses, including black people – also found room in the journal. This is a case study of obvious interest due to the influence

---

43 *The Big Picture*, a special issue of the *British Journal for the History of Science*, 26/4 (1993); John V. Pickstone, *Ways of Knowing: A New History of Science, Technology and Medicine* (Manchester, 2000); Lewis Pyenson, 'Comparative History of Science', *History of Science*, 40 (March 2002): 1–33.

44 Rossiter, *Women Scientists in America*, p. 3; Ann B. Shteir, 'Green-Stocking or Blue? Science in Three Women's Magazines, 1800–50', in L. Henson et al. (eds) *Culture and Science in the Nineteenth-Century Media*, pp. 3–13; Ann B. Shteir, '"Let Us Examine the Flower": Botany in Women's Magazines, 1800–1830', in Cantor and Shuttleworth (eds), *Science Serialized*, pp. 17–36.

it had on other countries[45] and the unique success of its formula. It was a formula that – with adjustments – still works today, based on a tricky balance required for addressing several different audiences: scientists and general readers, civil servants and policymakers, business people, science writers and journalists, politicians, historians of science.

Let me also stress the importance of a *longue durée* history of the popularization of science which could compare different national contexts. Studies adopting this approach should of course combine the 'close-ups with the long shots',[46] and could perhaps provide additional perspectives on elusive categories like 'centre' and 'periphery'. These were socially constructed during the nineteenth and twentieth centuries, but when examining the circulation of scientific knowledge, these same categories should be re-examined each time according to circumstance and context. For example: the Italian production of works on eugenics by nineteenth-century authors such as Cesare Lombroso[47] and Paolo Mantegazza – at both a research and a popular level – gave Italy a 'central' role in the international scene until the 1910s and 1920s, when many other countries were building the most extreme forms of nationalism based on racial issues. We do need big pictures, long-term histories and, above all, comparative histories if we want to understand phenomena like these. The tool enabling us to understand popular science – and its evolution over time – in individual national realities, is above all history, a trans-national, trans-disciplinary and trans-historiographical history of science.

On the Importance of Diversity

My own reconstruction of the Italian case seems to confirm that, between the nineteenth and twentieth centuries, the models of scientific popularization migrated from one country to another thanks to publishers, writers and readers

[45] Paola Govoni, '*Nature* à l'Italienne. La Presse de Science Populaire en Italie à la fin du XIXe Siècle', in Bensaude-Vincent and Rasmussen (eds), *La science populaire dans la presse et l'édition*, pp. 175–89; Paola Govoni, *Un pubblico per la scienza*, chapter 6.

[46] I am referring to the important historiographical reflections of Gianna Pomata in 'Close-ups and long shots: Combining particular and general in writing the histories of women and men', in Hans Medick and Anne-Charlott Trepp (eds), *Geschlechtergeschichte und Allgemeine Geschichte: Herausforderungen und Perspektiven* (Göttingen, 1998), pp. 101–24.

[47] For recent translations of two of Lombroso's books, see Cesare Lombroso and Guglielmo Ferrero, *Criminal Woman, the Prostitute, and the Normal Woman*, translated and with a new introduction by Nicole Hahn Rafter and Mary Gibson (Durham, NC, 2004); Cesare Lombroso, *Criminal Man*, translated and with a new introduction by Mary Gibson and Nicole Hahn Rafter (Durham, NC, 2006). On Lombroso see Mary Gibson, *Cesare Lombroso and the Origins of Biological Criminology* (London, 2002).

aware of innovative publishing in the sciences. Science itself, it is often said, is an enterprise that largely goes beyond national frontiers; yet the way the public receives it and the way science works its way into diverse social contexts are obviously linked to national (and even regional) cultural, economic and institutional traditions. To fail to take this into account may lead those trying to reconstruct the history of science popularization to fall into the same errors as the Italian popularizers and publishers I have been studying. Applying models originated abroad to a certain national context may mean carrying out an operation comparable to that attempted by science popularizers in post-unification Italy when, taking French or English popularization as a model, scientists and publishers attempted to transplant it into Italy, often ignoring the social context in which 'science for all' would act.

The STEP Minorca meeting where this book originated reflects a wealth of samples of research on the popularization of science, technology and medicine in national contexts, which only a few years back would have been unimaginable.[48] Now – after the end of the 'Science Wars'[49] and the tallies of the fallen – I think we are ready to admit that the history of science is rich in an extraordinary variety of points of view, even on a long-neglected subject like popular science. By emphasizing the different historiographical approaches, I mean that – just as in evolutionary biology – *diversity* is a strength rather than a weakness of the history of science. As I see it today, the history of science analyses with equal legitimacy intellectual and technical questions, laboratory practice and field work, as well as the social and public dimensions of science. Over recent years the history of science has been increasingly characterized by a plurality of voices, leading us away from the notion of a discipline relying on a single dominant model, but rather on the (difficult, but possible) coexistence between different research traditions.[50] The history of science has thus achieved both institutional status and scientific maturity in many countries. The history of the popularization of science seems

[48] Recalling the expression 'in context', I would like to mention the important studies edited by Roy Porter and Mikuláš Teich, starting with *The Enlightenment in National Context* (Cambridge, 1981). For a complete list of the books as well as the history of the enterprise, see Mikuláš Teich, 'How it all began: From the Enlightenment in national context to revolution in history', *History of Science*, 41 (2003): 335–43.

[49] A useful anthology providing an overview of the so-called Science Wars is Keith Parsons (ed.), *The Science Wars: Debating Scientific Knowledge and Technology* (New York, 2003).

[50] Proposals allowing us to overcome some of the tensions have been indicated in Lorraine Daston, 'Introduction: The coming into being of scientific object', in Daston (ed.), *Biographies of Scientific Objects* (Chicago, 2000), pp. 1–14; Giuliano Pancaldi, *Volta: Science and Culture in the Age of Enlightenment* (Princeton, 2003), pp. 273–89; Terry Shinn and Pascal Ragouet, *Controverses sur la science. Pour une sociologie transversaliste de l'activité scientifique* (Paris, 2005), chapter 3.

like an ideal topic for achieving another goal sorely needed: to set up a dialogue between professional historians of science and the general reader, while at the same time opening up new avenues of communication with scientists in the post-Public-Understanding-of-Science age.[51]

[51] For different points of view on the subject, see: David P. Miller, 'The "Sobel Effect": The Amazing Tale of How Multitudes of Popular Writers Pinched All the Best Stories in the History of Science and Became Rich and Famous while Historians Languished in Accustomed Poverty and Obscurity, and How This Transformed the World. A Reflection on a Publishing Phenomenon', *Metascience*, 2 (2002): 185–200; Paola Govoni, 'Historians of Science and the "Sobel Effect"', *Journal of Science Communication*, 4 (2005): 1–17; Peter Bowler, 'Experts and Publishers: Writing Popular Science in Early Twentieth-Century Britain, Writing Popular History of Science Now', *British Journal of the History of Science*, 39 (2006): 159–87; Frederic D. Schwarz, 'We should all be friends', *Technology and Culture*, 48 (2007): 407–10.

Chapter 3
Women and the Popularization of Botany in Early Nineteenth-Century Portugal: The Marquise of Alorna's *Botanical Recreations*

Palmira Fontes da Costa[1]

Poets want either to please or to instruct;
And mostly please and instruct together.

Marquise of Alorna, *Poetical Works*, vol. 5, 1844

A number of studies have shown that during the late eighteenth century and the first half of the nineteenth century, botany become popular in Britain and France.[2] The historiography on the subject has pointed out that this field of knowledge became particularly fashionable among the upper classes. However, some studies have revealed that, at least in Britain, interest in botany extended to an extremely wide range of participants from all strata of society.[3] It has also been shown that it particularly attracted the attention and involvement of women, not only as readers

[1] I would like to thank Fernando Mascarenhas for his interest in this project, and his generosity in allowing me to use some of the illustrations presented in this work and consult some of the original manuscripts at the Fundação das Casas de Fronteira e Alorna, Lisbon. My thanks also to Vanda Anastácio and Teresa Sousa de Almeida for their encouragement and very helpful suggestions; to Francesca Rayner for her linguistic revision of the chapter and translations of original quotations; to Christopher Auretta for his kind assistance with translations of excerpts from the poem *Botanical Recreations*; and to Fernando Reis for providing me with information on scientific periodicals. I would also like to thank the editors for their very helpful comments and suggestions. My thanks as well to the librarians of Arquivos Nacionais Torre do Tombo and the Biblioteca Nacional, Lisbon for their kind assistance.

[2] Ann B. Shteir, *Cultivating Women, Cultivating Science: Flora's Daughters and Botany in England, 1760 to 1860* (Baltimore, 1996); Greg Myers, 'Science for women and children: the dialogue of popular science in the nineteenth century', in J. Christie and S. Shuttleworth (eds), *Nature Transfigured: Science and Literature, 1700–1900* (Manchester, 1989), pp. 171–200; Roger L. Williams, *Botanophilia in Eighteenth-Century France. The Spirit of the Enlightenment* (Dordrecht, 2001); Pascal Duris, *Linné et la France* (1780–1850) (Geneva, 1993).

[3] David Allen, *The Naturalist in Britain: A Social History* (London, 1976); Anne Secord, 'Science in the Pub: Artisan Botanists in Early-Nineteenth Lancashire', *History of Science*, 32 (1994): 269–315.

of literature on the subject, but also as participants in the botanical activities of gardening, collecting and classification.[4] Moreover, women were also authors of important works aimed at promoting the cultivation of botany, and some of these works were specifically targeted at a female audience. One of the questions raised by the involvement of women, as both authors and a privileged audience, is the implication of this involvement in the kind of popularization of science that took place in the period. In particular, the literature written for women points to the problem of how scientific culture was presented to those presumed to exist outside it.

Did the general interest in botany reach countries usually considered to be outside the European centres of learning, such as Portugal? If so, were women also involved in its popularization? This chapter focuses on the poetical work *Botanical Recreations* [*Recreações Botânicas*], which was written at the beginning of the nineteenth century by the Marquise of Alorna, author of a significant body of poetry, as well as an essayist, translator and painter. The work was not only written by a woman, it was also dedicated to Portuguese women. Why did the author favour a female audience for her work, and what kind of female audience did she have in mind? In which ways was botanical knowledge appropriated in *Botanical Recreations*? Who was receptive to the work and how did it circulate?

Popularization processes do not simply transmit the main ideas of the scientific community; they also transform scientific knowledge as they relocate it in new textual forms and relate it to other elements from non-scientific culture. As I will show in this chapter, this is particularly evident in the poetical work *Botanical Recreations*. Furthermore, some authors have expressed doubts about the utility of the expression 'popularization of science' since it is difficult to conceive of this term as invoking a uniform or universal process.[5] Instead, as James Secord has remarked, what is crucial is to understand science as a form of communication and to analyse why and how scientific knowledge circulates, as well as to whom it is available.[6]

We also need to be responsive to a greater plurality of venues for the circulation of scientific knowledge. The analysis of the audience of *Botanical Recreations*, a work that was finished in 1813 but only published in 1844, after the author's death, points to other important locations for the circulation of natural knowledge besides the printed text, such as salons and personal correspondence.[7]

---

[4] On the various activities associated with botany in this period, see Nick Jardine, James Secord and Emma Spary (eds), *The Cultures of Natural History* (Cambridge, 1996).

[5] Roger Cooter and Stephen Pumfrey, 'Separate Spheres and Public Places: Reflections on the History of Science Popularization and Science in Popular Culture?', *History of Science*, 32 (1994): 237–67, p. 239; James Secord, 'Knowledge in Transit', *Isis*, 95 (2004): 654–72, p. 670.

[6] Secord, ibid.

[7] Marquesa de Alorna, *Obras Poéticas*, 6 vols (Lisbon, 1844–51). The work was compiled by the Marquise of Alorna's two daughters, Frederica d'Oeynhausen and Juliana, Countess of Stroganoff.

At the same time, there is also a need to be aware of the controversial nature of the bipolar distinction between centre and periphery. As Kostas Gavroglu has remarked, this division is

> incapable of capturing many salient details. There are, first of all, many centres and many peripheries. Moreover, and depending on the subject one is discussing, a place may be both centre and periphery. A centre may, over time, change into a periphery, and *vice-versa*. And a single country may contain both centres and peripheries, thereby making purely national distinctions of no use.[8]

Moreover, the existence of networks of social relationships between men and women from various countries resulted in an interchange of ideas and products that was not only restricted to one socio-geographic space. The distinction between centre and periphery is especially blurred if we consider the upper classes of the *Ancien Régime* who could travel abroad regularly and maintained regular contact with foreign members of their social network.[9]

Peripheries of learning can be defined not only in terms of their lesser role in the production of knowledge, but also in terms of the hindrance or even opposition to the circulation of new results, thoughts and ideas. For a long period of time, Portugal was subject to print censorship.[10] One of the consequences of this was that during the eighteenth century, the circulation of works by Enlightenment authors such as Voltaire (1694–1778), Julien Offray de La Mettrie (1709–51), Jean-Jacques Rousseau (1712–78) and Denis Diderot (1713–84), was strictly forbidden.[11] The dissemination of scientific works was also affected. Works arguing in favour of the heliocentric system were forbidden until almost the end of the eighteenth century.[12] Additionally, works that were openly in favour of Newtonianism only started to be authorized for publication in the latter part of the eighteenth century.[13] In addition,

---

[8] Kostas Gavroglu, 'The Centre from the Periphery: Appropriating Scientific Ideas during the Eighteenth Century', *Revue de la Maison Française d'Oxford*, 1 (2003): 11–32, p. 11.

[9] By upper classes, I mean noblewomen or women accorded noble treatment, the latter from a bourgeois background, seeking to buy or have access through marriage to titles that could safeguard their socio-economic position.

[10] Print censorship was introduced in Portugal in 1536 and lasted until 1821. It underwent some changes in structure in 1768 as a result of Pombal's nomination as Prime Minister, and when Queen Maria I became heir to the throne in 1777. On Portuguese literary censorship, see Maria Teresa Martins, *A Censura Literária em Portugal nos séculos XVII e XVIII* (Lisbon, 2005).

[11] Martins, *A Censura Literária em Portugal nos séculos XVII e XVIII*, pp. 393–540.

[12] Ibid., pp. 294–5.

[13] Hélio Pinto and Palmira Fontes da Costa, 'The Diffusion of Newtonianism in Eighteenth-Century Portugal', *Acta historiae rerum naturalium necon technicarum*, 8 (2004): 152–9.

some of the censors considered medical works very dangerous since they could easily accommodate and provoke materialistic ideas.[14] In the context of this article, it is important to stress that print censorship had a lower effect on the Portuguese upper classes. Not only could they buy some of the forbidden books when they travelled abroad, but some members of the clergy and the aristocracy were also able to secure special licences to acquire censored books.[15] With regard to the European upper classes, at least, this further calls into question a clear distinction between centre and periphery. As we shall see in the next section, this was of paramount importance in the education of the author of *Botanical Recreations*.

The Marquise of Alorna

Leonor de Almeida Portugal Lorena e Lencastre (1750–1839), the fourth Marquise of Alorna, had a long and turbulent life.[16] Daughter of a nobleman accused of conspiring to kill King José I (1714–77), her family was severely persecuted by the Marquis of Pombal (1699–1782), and she was imprisoned between 1758 and 1777 in the Chelas Convent near Lisbon, together with her mother and sister.[17]

This period of captivity coincided with her formative years, since she was eight years old when she was imprisoned and almost 27 when she was released. It was in this period of containment that she was able to acquire a solid education in classical and modern languages, literature and the sciences. The success of her education was certainly linked to the fact that, from an early age, Leonor was an avid and eclectic reader. Her letters to her father while she was in the Chelas Convent give an indication of her wide-ranging interests in literature, philosophy, natural philosophy and natural history.[18] Her innate inquisitiveness and her desire to improve herself beyond the restrictive education provided by the convent stimulated her various readings on these subjects. Her confinement was also important in the sense that, as she confessed to her father, 'Moderate study is the most certain delight one can choose in solitude'.[19]

---

[14] Graça Maria Rodrigues, *Breve História da Censura em Portugal* (Lisbon, 1980), p. 49.

[15] Vanda Anastácio, 'Perigos do livro', *Românica: Revista de Literatura*, 13 (2004): 125–42, p. 126.

[16] The only biographies of the Marquise of Alorna are: Hernani Cidade, *A Marquesa de Alorna: sua vida e obras* (Porto, 1930); Olga Morais Sarmento da Silveira, *Mulheres Ilustres: A Marquesa de Alorna* (Lisbon, 1941).

[17] During this period, her father was also imprisoned in the Junqueira Fort.

[18] On the letters sent by the future Marquise of Alorna while she was confined in Chelas, see Vanda Anastácio, 'D. Leonor de Almeida Portugal: As Cartas de Chelas', in V. Anastácio (ed.), *Correspondências. Usos da Carta no século XVIII* (Lisbon, 2005), pp. 45–53.

[19] Hernani Cidade (ed.), *Marquesa de Alorna, Poesias* (Lisbon, 1941), p. x.

Her letters to her father during this period also provide evidence that Leonor read and admired some of the most innovative and controversial authors of the period. She considered Voltaire the 'oracle of taste for wise men of all times' and, in her view, Rousseau was 'after Voltaire, the great philosophical genius'.[20] She found Jean-Baptiste d'Alembert (1717–83) 'the most amiable of characters', calling him 'reason itself'.[21] She was an admirer of Diderot and considered his works *The System of Nature* [*Système de la nature*] (1770) and *The Social System* [*Système social*] (1773) 'the two most celebrated works that have appeared this century'.[22] Leonor was also a reader of Georges Buffon (1707–88) and announced in one of her letters to her father that a supplementary volume of his *Natural History* [*Histoire naturelle*] had just been published.[23] She also read recent and innovative Portuguese publications, such as Luís António Verney's (1713–92) *True Method of Study* [*O verdadeiro método de estudar*] (1746), an influential work promoting pedagogical reform in Portugal in line with Enlightenment ideals, which attributed an important role to science.[24] There is also evidence that she read the first volumes of Teodoro de Almeida's (1722–1804) *Philosophical Recreations* [*Recriações filosóficas*] (1751–1800), an extremely important work in the popularization of science in eighteenth-century Portugal.[25]

As previously mentioned, several of these authors were still censored in Portugal. Nevertheless, although in prison, the future Marquise of Alorna was still able to gain access to them. It seems that Leonor's father had a licence to obtain censored literature, which enabled her to have access to these and other forbidden books. In letters to him, Leonor expressed a desire to obtain a licence for herself. At one point, she asked a relative, the Count of São Miguel, to acquire one on her behalf.[26] The Inquisition refused this request, remarking that no woman had ever dared to make such a request before.[27]

At the time, the teaching of Newton's ideas was still forbidden in Portugal, but this was not an impediment to Leonor's introduction to them. Moreover, in a

---

20 Marquês de Ávila e Bolama, *A Marquesa d'Alorna* (Lisbon, 1916), pp. 101–2.

21 Ibid., p. 102.

22 Both *The System of Nature* and *The Social System* were in fact works by Baron d'Holbach to which Diderot had contributed. On the Marquise of Alorna's assessment of the Encyclopedists, see Maria Helena Vilas-Boas e Alvim, 'A Marquesa de Alorna – De Defensora das Luzes a Agente Contra-Revolucionária', *Revista de História das Ideias*, 10 (1988): 265–76.

23 Bolama, *A Marquesa d'Alorna*, p. 101.

24 Cidade, *A Marquesa de Alorna, A sua vida e obras*, p. 22. On the importance of Verney in eighteenth-century Portugal, see António Alberto B. Andrade, *Verney e a projecção da sua obra* (Lisbon, 1980).

25 On the importance of Teodoro de Almeida in the popularization of science see, Francisco Contente Domingues, *Ilustração e Catolicismo: Teodoro de Almeida* (Lisbon, 1994).

26 Anastácio, 'Perigos do livro', p. 127.

27 Ibid., p. 126.

dispute with her confessor, she maintained that 'Newton and other heretics moved her in their careful search for the laws of nature', arguing that 'in science, the words of a wise heretic are worth more than those of an ignorant saint'.[28]

Leonor's desire to expand her intellectual horizons, combined with her independent and assertive personality and aristocratic origins, promoted her readings of forbidden literature. Nevertheless, during the period, there was still the belief that reading in itself could be dangerous to the health of women.[29] This view seems to have been shared by Leonor's father, who worried whenever he received a long letter from his daughter.[30] The same can be said of her doctor, who, each time she became indisposed, restricted the time she could spend reading.[31]

On special occasions, described as *Outeiros*, Leonor's confinement was attenuated. Various poets visited the Chelas Convent to read and compose verses dedicated to the nuns and other persons in the convent, who could hear and see them through the bars.[32] Leonor wrote her first poems when she was in the convent, and some of them become known through such events. It was during one of them that the distinguished poet Filinto Elísio (1734–1819), who often participated in the *Outeiros*, attributed to her the name 'Alcipe', which was to become associated with her poetical persona throughout her life.

With the death of King José I in 1777 and the succession of his daughter Queen Maria I, Pombal was dismissed and Leonor was finally released from prison. Afterwards, she frequented the court and enjoyed a close relationship with the queen. Following Leonor's marriage to the Count of Oeynhausen in 1779, she used her influence with the queen to obtain for her husband the post of Portuguese representative at the court of Vienna. On her journey to Austria, she travelled through Madrid and then Paris, where she had the chance to experience the atmosphere of the French salons and to meet Anne-Louise Germaine Necker (1766–1817), the future Mme de Staël, who became one of her acquaintances.[33] She lived in Vienna for four years, where she was in touch with various members of the aristocracy and cultural elite, including the Archduchess of Austria, Maria Theresa (1717–80). It was at this time that she sent her father her self-portrait 'Solitude' (Figure 3.1).

---

[28] Quoted in Cidade, *A Marquesa de Alorna, A sua vida e obras*, p. 25.

[29] Robert Darnton, 'History of Reading', in Peter Burke (ed.), *New Perspectives on Historical Writing* (Oxford, 1991), p. 152.

[30] Arquivo Palácio Fronteira (Lisbon), Mss. PAIALC101.

[31] Anastácio, 'Perigos do livro', p. 131.

[32] 'Biographical Note' in Alorna, *Obras Poéticas*, I, p. xvii.

[33] On the relationship between the Marquise of Alorna and Madame De Staël, see Maria Manuela Delile, 'Alcipe e Madame de Staël: entre admiração e discordância', in Anibal Pinto de Castro, José Esteves Pereira, Maria Manuela Delille and Teresa Sousa de Almeida (eds), *Alcipe e as Luzes* (Lisbon, 2003).

Figure 3.1 Self-portrait by the Marquise of Alorna entitled 'Solitude'. Palácio Fronteira e Alorna.

Leonor was supportive of the *Ancien Régime* and strongly anti-Napoleonic.[34] Fearing that the French would invade Portugal, she left for Britain in 1802.[35] The French invasions took place between 1807 and 1811, but she remained in Britain for two more years. Her sojourn in Britain gave the future Marquise of Alorna the opportunity to broaden her intellectual horizons by coming into contact with contemporary authors, some of whose works she translated, such as James Thompson's (1700–48) *The Seasons* (1726–30) and Alexander Pope's (1688–1744) *An Essay on Criticism* (1711).

When living in London, she frequented the house of the Portuguese ambassador, Domingos de Souza Coutinho.[36] It was during one of these meetings that she met

[34] Elias J. Torres Feijó, '*Ad maiorem gloriam ... feminae*: Enlightened Women and the Introduction of Models in Portugal during the Second Half of the Eighteenth Century', *Portuguese Studies*, 20 (2004): 73–88, pp. 80–81.

[35] She was a widow when she left for Britain, her husband having died in 1794.

[36] 'Biographical Note', p. xxxiv.

the poet Vicente Nolasco da Cunha, who at the time was translating Erasmus Darwin's poem *The Botanic Garden. Part I. Containing the Economy of Vegetation* (1791) into Portuguese. Botany was clearly the scientific subject most favoured by the future Marquise of Alorna. The importance she attributed to this field of knowledge was genuine and akin to her poetic inclination, in which proximity to nature played a prominent role. However, her interest in the subject seems to have increased during the time she spent in Britain, and it is probably no coincidence that she wrote *Botanical Recreations* in the country that was well known for the popularization of botany. The poem was written away from the turbulence of city life, as by then the future Marquise had moved to Gloucestershire, near Wales, where she rented a cottage.[37]

After the death of her brother on 2 January 1813, Leonor returned to Portugal to become heir to a title and the family properties. The Marquise of Alorna continued her translation projects, which included Claudian's *The Capture of Prosperine* and Hughes de Lamennais's (1782–1854) *Essay on Religious Indifference* (1817–23). She also wrote a *Paraphrase of the Psalms*. These latter two works testify to her strong religious convictions. During this period, she was also influential in Portuguese society through the salons that she organized in her palace in Lisbon, in which various poets and some scientists participated.[38]

The Marquise of Alorna's interest in scientific matters continued throughout her life. When she was almost ninety, she attended Luís Mousinho de Albuquerque's (1792–1847) courses on physics and chemistry at the Royal Mint [Casa da Moeda].[39] Her poem 'Epistle to Godefredo', written when she was elderly, is an apologia for the study of scientific doctrines 'which bring a hundred treasures to the human mind'.[40] It praises the scientific value and moral virtues of not just botany, but also geology, chemistry, physics, crystallography and mineralogy. Moreover, in this epistle, the Marquise asserts that all who ignore scientific knowledge 'are impious, rebellious and idiotic' and that science is crucial in the eradication of superstition, which will lead to 'the triumph of truth and pure reason'.[41]

---

[37] In one of the notes in the *Botanical Recreations*, the Marquise remarks that 'it was in the Wye-Cottage, in the confluence of the river Wye and the river Severn that she composed the *Botanical Recreations'*. See Marquesa de Alorna, 'Recreações Botânicas' [Botanical Recreations], in *Obras Poéticas*, vol. 4, *Canto* I, note 29, p. 131.

[38] 'Biographical Note', p. xxx. The naturalist Correia da Serra participated in the salons that the Marquise organized. See Bolama, *A Marquesa de Alorna*, p. 222. It is also very likely that the naturalist Félix de Avelar Brotero also participated in the salons organized by the Marquise of Alorna since he was a close friend of the poet Filinto Elísio, who was one of her closest relations.

[39] Silveira, *Mulheres Ilustres: A Marquesa de Alorna*, p. 116.

[40] Marquesa de Alorna, 'Epístola a Godefredo', in *Obras Poéticas*, vol. 2, pp. 54–61, p. 54.

[41] Ibid., pp. 59, 61.

Botanical Recreations as a Didactic Poem

The oldest known version of *Botanical Recreations* simply bears the title '*Botany*'.[42] Its later name was certainly more attractive and more in accordance with the interconnection between education, entertainment and the promotion of polite discourse which was common in the period.[43] It is no coincidence that other works which aimed to popularize science in the eighteenth century included the word 'recreations' in their titles, such as Teodoro de Almeida's aforementioned *Philosophical Recreations* and Cavaleiro de Oliveira's (1702–83) *Periodical Recreations or Dialogues on Natural Philosophy* [*Recriações periódicas ou diálogos de filosofia natural*] (1751). The success of these kinds of works is proof of the increased interest in scientific knowledge in eighteenth-century society.[44]

Botanical Recreations was written within the tradition of the didactic poem, a genre widely used during the late eighteenth and early nineteenth centuries in the popularization of botany. These poems aimed to combine utility and aesthetic pleasure. Their concept of utility was usually broad and implied not only the dissemination of botanical knowledge, but also the improvement of taste and moral virtues.[45] Thus, social manners and women's proper behaviour in society tend to have an important role in these poems. In *Botanical Recreations*, the description of the *Mimosa pudica* (the 'sensitive plant'), whose leaves close in upon themselves and droop when touched, is used as a model for the behaviour of young women. In fact, the plant's retraction when touched is used as a metaphor for women's necessary retraction from a gender-biased society:

---

[42] Arquivos Nacionais Torre do Tombo (Lisbon), Collection 'Casa de Fronteira e Alorna', Mss. 142. According to this manuscript, the first *Canto* was written in August 1810, the second *Canto* in 1806 and the third *Canto* in 1811.

[43] The title *Botanical Recreations* appears in the other three known manuscripts: Arquivos Nacionais Torre do Tombo (Lisbon), Collection 'Casa de Fronteira e Alorna', Mss 140 and 141 and an unclassified manuscript in Arquivo Casas de Fronteira e Alorna.

[44] Other forms of popularization that provide evidence of the importance of science in eighteenth-century society are public demonstrations and lecture courses. See Simon Schaffer, 'Natural Philosophy and Public Spectacle in the Eighteenth Century', *History of Science*, 21 (1983): 1–43; Jan Golinski, *Science as Public Culture. Chemistry and Enlightenment in Britain, 1760–1820* (Cambridge, 1992); Larry Stewart, *The Rise of Public Science: Rhetoric, Technology and Natural Philosophy in Newtonian Britain, 1660–1750* (Cambridge, 1992).

[45] Frances Rowden's (1780–1840) *A Poetical Introduction to the Study of Botany* (1801) combines scientific information with moral lessons about women's place in the world. Similarly, Charlotte Smith's (1749–1806) *Conversations Introducing Poetry: Chiefly on Subjects of Natural History* (1804) presents the theme of women and botany in order to explore 'proper' ways of being female. Sara Hoare's *A Poem on the Pleasures and Advantages of Botanical Pursuits* (1826) and *Poems on Conchology and Botany* (1831) advocate the study of nature as a useful pursuit and a source of spiritual knowledge. See Shteir, *Cultivating Women, Cultivating Science*, pp. 62–77.

What a modest maid, the *Sensitiva*
She recoils from the profane touch
Like a virgin, frightened by a simple touch,
She retracts, discolours and fades.
Crown yourselves with her, choose her as your guide:
To escape is to triumph for our sex.[46]

In the tradition of natural theology, a significant number of didactic poems dealing with botany also associate the study of nature with knowledge of God.[47] *Botanical Recreations* is no exception, and one of its verses is particularly illustrative of this message:

Daughter, walk with me: flowers, glades,
Rivers, mountains, cliffs and groves
These are all worthy of study and reason:
They impose silence upon all worldly corruption,
And bring us closer to Nature's Author.[48]

Poetry, as well as prose in the form of a dialogue, had the advantage of rendering scientific notions pleasant and agreeable. In particular, as John Mullan has noted, poetry and natural philosophy possessed a 'natural affinity', in that poetry best captured the wonderment and delight to be found in the study of the natural world.[49] Poetry and references to classical literature lent politeness to the science they illustrated by coupling this form of scientific study with literary study, and hence with polite learning and a liberal education.[50] In addition, the genre of the didactic poem was especially appealing to women; on the one hand, pedagogy was one of the few accepted pairings of woman and culture,[51] while on the other, poetry was considered to be part of the education and sensibility of women, especially among the upper classes.[52]

---

46 Alorna, 'Recreações Botânicas', *Canto* 4, p. 102. All the translations of excerpts from *Botanical Recreations* included in this chapter are by Christopher Damien Auretta.

47 Williams, *Botanophilia in Eighteenth-Century France*, p. 89.

48 Alorna, 'Recreações Botânicas', p. 27.

49 John Mullan, 'Gendered Knowledge, Gendered Minds: Women and Newtonianism, 1690–1760', in Marina Benjamin (ed.), *A Question of Identity: Women, Science and Literature* (New Brunswick, 1993), pp. 41–56, pp. 51–6.

50 Alice N. Walters, 'Conversation Pieces: Science and Politeness in Eighteenth-Century England', *History of Science*, 35 (1997): 121–54, p. 128.

51 Marina Benjamin, 'Elbow Room: Women Writers on Science, 1790–1840', in Benjamin (ed.), *Science and Sensibility: Gender and Scientific Enquiry, 1780–1945* (Oxford, 1991), pp. 27–59, p. 28.

52 Walters, 'Conversation Pieces', p. 128.

Although the most famous didactic poem on botany was written by a man, Erasmus Darwin's *The Loves of Plants* (1789), there were several female authors working within the same genre.[53] Frances Rowden's *A Poetical Introduction to Botany* (1801) combined an emblematic tradition associating women and nature with empirical knowledge and an interest in women's education.[54] Charlotte Smith's *Conversations Introducing Poetry: Chiefly on Subjects of Natural History* (1804) and Sara Hoare's *A Poem on the Pleasures and Advantages of Botanical Pursuits* (1826) are other examples of works by women writers who cultivated the genre.[55] Moreover, French authors such as Anne-Elizabeth Voiart (1786–1866) and Sabine-Casimire-Amable Tastu (1798–1885) were associate members of the Linnaean Society of Paris from its founding, and authors of publications falling within this genre.[56] Some of their works were even published in the *Memoirs of the Linnaean Society of Paris* under the title 'literature applied to the natural sciences'.

Botanical Recreations is more than a didactic poem on botany and moral virtues. It is also a personal narrative in disguise, especially in the last three cantos. As Maria Helena Pereira has remarked, in this work the Marquise of Alorna 'gave vent to all her woes, finding in several plants analogies with her persecutors and the innocence of her persecuted relatives'.[57] Thus, the description of the 'carvalho' (oak) tree in the poem functions as a pretext to denigrate her greatest enemy, Sebastião José de Carvalho e Melo, the Marquis of Pombal.[58] It can be said that what most distinguishes *Botanical Recreations* from other works of the genre is the fact that the work was written in exile and by an author who had suffered numerous trials and tribulations in her lifetime. In her letters to António Araújo de Azevedo, Count of Barca (1754–1817), she confessed how difficult and even tragic this period was, and how poetry was the balm with which she healed her wounds.[59]

---

[53] On Erasmus Darwin's taxonomic poem *The Botanic Garden II: The Loves of Plants*, see Janet Browne, 'Botany for Gentlemen: Erasmus Darwin and the Loves of the Plants', *Isis*, 80 (1989): 593–621 and Londa Schiebinger, 'The Private Life of Plants: Sexual Politics in Carl Linnaeus and Erasmus Darwin', in Benjamin (ed.), *Science and Sensibility*, pp. 121–43.

[54] On the association between nature and women, see Carolyn Merchant, 'Nature as Female', in Merchant, *The Death of Nature: Women, Ecology and the Scientific Revolution* (San Francisco, 1980), pp. 1–41.

[55] Shteir, *Cultivating Women, Cultivating Science*, pp. 61–77. From 1780 to 1830, English women writers also used letters and conversations as narrative forms in the popularization of botany.

[56] Duris, *Linné et la France*, p. 197.

[57] Maria Helena da Rocha Pereira, *'Utili Dulci' nas 'Recreações Botânicas' da Marquesa de Alorna* (Coimbra, 1983), p. 7.

[58] Alorna, 'Recreações Botânicas', *Canto* IV, p. 99.

[59] A.L. Vaz (ed.), *A Marquesa de Alorna, Cartas do exílio em Londres (1804–1814)* (Braga, 1974), p. 81.

Botanical Recreations and the Linnaean System of Classification

The Marquise of Alorna was not a mere amateur in botanical knowledge. The poem *Botanical Recreations*, together with her translation of *Letters on the Elements of Botany: Addressed to a Lady By the Celebrated J.J. Rousseau, Translated into English, with Notes, and Twenty-Four Additional Letters, Fully Explaining the System of Linnaeus* by Thomas Martyn (1785), testifies to the contrary. Martyn's (1735–1825) work went through eight editions over the next thirty years, and was one of the most popular writings on botany. The Marquise of Alorna's translation into Portuguese included additional notes which showed her interest in understanding Linnaeus' system of classification. In one of the notes, she comments on the popularization of Linnaeus' system of classification in Britain and Portugal:

> The English or Portuguese student can only encounter advantages in the possession of many elementary books which explain terms to them in their own language. Linnaeus is translated into English. There is also the flora of Hudson, and the arrangement of Dr. Wecthering which combine the English terms with those of Linnaeus, Dr. Wardelly, Professor Brotero, Gresley and the Count Hoffmannsegg it is they who have introduced Botany into Portugal. The Abbé Correia has written nothing on this subject as yet.[60]

Carl Linnaeus' (1707–78) work had given the study and popularization of botany a special impetus.[61] One of the advantages of Linnaeus' system of classification was its easy accessibility, something that rendered it open to both academics and novices, men as well as women.[62] It therefore comes as no surprise that it was promoted in many works which aimed to popularize botany, including the genre of the didactic poem. *Botanical Recreations* is no exception, and Linnaeus and his system of classification based on sexual characteristics have a central place in the poem:

> Astonished the Wise, who long Believe
> Linnaeus converses with the Deity:
> Or that sentient plants confide to him
> Their loves in florid eloquence[63]
> …

[60] Arquivos NacionaisTorre do Tombo (Lisbon), Collection 'Casa de Fronteira e Alorna', Mss. 146, note v, from the Marquise of Alorna's translation into Portuguese of *Letters on the Elements of Botany* (undated manuscript).

[61] On Linnaeus, see Frans A. Stafleu, *Linnaeus and the Linnaeans: The Spreading of their Ideas in Systematic Botany, 1735–1789* (Utrecht, 1971).

[62] Lisbet Koerner, 'Women and Utility in Enlightenment Science', *Configurations*, 2 (1995): 233–55, pp. 245–6.

[63] Alorna, 'Recreações Botânicas', Dedicatory Epistle, p. 9.

Linnaeus is the Polymarch of their army
Who knows the laws of castamentration:
Like the General whose soldiers
Are placed on the battle-field; the orders and species of plants
Linnaeus glimpsed without confusion
That ray of light that all things illuminates
And *arcanas* perceived in benign revelation.[64]

Nevertheless, she stated her awareness of the limitations of Linnaeus' system of classification, especially its degree of artificiality:

The Species is certain and inalterable
...
It is within Nature;
But further divisions are mind designed
Methods diverse which memory holds
In orders, *genera* and systems
Of these species Botanists composed
New groups, designating them with the name
Of *genus* and *genera* they sought
(even Linnaeus) to discover them in Nature
But it is time and observation which all discloses
And it is before truth that conjecture ceases.[65]

She further asserted in the poem that there is more to botanical knowledge than classification:

To know the name of plants does not suffice
...
Analogies, habit, character
Lead Science to the greatest heights.[66]

Nor is Linnaeus the only foreign naturalist mentioned in *Botanical Recreations*. She also stresses the contributions of John Ray (1627–1705), Joseph Pitton de Tournefort (1628–1705) and Antoine Laurent de Jussieu (1748–1836) in progress on the subject. Additionally, Rousseau's *Elementary Letters on Botany* is presented as one of the most useful works in the improvement of botanical knowledge.[67] The *Letters* were written between 1771 and 1774 and published posthumously

[64] Ibid., *Canto* II, p.46.

[65] Ibid., *Canto* III, p. 54.

[66] Ibid.,*Canto* I, p. 21.

[67] In her poem, Alorna alludes to the 'enchanting laws of plants' and how a genial mind has beautifully presented them in a work that she recommends to her readers. In a footnote

in Geneva in 1781. They enjoyed great popularity in the period and contributed greatly, especially in France, to the dissemination of the Linnaean system of classification.[68]

The Portuguese naturalist Félix de Avelar Brotero (1744–1828) and the Abbé José Correia da Serra (1751–1823) are also praised in the poem:

> If it is Science you seek, seek him [Brotero]
> Who planted the nursery of botanists;
> And, on the banks of our Mondego studies Nature
> Call that Lusitanian genius [Correia da Serra] in patriotic chorus
> From that long exile to which
> Our nation has abandoned him.[69]

Together with Domingos Vandelli (1732–1815), Brotero was the main naturalist behind the dissemination of the Linnaean system of classification in Portugal.[70] He published his *Compendium of Botany* [*Compêndio de Botânica*] in 1788.[71] In this work, Brotero acknowledges the didactic value of the Linnaean system, as well as its broad acceptance within Europe, something that he presents as sufficient reason to provide a detailed presentation of this method of classification. However, he is not a strict follower of Linnaeus, and expresses his awareness of the limitations and disadvantages of the Linnaean system. Thus, he also presents the views of other famous botanists on the subject to ensure that 'the reader is made aware not only of its advantages, but also its limitations, and therefore can attribute to it its real value'.[72] Brotero was also the author of *Flora Lusitanica* (1804), in which 1,885 species from Portugal were for the first time described and classified according to the Linnaean method of classification.

---

presented on the same page she refers specifically to Rousseau's *Elementary Letters on Botany*, ibid., *Canto* II, p. 38.

68 Williams, *Botanophilia in Eighteenth-Century France*, pp. 103–7.

69 Alorna, 'Recreações Botânicas', Dedicatory Epistle, p. 8. On Correia da Serra's life and work, see Ana Simões, Maria Paula Diogo and Ana Carneiro, *Cidadão do Mundo, Uma Biografia Científica do Abade Correia da Serra* (Porto, 2006).

70 Correia da Serra was also an influential Portuguese naturalist from the period. In contrast to Brotero and Vandelli, he distanced himself from the Linnaean tradition of classification and advocated the supremacy of natural systems over artificial ones. On Correia da Serra's views on natural historical classification, see Maria Paula Diogo, Ana Carneiro and Ana Simões, 'The Portuguese naturalist Correia da Serra (1751–1823) and his impact on early nineteenth-century botany', *Journal of the History of Biology*, 34 (2001): 353–93.

71 On the life and work of Brotero, see Cristina Castel-Branco, *Félix de Avelar Brotero: botaniste portugais (1744–1828)* (Paris, 2004).

72 Felix de Avelar Brotero, *Compêndio de Botânica* (Lisbon, 1788), p. vii.

Vandelli was one of Linnaeus' correspondents, and the author of the *Dictionary of Technical Terms in Natural History taken from the Works of Linnaeus* [*Diccionario dos termos technicos de Historia Natural, extrahidos das obras de Linneo*] (1788). He was influential in the education of naturalists who were later involved in scientific trips to Africa and Brazil, and believed that the international success of these trips could only be guaranteed through the use of the Linnaean system of classification.[73] Despite his contribution to the spread of botany in Portugal, his name is absent from Botanical Recreations. Every form of communication includes, just as it excludes, and the reason for this absence is likely related to the fact that Vandelli collaborated with the French during the French invasions and that the Marquise of Alorna was strongly anti-Napoleonic.

The content of the printed version of *Botanical Recreations* differs from the manuscript on which it was based in the notes added by the naturalist Caetano Bordão, together with the notes written by the editor, Carlos Manuel Soye.[74] These notes extend the botanical knowledge provided by the work, and include two tables on the Linnaean system of classification that were taken directly from Brotero's *Compendium of Botany*.[75]

[73] On Vandelli, see José Luís Cardoso, 'From natural history to political economy: the enlightened mission of Domenico Vandelli in late eighteenth-century Portugal', *Studies in the History and Philosophy of Science*, 34 (2003): 781–803.

[74] The printed version was based on manuscript 140 in the Arquivos Nacionais Torre do Tombo (Lisbon), Collection 'Casa de Fronteira e Alorna'. The first three *cantos* of the oldest known manuscript of *Botanical Recreations* (Mss 142, Arquivos Nacionais Torre do Tombo (Lisbon), Collection 'Casa de Fronteira e Alorna') are very similar to the printed version, but the last three *cantos* seem to only exist in their early stages. They have fewer pages than the printed version, and several of the verses were corrected or omitted in the final version. Besides, the notes on the work included in this version are very brief. The other two known manuscripts of the work (Mss 142, Arquivos Nacionais Torre do Tombo (Lisbon), Collection 'Casa de Fronteira e Alorna' and an unclassified manuscript of the *Botanical Recreations* in the archive of the Casas de Fronteira e Alorna) are closer to the aforementioned manuscript 140, but it still contains several amendments, especially in the last three *cantos* which seem to have been the most difficult for the Marquise of Alorna to compose. These last three cantos are more biographical.

[75] Only the notes of the first *Canto* were written by the Marquise of Alorna. These notes include information on mythological and scientific figures such as Pliny, Tournefort, John Ray, Jussieu and Linnaeus. On Soye's contribution to the printed version of the *Botanical Recreations*, see Luís de Pina, *A Botânica na Poesia da Marquesa de Alorna* (Porto, 1953).

The Audience for Botanical Recreations

Botanical Recreations was dedicated to Portuguese women [my patricians], and especially to the younger female generations:

> For you, in whom the flower of youth,
> An unknown love your spirit moves;
> And subtle impulse leads to glades;
> For you, Nymphs gentle, I write these verses.
> That an ardent love for flowers your soul reaches,
> Such sweet passion, sovereign and triumphant,
> That opens up the Temple of Peace and Delight.[76]

The Dedicatory Epistle urged them to study nature and develop their passion for botany instead of spending their time in futile activities. The intended audience for *Botanical Recreations* was certainly women; however, the readership was probably not restricted to a female audience. In any case, the work was especially intended for the upper classes, those who, according to the Marquise, cultivated the pleasures of botanical knowledge: 'Botany, origin of pleasures that the turbulent masses never know'.[77]

Throughout her life, the Marquise of Alorna was concerned with the education of the less fortunate. The 'Biographical Note' published in her *Poetical Works*, for example, mentions her dedication to instructing the 'poor girls' from the village of Almeirim in the Santarém district, where the Marquise had a country residence. For this task, she employed a teacher to educate them in 'reading, sewing and more tasks suitable to their sex' without any reference to initiation into the study of nature.[78] These activities can be contrasted with the recommendation in a letter she wrote to one of her nieces to read Buffon and:

> not to treat plants with contempt. Become accustomed to their physiognomy as one becomes accustomed to distinguishing different faces in a crowd. Soon you will discover pleasant countryside delights, even before you know their names, which are always arbitrary and never bestowed by Nature.[79]

As mentioned above, *Botanical Recreations* was not published during the Marquise of Alorna's lifetime. The usual explanation given for this is related to

---

76 Alorna, 'Recreações Botânicas', Dedicatory Epistle, p. 7. Moreover, each of the first three *cantos* was specifically dedicated to one of the Marquise's three daughters.

77 Alorna, 'Recreações Botânicas', *Canto* I, p. 18.

78 'Biographical Note', pp. xxix–xxx.

79 Quoted in Cidade (ed.), *Marquesa de Alorna, Poesias*, pp. xxx–xxxi.

the confessional nature of the work.[80] It is likely that this contributed to the delay in publication, but is probably insufficient in explaining it completely. Indeed, the careful presentation of the history of botany and the Linnaean system of classification, together with the notes provided by the author in the first canto, indicate that the Marquise of Alorna originally thought of publishing *Botanical Recreations*. We do not know, however, why she abandoned the project.

It is true that the Marquise carefully crafted the relatively few texts that represented her public face to the literate world.[81] At the same time, during that period there was an elitist and aristocratic attitude that made some writers prefer recognition by a small group of peers to widespread public success.[82] We should also take into account the fact that during this period women who wrote were aware that publication could damage them and be interpreted as a sign of vanity.[83] As Mary Terral has pointed out, numerous eighteenth-century scholars defined glory in opposition to femininity. For women, the quest for glory also entailed provoking potential ridicule as opposed to admiration. Ridicule was frequently used as a weapon against all sorts of intellectual pleasures, and women who aspired to wisdom and rationality were especially vulnerable to dismissal by ridicule.[84]

However, the fact that *Botanical Recreations* was only published in 1844 does not imply that the work, or at least part of it, did not have an audience before this date. It is known that several letters circulated publicly among the upper classes.[85] Excerpts from *Botanical Recreations* might have been spread in this way. The circulation of copies of the work among a restricted group of friends was probably

---

80 'Biographical Note', p. xxxiii. See also Pereira, *'Utili Dulci' nas 'Recreações Botânicas' da Marquesa de Alorna*, p. 7.

81 During her lifetime, the Marquise of Alorna only published under her name the poetical work *Paraphrase dos Salmos* em duas partes [*Paraphrase of the Psalms* in two parts] (Lisbon, 1817 and 1833) and her translation into Portuguese of Félicité Robert de Lamennais' *Essai sur l'indifférence en matière de religion* [*Essay on Indifference in Religious Matters*] (Lisbon, 1820). Her translations of François Auguste de Chateaubriand's *De Buonapart et des Bourbons* [*On Bonaparte and the Bourbons*] (Lisbon, 1814), and of Horace's *Poetical Art* [*A Arte Poética de Horácio*] (London, 1812) and Alexander Pope's *Essay on Criticism* [*Ensaio sobre a Crítica de Alexandre Pope*] (London, 1712) were published anonymously.

82 Feijó, *Ad maiorem gloriam*, p. 82.

83 Vanda Anastácio, 'Cherchez la femme (À propos d'une forme de sociabilité littéraire à Lisbonne à la fin du XVIIIe siècle)', *Arquivos do Centro Cultural Calouste Gulbenkian*, 49 (2005): 93–101, p. 94.

84 Mary Terral, 'Émile du Châtelet and the Gendering of Science', *History of Science*, 33 (1995): 283–310, p. 299. See also Raquel Bello Vásquez, *A Condessa do Vimieiro* (Lisbon, 2006), p. 18.

85 See Raquel Bello Vázquez, 'Lisbon and Vienna: The Correspondence of the Countess of Vimieiro and her Circle', *Portuguese Studies*, 20 (2004): 89–107 and Anne Goldgar, *Impolite Learning: Conduct and Community in the Republic of Letters 1680–1750* (New Haven and London, 1995).

yet another way in which the work was disseminated. In a letter to the Count of Barca dated 8 February 1815, the Marquise of Alorna expressed her desire to send him some of her works and mentioned that she was already in the process of copying them.[86] Furthermore, in one of the letters from the poet Filinto Elísio to the Marquise, he explicitly mentions receiving a copy of *Botanical Recreations* and that he read the poem with great pleasure.[87]

During the eighteenth century, new forms of sociability emerged in a variety of urban centres, which included coffee houses, salons and Masonic lodges.[88] These meeting places minimized the distinctions created by wealth or inherited rank, and offered a space where participants could meet and exchange ideas as equals, at least in theory. In particular, the salons offered women the possibility of playing a central role in facilitating and promoting intellectual exchange. Moreover, salon culture portrayed itself as replacing the hierarchical status relationships in court life with an egalitarian ideal of circulating individuals and ideas.

Salons, or *assembleias*, become fashionable in Portugal shortly after the 1755 Lisbon earthquake, and were frequently organized by upper-class women until the 1840s. According to Vanda Anastácio, during this period these meetings constituted one of the most efficient channels for the dissemination of unpublished works, and they were true spaces of legitimization in terms of authorship.[89] At these *assembleias*, women not only exerted their influence and organizational skills in attracting writers, scholars and government officials, but some of them also revealed their literary talents. The important role of the *assembleias* in society of the time demonstrates that although women writers from this period published very few works, this should not conceal their role as intellectual producers and mediators.[90]

The *assembleias* that the Marquise of Alorna held in her palace at Benfica until she was almost 90 were some of the most famous from that time. Hence, it is very likely that they contributed to the circulation of *Botanical Recreations*. The importance of these *assembleias* is also helpful in understanding why her literary skills were known during her lifetime, even abroad, although she had published very few works by then. In his *Statistical Essay on the Kingdom of Portugal and Algarve* [*Essai statistique sur le Royaume de Portugal et d'Algarve*] (1826), Ferdinand Denis asserted that

---

86 Vaz (ed.), *A Marquesa de Alorna, Cartas do exílio em Londres*, p. 129.

87 Biblioteca Nacional Mss 251 (1). A facsimile of the manuscript is published in Bolama, *A Marquesa de Alorna*, p. 67.

88 See Thomas Broman, 'The Habermasian Public Sphere and Science in the Enlightenment', *History of Science*, 26 (1998): 123–49.

89 Vanda Anastácio, 'Mulheres Varonis e Interesses Domésticos' (Reflexões acerca do discurso produzido pela História Literária acerca das mulheres escritoras da viragem do século XVIII para o século XIX)', *Ariane: Revue d'études littéraires françaises*, 18–20 (2003–05): 537–56, p. 541.

90 Anastácio, 'Cherchez la femme', p. 100.

> it has been lamented that in Portugal the education of women leaves much to be desired; yet it seems that the same reproach can no longer be made, and that many ladies are famous for their works. One of the most important is the Countess of Oeynhausen [Marquise of Alorna], who has written works of all kinds and who combines a knowledge of languages with a remarkable talent for writing poetry.[91]

Botanical Recreations was published not independently, but together with other poetry in one of the six volumes of the Marquise of Alorna's *Poetical Works* (1844). This probably contributed to its lesser impact in terms of the reception of the work. Moreover, the *Poetical Works* was an expensive publication, and the editors specifically intended it for 'an Enlightened public'.[92] The work did not enjoy a second edition, in contrast with similar works that were translated into Portuguese such as *The Marriage of the Flowers* [*Le marriage des fleurs*] by Demetre Lacroix, which went through a first edition in 1801 and a second in 1811.[93] The circulation of these works shows that there was an audience for this kind of poetry in Portugal.

Independently of the degree of success of her *Botanical Recreations*, the Marquise of Alorna was certainly influential in Portuguese society and, in particular, in the career of the poet Barbosa du Bocage (1800–75) and the writer Alexandre Herculano (1810–77), who described her as an extraordinary woman 'who only lacked another country, which was not this poor, forgotten land of Portugal, to be one of the most brilliant refutations of the vain pretences to excessive superiority of our sex'.[94]

---

91 Quoted in Silveira, *Mulheres Ilustres: A Marquesa de Alorna*, p. 111. Silveira mentions also the views of António Feliciano Castilho, of Pinheiro Chagas, of Romero Ortiz and of Adrien Balbi on the Marquise of Alorna; ibid., pp. 109–11.

92 'Biographical Note', p. xliii.

93 René Richard Castel's *The Plants* [*Les plantes*] (1797) and Jacques Delille's *The Gardens* [*Les jardins*] (1782) were also available in Portuguese. The Fundação das Casas de Fronteira e Alorna is in the process of publishing a critical edition of the *Complete Works* of the Marquise of Alorna, edited by Vanda Anastácio.

94 Quoted in Silveira, *Mulheres Ilustres: A Marquesa de Alorna*, p. 105. The Marquise of Alorna was portrayed by some authors as having a masculine mind. Such is the case of Maria Amélia Vaz de Carvalho, the first Portuguese woman to be elected to the Lisbon Academy of Sciences, who characterizes her as having 'an almost virile vigour of mind acquired and refined during her long years in prison'. See Anastácio, 'Mulheres Varonis e Interesses Domésticos', pp. 543–6.

Conclusions

The recent attention focused on the history of women in science has led to an increased interest in the history of scientific communication.[95] This interest has found expression in a growing number of case studies which demonstrate that during the European Enlightenment and the early nineteenth century, scientific authorship constituted an important and hitherto neglected way that women could participate in science, yet equally interestingly, that women science writers were active interpreters of specialist knowledge.[96]

However, the study of the history of women in science has generally focused on European centres of learning. The analysis presented in this chapter has shown that, in Portugal, women seem to have been virtually absent from the popularization of the sciences. In fact, with the exception of the Marquise of Alorna's *Botanical Recreations*, there are no traces of works dealing with scientific subjects written by women or aimed specifically at a female audience. At the same time, the Portuguese periodicals in which scientific news and articles were published did not publish any articles authored by women.[97] It should also be stressed that all the didactic poems dealing with botany that were published in Portugal were authored by men. Of course, this situation does not mean that Portuguese women were not consumers of scientific knowledge; but it certainly does indicate that their visibility in activities related to the popularization of science was indeed very limited. The difference between this situation and that of other European countries such as Britain and France can be explained at least in part by the highly conservative ideology of Portuguese society during this period, in which the power of the Catholic Church bore a significant influence. This situation contributed to relegating the role of women to the private sphere of domesticity.

The study presented in this chapter has shown that the dichotomy between centre and periphery is blurred for the upper echelons of Portuguese society. In terms of access to knowledge, these classes suffered less from the fact that they lived in a peripheral country dominated by religious, political and intellectual censorship. As we have seen in the case of the Marquise of Alorna, although she was born and lived for a considerable time in a peripheral European country like

[95] See, among others, Barbara T. Gates and Ann B. Shteir (eds), *Natural Eloquence: Women Reinscribe Science* (Madison, WI and London, 1997).

[96] Gates and Shteir (eds), *Natural Eloquence*. See also Barbara T. Gates, 'Ordering Nature: Revisioning Victorian Science Culture', in Bernard Lightman (ed.), *Victorian Science and Victorian Culture* (Chicago and London, 1993); Benjamin (ed.), *A Question of Identity*; Shteir, *Cultivating Women, Cultivating Science*.

[97] This information was personally supplied by Fernando Reis and is based on his research in the *Jornal Enciclopédico Dedicado à Rainha* (1779; 1788–93; 1806), *Semanário de Instrução e Recreio* (1812–13), *Jornal Enciclopédico de Lisboa* (1820), *Correio Brasiliense* (1808–22), *O Investigador, Português em Inglaterra* (1811–18), and *Annaes das Ciências, das Artes e das Letras* (1818–22).

Portugal, she also lived in, and was in active contact with, European centres of knowledge. In fact, it was while she was living in Britain that she wrote *Botanical Recreations*. Even when the Marquise was living in Portugal, her upper-class origins enabled her to have access to books that were forbidden through censorship, and her network of correspondents and influences gave her privileged access to knowledge and ideas from the so-called 'centres' of Europe.

Moreover, the upper classes could enjoy other places of sociability, such as the salons, where women played a prominent role as social and cultural mediators, and where they could freely present their ideas and literary works. It was in these kinds of places that excerpts from *Botanical Recreations* probably enjoyed their widest circulation. In comparison with European countries from the so-called European centres of learning, therefore, in Portugal there was a greater gap between the upper classes and the middle classes in terms of access to knowledge, and this breach was probably wider in the case of scientific knowledge. As we have seen, the intended audience of *Botanical Recreations* was the upper classes, and the circulation of the work before it was published was confined to certain social networks and places of sociability of the Portuguese upper classes.

This more privileged access to knowledge enjoyed by the upper classes does not mean, however, that at least some of them, such as the Marquise of Alorna, were not profoundly aware of the marginal situation and ignorance of their homeland:

> In this my birthplace, this wretched western clime
> All we know is masked (Numen, reason) …
> Fettered ideas in exiled confinement
> To no freedom of flight consent;
> And, collapsing upon its own central point,
> Appears irremediably stagnant
> …
> Think not, then, that I am guided
> By that cruel ignorance which affects our Nation;
> In grievous lament I see our errors
> With the honest envy of one who sees
> The advantages which study in other nations confers,
> And ever watchful of her own misconceptions
> Shall faithful to your precepts be.[98]

98 Alorna, 'Epístola', *Obras Poéticas*, vol. 2, pp. 13–14.

Chapter 4
Science for the People: The Belgian *Encyclopédie populaire* and the Constitution of a National Science Movement

Geert Vanpaemel and Brigitte Van Tiggelen

In 1872 the Belgian Royal Academy of Science, Literature and Art celebrated its first centenary. It was a major public event, honoured by the presence of King Leopold II and many political and scientific personalities, both from Belgium and from other countries. The celebrations lasted for two days, during which numerous speeches were held on the subject of Belgian science. There were musical performances by members of the Academy, and several banquets. The proceedings of the celebrations were published in two volumes, setting the tone for many historical accounts in later years.[1]

The rhetoric of the celebrations highlighted the importance of science for the nation. Looking back on the development of science in Belgium, the palaeontologist Pierre-Joseph Van Beneden remarked: 'At the time of the creation of the Academy, Belgium was not in a favourable situation with regard to [scientific] studies'. But now, 'Belgium can feel proud of its scientific situation; it has representatives in every field of science and no discipline has been completely neglected'.[2] The appreciation of the scientific movement was further extolled by the discourse of King Leopold II. 'In our free and independent country, intellectual activities ought to be executed without restraint. The field that you have cultivated is a neutral territory where people can meet, hand in hand, who in their active lives may move in quite different directions.'[3] The words of King Leopold II were true. Science had been able to steer a neutral course between the opposing political factions which had marked Belgian politics since the country's independence in 1830. Catholics and liberals were in constant opposition, which extended towards the realm of education. Although Belgium already had two state universities (Ghent and Liège), both Catholics and free-thinking liberals created their own universities

---

1 *Centième Anniversaire de Fondation (1772–1872)*, 2 vols, (Brussels, 1872).

2 Ibid., vol. 1, p. 25: 'Lors de la création de l'Académie, la Belgique ne se trouvait pas dans une situation favorable aux études'; p. 51: 'la Belgique peut, à juste titre, s'enorgueillir de sa situation scientifique; elle compte des représentants dans toutes les branches des sciences; aucune d'elles n'est complètement négligée'.

3 Ibid., p. 8.

viz. in Leuven and Brussels. This led to tensions and controversies, but natural science in particular was able to stay out of party politics. Only after, and partly as a result of, the emergence of Darwinist debates did further polarization of Belgian society more strongly affect the institutional basis of science. But in 1872, all still seemed well.[4]

The relationship between science and government in Belgium was very much determined by the effort of scientific leaders to maintain that neutral position. Belgian scientists were well aware of the fact that the success of Belgian science and its growing cultural importance was closely linked to the support of Belgian political authorities. Although the beginning of the revival of Belgian science actually preceded Belgian independence in 1830, the Belgian government had always strongly and deliberately supported the national scientific movement, either through the creation of scientific institutions or by supporting individual research projects. If scientists before 1830 had often complained that their efforts were scarcely noted by the public in general, by the end of the century scientists had acquired the status of intellectual and cultural leaders of the nation, and were heralded as bearers of moral authority. Expansion of the scientific community was made possible by the creation and promotion of national institutions, around which a varied universe of scientific societies and private schools flourished. The centre of this universe was the Academy, situated in the capital of the nation, in close contact with the political and cultural elite.

The relationship between science and national prestige was carefully embedded in public discourse. In popular accounts the contemporary development of Belgian scientific life was heralded as a mark of national maturity, in particular as a beneficial consequence of the intellectual freedom brought about by national independence. The scientific achievements of such sixteenth-century pioneers as Vesalius, Mercator, Dodoens and Stevin were appropriated as the historical heritage of the modern nation.[5] A coherent historical analysis was elaborated and divulged by the statistician Adolphe Quetelet, who served for 40 years as the secretary and spokesman of the Academy. Quetelet maintained that the science of a nation was the most trustworthy indicator of its cultural development. It was thus an obligation of national importance to the people to promote and support science. In this respect Belgian independence had indeed created the necessary conditions for Belgian science to grow. Quetelet was the personal incarnation of his own views. From the beginning his career was marked by his clever enlisting

---

4 The most recent overview of scientific life in Belgium can be found in R. Halleux., J. Vandersmissen, A. Despy-Meyer and G. Vanpaemel (eds), *Histoire des Sciences en Belgique 1815–2000*, 2 vols (Tournai, 2001).

5 G. Vanpaemel, 'Omtrent Dodoens: Wetenschapsgeschiedenis en de cultuur van het nationale erfgoed', in *Liber Amicorum Roger-A. Blondeau* (Poperinge, 1999), pp. 307–19; G. Vanpaemel, *Een standbeeld voor Stevin* (Nijmegen, 1995); Jo Tollebeek and Tom Verschaffel, 'Group portraits with national heroes: the pantheon as a historical genre in nineteenth-century Belgium', *National Identities*, 6 (2004): 91–106.

of government support for his many projects. The scientific movement in Belgium was largely the result of his indefatigable efforts to create and shape it.[6]

The simultaneous emergence of the Belgian nation state and its national scientific community suggests a process of co-production in which the creation of both political and intellectual institutions is closely intertwined.[7] Inasmuch as the celebration of the Academy, honoured by the presence of the political elite of the country, was a vindication of the nascent scientific community, at the same time it can be interpreted as a pronouncement of Belgium's right to independence and the result of enlightened government. The problem is not whether the scientific movement was making use of an already existing nationalist idiom, or whether the nation-building process followed the rise of the scientific community. Both processes were coexistent, each becoming more articulated by the advance of the other. Co-production thus blurs the distinction between the natural and the social order, while directing the attention of scholars to the creative interaction between science and its socio-political environment. This approach, originating from the field of science studies, is easily integrated into the new concepts in the historiography of nineteenth-century nation building, which emphasizes the symbolic construction of the nation based on actual, but also very much on 'invented' representations of a community.[8]

The framework of co-production makes it possible to understand the often contradictory strategies visible in the 'nationalization' of science. Indeed, two different strategies can be discerned, depending on the pre-eminence given to science or to nationalism. The nationalization of science in the nineteenth century has been studied as an appropriation of nationalist feelings and values by the scientific movement.[9] Typically, nationalist science would highlight the contributions of the

---

[6] Adolphe Quetelet, *Histoire des sciences mathématiques et physiques chez les Belges* (Brussels, 1864) and *Sciences mathématiques et physiques au commencement du XIXe siècle* (Brussels, 1867). On the historiography of science in Belgium, see Geert Vanpaemel, 'Wetenschapsgeschiedenis in België', in R. Halleux, C. Opsomer and J. Vandersmissen (eds), *Geschiedenis van de wetenschappen in België van de Oudheid tot 1815* (Brussels, 1998), pp. 419–32; Kaat Wils, 'Een zaak van nationaal belang? Wetenschapsvulgarisering in België', in Els Witte (ed.), *Natie en Democratie* (forthcoming).

[7] On the concept of co-production, see Sheila Jasanoff (ed.), *States of Knowledge: The Co-Production of Science and Social Order* (London and New York, 2004).

[8] The foundational text of this new historiography is Benedict Anderson, *Imagined Communities; Reflections on the Origins and Spread of Nationalism* (London, 1983, 1991^2). For the history of science relating to these concepts, see Ludmilla Jordanova, 'Science and Nationhood. Cultures of Imagined Communities', in Geoffrey Cubitt (ed.), *Imagining Nations* (Manchester, 1998), pp. 192–211; Ralph Jessen and Jakob Vogel (eds), *Wissenschaft und Nation in der Europäischen Geschichte* (Frankfurt and New York, 2002).

[9] Traditional discussions of the nationalization of science can be found in Brigitte Schroeder-Godehus, 'Nationalism and Internationalism', in R.C. Olby, G.N. Cantor, J.R.R. Christie and M.J.S. Hodge (eds), *Companion to the History of Modern Science* (London, 1990), pp. 909–19; Elisabeth Crawford, *Nationalism and Internationalism in Science:*

nation to the progress of scientific knowledge; it draws attention to the national talents of countrymen for doing science, and it emphasizes the importance of science for national prosperity. The special skills and expertise of the scientist are needed for the survival and development of the nation in a rapidly evolving world of scientific and technical progress. Nationalist science thus lends scientists a social function as servants of the nation, for which they are to be appreciated and admired, and which encourages them to enlist public support.

Yet another, contradictory, strategy consists of deliberately and consciously *not* adapting science to the frame of a nation's own cultural resources. By showing science to be universal and therefore *not* confined to national boundaries, science can be presented as an objective body of knowledge open to anyone, from which any well-educated nation can and ought to benefit, but which no nation can consider its own. In this strategy, the existence of a national science movement is regarded as proof of the maturity of a nation and of its willingness to compare itself with other nations. The scientist is a prime protagonist for the nation, a defender of national honour, a representative of their country. However, science itself escapes national labels and cannot be nationalized.

Obviously, both strategies can contribute equally to the production of a nationalized representation of science. Some scientists may adopt one strategy, while others pursue a different one. Strategies may be adapted easily to the particular audience for which they are destined. Through both of these strategies scientists can determine their position with regard to the cultural landscape, soliciting support or taking on a leadership role. The strategy chosen may depend on the social power of the scientist or the scientific discipline involved: the more recognized scholar will probably be less inclined (or compelled) to enlist in a nationalist perspective. Scientific disciplines close to industry will point rather to national inventiveness, whereas theoretical research will make primary use of its international recognition.

Inasmuch as the co-production of nation and science is based on symbolic representations of scientific research, scientific institutions and individual scientists, special attention should be given to the selected audiences which legitimize the images produced. The possibility for scientists to address elite audiences or to be heard by political leaders or captains of industry will contribute most effectively to the social power of the scientific establishment. But equally important is the command of large audiences, the general public or the people of the nation. In this respect the popularization of science takes a central place in the study of the scientific movement in a country.[10] Popular representations of science

1880–1939 (Cambridge, 1992); Alan J. Rocke, *Nationalizing Science: Adolphe Wurtz and the Battle for French Chemistry* (Cambridge, MA, 2000).

[10] Bernadette Bensaude-Vincent, *La science contre l'opinion: histoire d'un divorce* (Paris, 2003); Andreas W. Daum, *Wissenschaftspopularisierung im 19. Jahrhundert: Bürgerliche Kultur, naturwissenschaftliche Bildung und die deutsche Öffentlichkeit, 1848–1914* (Munich, 1998).

are a powerful means by which to claim credibility and legitimation. By looking at popularization from the perspectives of both science and nationalism, the co-productive processes are likely to become visible.

In this chapter we will examine two different popularization strategies that lend themselves to close comparison. From 1846 on, two series of scientific books were published which were influential in promoting the scientific movement in Belgium. Both these series were published around the same time, and by the same publisher; yet the involvement of scientists in the production of these books was very different. Whereas the first series shows the adaptation of science to nationalist feelings, the second becomes a vehicle for the promotion of universal, scientific values, at the service of an emergent local scientific community.

Science Nationalized: The *Bibliothèque Nationale*

In January 1846 the Brussels-based publisher Alexandre Jamar (1821–88) issued a prospectus[11] for the publication of a *Bibliothèque nationale*.[12] The *Bibliothèque* would consist of 48 beautiful volumes in-8 (the actual books would be in-12), on satin paper, illustrated by the best artists. Each month one volume would be published. Subscriptions were only taken for the complete series. The price was fairly modest: 1.50 francs per volume, with a discount of 0.25 francs for early subscribers.

Alexandre Jamar was the son of a Brussels registrar (Figure 4.1). At 17, he entered his older brother's publishing house and soon became manager of the company.[13] Like many Belgian publishers of the time, the Jamar press made a fortune publishing cheap (illegal) editions of French books, which was possible under Belgian law until 1852.[14] But what set Jamar apart from other publishers was his very pronounced political and patriotic idealism, which inspired him to launch several series of books destined for the education of the people. In 1840 he started the *Panthéon national*, which focused on great moments in Belgian

[11] This prospectus can be found in the papers of Adolphe Quetelet, preserved in the Archives of the Belgian Royal Academy in Brussels: Archives de l'Académie Royale de Belgique (ARB), *Correspondance A. Quetelet*, 2784. Liliane Wellens-De Donder, *Inventaire de la correspondance d'Adolphe Quetelet déposée à l'Académie royale de Belgique* (Brussels, 1966).

[12] P. Kauch, *Alexandre Jamar: Quatrième gouverneur de la Banque Nationale de Belgique, 1821–1888* (Brussels, 1954), repr. from *BNB, revue du personnel de la Banque Nationale de Belgique*, 10 (1954): 11; P. Kauch, 'Jamar (Alexandre)', *Biographie nationale*, XXIX (1957), cols 719–22.

[13] On the Jamar press, see Thomas Decreus, 'Het verleden heruitgegeven. Geschiedenis van de uitgeverij Jamar' (master's thesis, Catholic University of Leuven, 2006).

[14] H. Dopp, *La contrefaçon des livres français en Belgique* (Leuven, 1932).

history.[15] The books were not copied from French or English books, but written by well-known Belgian scholars and lavishly illustrated. The success of the series inspired Jamar to adopt the same format for the *Bibliothèque nationale* (1846–51) and the *Encyclopédie populaire* (1849–56), although these series were not quite so glamorous. In the 1850s, Jamar published another landmark book, the *Biographie nationale. Vie des hommes et des femmes illustres de la Belgique depuis les temps les plus reculés jusqu'à nos jours*, edited by André Van Hasselt.

Figure 4.1 Alexandre Jamar (1821–88). From a brochure (1954).

As a successful businessman, Jamar was well introduced to the elite circles of the Belgian capital, where his liberal and patriotic views found a sympathetic ear among the Brussels bourgeoisie. From the middle of the nineteenth century, Brussels had been the cradle of progressive liberalism, popular among the middle classes. This liberalism was mildly critical of government policies, controlled by conservative Catholics or liberals, but in general was aimed at a harmonious society where the working classes were not dominated or controlled, but integrated into more civilized society. Much of this movement was of a moralist nature, hoping to educate the masses rather than force them by imposed rules. Progressive liberalism became fairly successful, although it was also blamed for stirring up discord within the liberal faction.[16] Jamar was obviously influenced in his social views and professional work by the views of progressive liberalism.

In 1859 Jamar quit his printing company and was elected a Member of Parliament for the Liberal Party. He also became a financial expert, participating in the foundation of several financial institutions. From 1868 to 1870 he was briefly minister of public works, and ended his career as the fourth governor of the National Bank, where he could use his expertise as a printer to lead the battle against counterfeit money.

---

15 The *Panthéon national* produced three books: Th. Juste, *Histoire de la Belgique, illustrée par l'élite des artistes belges* (Brussels, 1840); F. Bogaerts et al., *La Belgique monumentale, historique et pittoresque* (Brussels, 1844); *Les Belges illustres*, 3 vols (Brussels, 1844–45).

16 Els Witte and Jan Craeybeckx, *Politieke Geschiedenis van België sinds 1830* (Antwerp, 1985), 4th edn, pp. 79–86.

The *Bibliothèque nationale*, announced in 1846, was an ambitious project to stir up the flame of nationalist feelings among the Belgian population. It would form a vast collection of works on varied topics, which together would constitute a true Belgian encyclopedia. In the *Bibliothèque nationale* the reader would find complete historical and theoretical instruction in all sciences, each considered from a national point of view. The *Bibliothèque* was divided into five thematic sections: history, science, literature, religion and art. Almost every single book carried the word 'Belgium' or 'Belgian' in its title: 'History of French literature in Belgium', 'History of the introduction of Christianity in Belgium', 'The Belgians during the Crusades', 'Belgian Hagiography, or the life of saints belonging to Belgium', 'Biography of Belgian artists, with a catalogue and appreciation of their works' and so on.

The short prospectus emphasized the nationalist spirit in which the series was conceived. The newly won Belgian independence, although guaranteed by the nations of Europe, could only be maintained by 'inspiring a firm and enlightened patriotism' among the population.[17] This patriotism had to be based on the forging of an identity for the new nation, which throughout its history had rarely been independent. The Belgian people, the prospectus read, were one of the oldest in Europe, and had retained their homogeneity since the times of Caesar; but they were also among the youngest children of the European family. Belgium had often been 'united by intention, but divided by fact'.[18] The patriotism of Belgians was exalted. They loved their country, but they did not know it well enough, and because of that, they were not particularly self-conscious of their identity. Belgians were wary of foreigners, and yet prone to adopt their ideas and views, even with regard to themselves. Their nationalism, the prospectus concluded, was both excessive and defective.

The *Bibliothèque nationale* was destined for a wide audience, but was certainly too expensive to be read by workers or farmers. The price of the volumes was too high for ordinary workers, in particular because subscribers were required to subscribe to the whole series.[19] Moreover, the average literacy level was extremely low, especially in rural areas, so the vast majority of workers and farmers would probably not have been able to read scientific books at all.[20] Early subscribers would have had the honour of having their names published on the front page, although no such lists have been found. Such an honour clearly catered for

---

[17] *Bibliothèque nationale* prospectus (hereafter Prospectus), p. 3: 'inspirer un patriotisme à la fois ferme et éclairé'.

[18] Ibid.: 'le plus souvent uni d'intention, et divisé de fait'.

[19] Around 1850, the average wage for a factory worker was 1.50 francs a day. See J. Art and R. van Eenoo, '1850–1914. Nieuwste Tijd', in *Geschiedenis van de Kleine Man* (Brussels, 1983[3]), pp. 129–54, see p. 136.

[20] In 1866, 59 per cent of the population above eight years old could read and write. Ibid., p. 140; Chris Vandenbroeke, *Sociale Geschiedenis van het Vlaamse volk* (Leuven, 1984[2]), pp. 275–82.

bourgeois audiences, although the commercial setup shows that other audiences were also targeted. The prospectus pointed out that this prestigious and patriotic series should not be omitted from any public libraries, army libraries, schools or colleges. The *citoyens* were called upon to support this important work of popularization, and anyone subscribing to six copies would get the seventh free. The king himself ordered 100 copies.[21]

In this patriotic perspective, science was properly nationalized. The announcement of the scientific section read:

> The sciences, of course, are of all ages and of all nations. We know very well that most of them represent eternal and universal truths, and that there is no such thing as Belgian physics, chemistry, algebra, as much as these sciences are not French. But science too can often become nationalized in its application, either by looking only at those sciences which in their invention or development have given a particular glory to their fatherland, or by studying those sciences with relation to the sky, the land, the needs and the habits of a particular nation. In this way political economics, jurisprudence, medicine, geography and natural history present really national specialties.[22]

Not all the sciences could be nationalized. A list of eleven titles was mentioned in the prospectus, of which four were connected to the natural history of the country (climate, zoology, botany, geology); two titles mentioned the history of science and technology and two others presented a statistical and geographical description of the country. Other volumes gave summaries of the Belgian economy and its political constitution.

- Inventions and discoveries made by Belgians
- History of mathematical and physical sciences in Belgium
- Physical, political and picturesque geography of Belgium
- Statistics of the productive and commercial force of Belgium
- Social economy of Belgium, its agricultural resources, mines, coalmines, salaries, relations among masters and workers, prisons, hospitals, social care etc.

---

21 Mentioned in Tollebeek and Verschaffel, 'Group Portraits', p. 94.

22 Prospectus, p. 5: 'Les sciences, sans doute, sont de tous les siècles et de tous les pays. Nous savons fort bien que la plupart d'entre elles présentent des vérités éternelles, universelles, et qu'il n'y a point de physique, de chimie, d'algèbre belge, non plus que française. Mais souvent les sciences peuvent, dans leur application, se nationaliser aussi, en quelque sorte, soit qu'on se borne à celles qui, dans leur invention ou leurs developpements, ont donné à la patrie un éclat particulier, soit qu'on en étudie d'autres sous le point de vue du ciel, du sol, des besoins, des habitudes de la nation. C'est ainsi que l'économie politique, la jurisprudence, la médecine, la géographie, les sciences naturelles présentent des spécialités tout à fait nationales.'

- The communities of Belgium, their history, their constitutions
- Elements of constitutional, civil and commercial law, the custom law and the transient legislations in Belgium
- Hygiene and popular pathology, accommodated to the climate of Belgium
- Natural history of mammals, birds, fishes, reptiles, insects of Belgium
- Belgian flora
- Geology and mineralogy of Belgium

Not all of the books on this list seem to have been actually published. It is, however, quite difficult to retrace the publishing history of this series. The books had cheap paper covers and were often bound together with other volumes. The title of the series on the cover would thus be lost. The editor also appears to have used different layouts for the covers, sometimes even without any mention of the title of the series. Library catalogues therefore rarely mention the title of the series, and when they do, they often confuse it with other series (such as the '*Panthéon national*', the '*Bibliothèque populaire*' or the '*Collection nationale*'). It is also possible that the editor deliberately created some confusion in producing second editions (never mentioning a publication date) either in response to the market, or as a means of circumventing his own restriction that subscriptions could only be taken for the whole series. By using different names, he may have found a way to sell books outside the series. In any case, a search through libraries and bibliographic dictionaries has produced at most 22 titles for the whole series, only two of which are on science.[23] Many of the titles consist, however, of two or more volumes, so most of the 48 volumes announced in the prospectus may indeed have been printed, albeit not completely according to the proposed scheme.

The science books were not very well represented. Only the books on botany and natural history still exist. Joseph Désiré Hannon published a three-volume book on Belgian flora, while Julien Deby wrote a two-volume treatise on the natural history of mammals.[24] But these already give a fair indication of how science was to be nationalized according to the guidelines of the *Bibliothèque nationale*.

Julien Deby (1826–95) was a young teacher of agriculture and natural history at the Brussels Central School of Commerce and Industry.[25] He was at the time still in the early years of his career, and the natural history volume was probably his

---

[23] Decreus, 'Het verleden', pp. 38–40. Decreus has searched the *Bibliographie nationale. Dictionnaire des écrivains belges et catalogue de leurs publications, 1830–1880*, 2 vols (Brussels, 1886–92). He lists at least 34 volumes in all. My search through library catalogues has only yielded 16 clearly identified titles.

[24] J.D. Hannon, *Flore belge. I. Phanérogamie. II. Cryptogamie*, 3 vols (Brussels, n.d. [1849]). J. Deby, *Histoire naturelle de la Belgique*, 2 vols (Brussels, 1848).

[25] 'Deby, Jules', *Bibliographie Nationale*, vol. I (1886), p. 360. 'Deby, J.', *Le livre d'or de l'ordre de Léopold et de la croix de fer* (Brussels, 1858), vol. 2, p. 167. Deby was not included in the 44 volumes of the *Biographie Nationale* (1866–1986).

first publication. He would later publish on the culture of vegetables and became known as a collector of diatoms (single-celled algae). He went on several scientific expeditions, notably in 1852 to Guatemala, in 1876 to the United States and in 1883 to Cyprus. In 1876 he represented Belgium at the Philadelphia Centennial Exhibition. He was an active member of the Société Belge de Malacologie (the Belgian Malacologocial Society) and the Société Belge de Microscopie (the Belgian Society for Microscopy).

Deby's career typifies him as an amateur scientist with diverse interests and genuine talents, but without a professional profile. He probably never held a university position, and throughout his writings he freely mixed scientific observations with historical anecdotes and personal remarks – a style which is already apparent in his *Histoire naturelle de la Belgique*. Although he acknowledged the support of the leading Belgian botanists Charles Morren and Edmond de Sélys-Longchamps, his work holds virtually no scientific interest. But it does fulfil beautifully the aims of the *Bibliothèque nationale*. It gives basic information on common animals, focusing on the varieties which could be found in Belgium. Animals are designated by their Latin names, with French (and sometimes Dutch) translations. In the chapter on whales, Deby gives an overview of all the whales that had been beached on the Belgian coast, with historical descriptions of them. In the chapter on farm animals, he adds quantitative details on Belgian agriculture, including the information that an ordinary dairy cow annually yields 2,520 litres of milk, worth 504 francs on the Brussels market. He even adds a chapter on '*L'homme belge*', a study of the races composing the Belgian people.[26]

The second science book is more focused on 'pure' science, with most of the book devoted to a very dry exposition of taxonomic classes, families and species. It was written by Joseph Désiré Hannon (1822–70),[27] still a medical student at the time, but already the author of a book on domestic dogs. Hannon was less of an amateur than Deby, and he did indeed pursue an academic career, having been appointed professor at the Université Libre de Bruxelles in 1850.[28] However, he would not become an important scientist, and his publications are quite limited in number. His contribution to the nationalization of science consists mainly of the 60-page essay 'Histoire de la botanique en Belgique' at the end of the *Flore belge*.[29] Hannon opened his essay with the proud observation

26 Deby, *Histoire naturelle*, vol. 1, chapter 3: 'L'homme belge', pp. 35–45.

27 'Hannon, Joseph', *Bibliographie Nationale*, vol. 2 (1892), p. 206. E. De Wildeman, 'Hannon (Joseph-Désiré)', *Biographie Nationale*, XXIX (1956–57), cols 643–4.

28 The Université Libre de Bruxelles was not a state university, but a private initiative. Although by the end of the nineteenth century it had become one of the centres of scientific research in Belgium, around 1850 it was not very prestigious. Most scientists preferred an appointment at a state institution, rather than at the free universities of Brussels and Leuven.

29 Hannon, 'Histoire de la botanique en Belgique', vol. 3, pp. 131–95.

that 'one of the main characteristics of the Belgian people is its agriculture. ... In proportion to its surface area, no other country can boast as many great botanists.'[30] His historical account also gave him the opportunity to elaborate on the efforts of contemporary Belgian botanists, in particular in relation to the study of the national flora.

It was perhaps no coincidence that both authors were young and relative outsiders in their fields. As an amateur, Deby would always remain on the margins of Belgian scientific life. His book on mammals was not the starting point for a research career. His later activities meander freely according to his personal interests. Neither can Hannon be taken as a representative member of the scientific community. Andreas Stynen has shown that among Belgian botanists the idea of writing a national flora was rekindled after 1830, after a new appeal had been launched by Ghent professor Jean Kickx in 1835.[31] Yet, probably due to lack of organization, government support and professional interest, Hannon's book would be the first to effectively do so. Another attempt to write a national flora was even sharply criticized by the botanical community.[32] The nationalist imperative in Belgian botany only made a definitive start after 1860, with the foundation of the Société Royale de Botanique and the work by François Crépin. When Hannon's book was published, the nationalization of botany was not a widespread preoccupation of Belgian botanists.

The contributions by Deby and Hannon to the *Bibliothèque nationale* were thus inspired more by Jamar, and less by the authors themselves. As young and as yet unknown scientists, they were probably motivated by a desire to start a scientific career by putting a popular book on the market. The nationalist perspective suited their objectives very well. Through the *Bibliothèque nationale* they were able to reach a wide, bourgeois audience, to whom their work was presented as part of the patriotic effort of Belgian intellectuals. The legitimation of their work depended on the widespread acceptance of nationalist feelings. National pride and patriotism thus offered suitable opportunities for the scientific movement to make itself known and to catch the public eye. A rather different use of nationalism was visible in the second scientific series Jamar produced, the *Encyclopédie populaire*.

---

30 Ibid., p. 131.

31 Andreas Stynen, 'Vaderlandse weelde op de kaart gezet. Belgische botanici, wetenschappelijke ijver en nationale motieven', *Bijdragen en mededelingen betreffende de geschiedenis der Nederlanden*, 121 (2006): 680–710, in particular pp. 689–92.

32 C. Mathieu, *Flore de Belgique, contenant la description de toutes les plantes qui croisent dans ce pays* (Brussels, 1853).

The Autonomy of Science: the *Encyclopédie populaire*

Three years after Alexandre Jamar had started his *Bibliothèque nationale* he took the initiative for a new series of scientific popularization. The new prospectus,[33] dated August 1849, was written under the aegis of the *Comité de publication* of the otherwise completely unknown *Société pour l'émancipation intellectuelle*. The *Société* was, according to the prospectus, founded in 1836 with the aim 'to propagate education among all social classes [...] and to bring scientific knowledge within reach of everyone'.[34] During the first years of its existence, the *Société* had published some very cheap booklets for 15 centimes, of which all 100,000 copies had been sold. But the recent developments in Europe, in particular the social unrest of 1848, made new initiatives urgent. The *Encyclopédie populaire* was to be a dyke against 'those people, who call themselves friends of the people, but in reality are behaving egoistically or without proper reflection'.[35] The prospectus for the *Encyclopédie* is signed by the Bruges historian Octave Delepierre (1802–79) as secretary of the *Société*.[36] Delepierre was well known for his nationalist historical writings. In particular in 1836, he had published a book on the history of Belgian science, with a second edition in 1840.[37] Delepierre, however, had moved to London in 1843, so it is unlikely that he was effectively still involved in the *Société*'s activities. Decreus states that Ghent historian Henri Moke was president of the *Société*, with the Brussels archivist Antoine Schayes as its secretary and Jamar as editor-administrator.[38] Both Moke and Schayes had contributed to the *Bibliothèque nationale*, and were promoters of a nationalist historiography.[39] There is no further information on the *Société*. One of the books

[33] Société pour l'émancipation intellectuelle, *Encyclopédie populaire ou répertoire universel et élémentaire des connaissances humaines* 'Prospectus' (Brussels, 1849), hereafter referred to as *Répertoire*. This prospectus can also be found in the papers by Adolphe Quetelet, preserved in the Archives of the Belgian Royal Academy in Brussels: (ARB), *Correspondance A. Quetelet*, 2784.

[34] *Répertoire*, pp. 4–5.

[35] *Répertoire*, p. 5: '... la digue la plus puissante à opposer à ceux qui, en se disant les amis du peuple, n'obéissent réellement qu'à une ambition égoïste ou à un entraînement irréfléchi.'

[36] J. Cuvelier, 'Delepierre (Octave)', *Biographie Nationale*, XXIX (Brussels, 1956–57), cols 535–7.

[37] Octave Delepierre, *Aperçu historique et raisonné des découvertes, inventions, innovations et perfectionnemens en Belgique dans les sciences, les arts, l'industrie, etc. depuis les Romains* (Bruges, 1836); the second, slightly expanded edition appeared as *La Belgique illustrée par les sciences, les lettres et les arts* (Brussels, 1840).

[38] Decreus, 'Het verleden', pp. 34–5. Letters with a letterhead, mentioning the administration of the *Société*, are preserved in the Archives of the Belgian Royal Academy: ARB, *Correspondance A. Quetelet*, 1380.

[39] On the wave of nationalist historiography, in which Moke played an important role, see Jo Tollebeek, 'Historical Representation and the Nation-State in Romantic Belgium

in the *Encyclopédie populaire* contains a list of 'honorary members', possibly simply a list of subscribers.[40] The list features politicians, businessmen, students and (in particular) military men.

The prospectus of the *Encyclopédie populaire* put less emphasis on its patriotic goals than the *Bibliothèque nationale*, and focused more on popularization and bringing enlightenment to the general public. Its model was the highly successful *Information for the people* series, published by Scottish publishers William and Robert Chambers. This series of booklets touched on a wide range of topics, including mathematics, science, geography, history and literature.[41] Octave Delepierre had possibly brought this initiative to the attention of Jamar while in London. Yet, whereas the *Information* appears to have been a periodical sheet with short articles of (densely printed) 16 pages, the *Encyclopédie populaire* was to become a series of treatises of several hundreds of pages.

As the *Encyclopédie*'s prospectus explained, 'popular' would refer to the whole of the Belgian people, and in particular those persons who were not able to follow a proper education. A knowledge of scientific principles would enlighten these people and save them from political agitators. Science would thus bring progress and prosperity, in particular by promoting conservative action. By bringing science to the people, the people would be empowered and would be put in a position to become prosperous. Whereas in earlier times wealth was the source of power and knowledge, in modern society knowledge had become the true foundation of both power and wealth: 'Savoir c'est pouvoir, et pouvoir c'est avoir' (Knowledge is power and power is wealth).[42] Once people started to feel and to regulate their own power, they would be less inclined to fight the existing powers of the privileged classes. Both the lower and the upper classes would always need each other:

(1830–1850)', *Journal of the History of Ideas*, 59 (1998): 329–353. On Schayes, see Ellen Van Impe, 'A National Architectural History in Word and Image. Antoine Schayes and the *Histoire de l'architecture en Belgique* (1849–1852)', paper presented at ICAM 13 (International Council of Architectural Museums) Conference, Athens, 4–8 June 2006.

40 J. Thonissen, *Le socialisme et ses promesses* (Brussels, 1850).

41 The publishing history of the *Information for the people* series is puzzling. It is hardly mentioned in any article on the popular science press in Victorian Britain. Some libraries contain bound series of separately published sheets. The publishing ideology of the Chambers brothers is usually related to their *Edinburgh Journal*. This journal also contained a regular feature, 'Popular information on science'. See Joel S. Schwartz, 'Robert Chambers and Thomas Henry Huxley, Science Correspondents: The Popularization and Dissemination of Nineteenth Century Natural Science', *Journal of the History of Biology*, 32 (1999): 343–83. On Robert Chalmers, who was also the author of the anonymous *Vestiges of the Natural History Creation* (1844), see among others, M.J.S. Hodge, 'The Universal Gestation of Nature: Chambers' "Vestiges" and "Explanations"', *Journal of the History of Biology*, 5 (1972): 127–51.

42 *Répertoire*, p. 1.

'everyone has to recognize that the interest of the lower classes is to be instructed; that of the upper classes to instruct them'.[43]

To put these views into practice, the sciences were still to be approached from a nationalist perspective. In fact, the *Encyclopédie*'s prospectus used almost the exact words of the *Bibliothèque nationale*'s prospectus to emphasize that this series, too, was popular and Belgian. One of the criteria for the inclusion of topics in the programme was their 'patriotic utility'. The whole series was placed under the clearly nationalist motto 'Tout pour le pays', while the editors encouraged citizens not to judge the undertaking along the dividing lines of party politics. The prospectus also contained an engraving by Henri Hendrickx (1817–94) and Edouard Vermorcken (1820–1906) illustrating a temple worshipped by people representing the whole nation. The temple carried the words 'A la science – au travail' (To science – to work), indicating the similar but complementary position of intellectuals and workers (see Figure 4.2).[44]

The series would also be larger than the *Bibliothèque nationale*, consisting of no less than 125 small volumes. They were priced at 1.25 francs, with reductions for early subscribers. Still, single volumes were not sold separately, thus making it very difficult for individual poorer people to subscribe, but Jamar probably had the same policy in mind as with the *Bibliothèque* – that is selling the series to the libraries of schools, communities, the army etc. or urging rich philanthropists to distribute copies. As with the *Bibliothèque* it is not easy to track all the existing volumes, as they are listed under quite different denominations in library catalogues;

---

43 Ibid., p. 6. It is interesting to compare the programme of the prospectus with the more far-reaching programme, which was implied in the creation of the British Mechanics' Institutes. Whereas, according to Shapin and Barnes, the Mechanic's' Institutes wished to replace the debauched mind of the workers with solid knowledge, the prospectus did not give a negative assessment of the people's mind. Its programme was aimed at integrating the lower classes into society by showing them how the knowledge of the 'teaching classes' could help them with their interests. Also, the prospectus did not target specific audiences (e.g. factory workers, day labourers, mechanics, etc.), but rather the Belgian population as a whole. It is interesting, however, that the prospectus in style, tone and content does not seem to have been written directly for these audiences, but for the wealthy and liberal supporters of the good cause. The programme of the prospectus also seems different from the concept of 'useful knowledge' which Susan Sheets-Pyenson identifies in the emergence of the popular science periodicals in London and Paris. In any case, Jamar's efforts were too short-lived, and probably too idealistic, to warrant overly general conclusions about the social impulses behind the publication project. See S. Shapin and B. Barnes, 'Science, nature and control: interpreting mechanics institutes', *Social Studies of Science*, 7 (1977): 31–74; S. Sheets-Pyenson, 'Popular Science Periodicals in Paris and London: The Emergence of a Low Scientific Culture, 1820–1875', *Annals of Science*, 42 (1985): 549–72.

44 Hendrickx and Vermorcken were among the most popular romantic illustrators of the period, often engaged in the production of nationalist historical pictures. They both contributed to many of Jamar's publications. See Tom Verschaffel, *Beeld en geschiedenis. Het Belgische en Vlaamse verleden in de romantische boekillustraties* (Turnhout, 1987).

Figure 4.2 Engraving for the *Encyclopédie populaire* prospectus (1849). Archives de l'Académie Royale de Belgique: ARB, *Correspondance A. Quetelet*, 2784.

but a preliminary count, based on a search through digital catalogues, yields 62 titles, most of which consisted of more than one volume.[45] This means that Jamar succeeded in bringing his ambitious *Encyclopédie* fairly well to completion.

From the start, Jamar had divided the *Encyclopédie* into ten separate series: religion and moral philosophy, economics, education, legislation and administration, mathematics and physics, natural history and medicine, history and geography, agriculture, industry and literature. Science and technology were well represented: five series discussed scientific topics. For every series, one director was appointed, who had to take care of the publication of volumes. Clearly, Jamar had good connections in the Brussels intellectual circles. His directors were chosen from the Royal Academy of Science, Letters and Arts, and from various government agencies. All directors had an obvious link to the capital, which made the whole undertaking very centralized.

The general management of the series was entrusted to Auguste Baron (1794–1862).[46] Baron was a French immigrant who had studied at the École Normale in Paris and settled in Brussels in the early 1820s. He soon established himself as a leading man of letters in the Brussels *salons*, and became a well-appreciated teacher and later a member of the Freemason's lodge 'Les Amis Philanthropes'. In 1827 he was appointed to teach literature at the *Musée des sciences et des arts*, a state institution where several courses were offered to the general public for free.[47] Baron was invited to make the opening address; his course was an overwhelming success. When the future of the *Musée* was in danger as a result of Belgian independence, Baron worked to create a university in the capital, which was finally founded in 1834. For 15 years he taught literature at Brussels University before being nominated to the State University of Liège. In 1847, Baron entered the Royal Academy, in a sense making his own 'nationalization' process complete.[48] The choice of Baron as general director of the *Encyclopédie populaire* was apposite. Not only was he an influential figure in Belgian cultural life, with many friends in political and intellectual circles, but he also shared Jamar's idealism in promoting scientific instruction as a means to social improvement. Furthermore, Baron had already cooperated with Jamar on an earlier series, writing some chapters for *Les Belges Illustres* (1845). Baron directed two series in the *Encyclopédie*, viz. on history and geography, and literature and art.

[45] Decreus, 'Het verleden', pp. 40–43 lists 69 titles but included some double counts.

[46] G. Charlier, 'Baron (Auguste)', *Biographie Nationale*, XXIX (1956–57), cols 204–12.

[47] The *Musée* was an expansion of former public courses on science given by Quetelet and Drapiez. The courses combined scientific vulgarization with the encouragement of nationalism. Geert Vanpaemel, 'Onderwijs voor "de meer beschaafde klasse". Het Museum voor Wetenschappen en Letteren te Brussel (1826–1834)', *Scientiarum Historia*, 23 (1997): 3–19.

[48] In 1830 Baron had taken the side of the Belgian insurrection against Holland, but he retained his French nationality for some years before finally acquiring Belgian nationality.

The different sections of the *Encyclopédie* provided very different opportunities for vulgarizing science and putting it into a nationalist perspective. The section on technology, directed by Jules Kindt, concentrated on the description of industrial processes, which provided ample opportunity to discuss Belgian industries. The series on agronomy was rather one-sided. Most of the volumes in this series were written by one author, Alexandre Ysabeau, who to a considerable extent based his work on English treatises. The authors of these two series of booklets on technology demonstrate the intention of being useful in providing less-educated people with the basics of the techniques and craft of knowledge. The series on history, geography, art, education and politics highlighted national issues or provided elementary expositions of basic knowledge. Many of these volumes could just as well have been published in the *Bibliothèque nationale*, representing the same mixture of nationalism and popularization. But a marked difference emerges regarding the series on science.

The two sections on science – mathematics and physics, and natural science and medicine – were directed by Adolphe Quetelet (1796–1874).[49] Quetelet was the most important figure in the Belgian science movement for much of the nineteenth century. He graduated as a mathematician from the University of Ghent in 1819 and soon moved to Brussels, where he secured an appointment as a mathematics teacher at the *Athenée* and managed to be elected a member of the Royal Academy. His career made rapid progress when the government accepted his plan for the creation of an observatory, of which he was to become the first director. He was sent on a tour of the major European observatories to establish contacts with all the important astronomers and mathematicians. After Belgian independence, his position grew even stronger. He became a professor of mathematics at the newly founded Military School, and he was made permanent secretary of the Royal Academy. On top of that he assumed the presidency of the *Commission centrale de statistique*, which organized the population census and prepared statistical reports.

Quetelet was a great promoter of science in Belgium. As a young professor, he organized public lectures on mathematics and probability theory, which he continued at the *Musée*. He published several popular books on science and the history of science, in which he freely compared historical and contemporary situations. He was well aware of the fact that science needed public esteem to be able to grow, and he had a keen sense of how to address lay audiences. When in the course of 1849 Jamar and Baron approached him to collaborate on the *Encyclopédie*, the whole idea must have been very close to his heart. Indeed, in

[49] There is extensive literature on A. Quetelet, mainly connected to his statistical work. See, for example, S.M. Stigler, *The History of Statistics before 1900* (Cambridge, 1900); T.M. Porter, T*he Rise of Statistical Thinking 1820–1900* (Princeton, 1986). There is still no full-length biography. Most accounts are based on E. Mailly, 'Essai sur la vie et les ouvrages de Lambert-Adolphe-Jacques Quetelet', *Annuaire de l'Académie royale de Belgique*, XLI (1875): 109–297; Joseph Lottin, *Quetelet: Statisticien et sociologue* (Louvain, 1912); E. Waxweiler, 'Quetelet (Adolphe)', *Biographie Nationale*, XVIII (1905), cols 477–94.

December 1849, barely a few months after the appearance of the Prospectus, he wrote that 'somewhat more than twenty years ago, I had conceived a project to publish a *Bibliothèque populaire*, a series of small treatises, simple and interesting, which would give a general idea of the large field of science to ordinary people, and which would allow them to appreciate the results that have been reached.'[50] Indeed, with friends (E. Jacquemyns and E. Lefrançois), he had set up in the late 1820s a series of popular books but, according to Quetelet, only four treatises were ever written, including his own books on astronomy and on the calculus of probabilities.[51] The enterprise was then interrupted and finally halted by the Belgian war of independence in 1830.

In 1846 Quetelet had declined Jamar's invitation to collaborate as one of the main promoters of the *Bibliothèque nationale*.[52] He accepted the directorship of the two scientific series for the *Encyclopédie populaire* on the condition that he alone would decide what kind of work would be published and which writers would be appointed.[53] This forced Jamar to somehow modify his initial plan, but Quetelet immediately seized upon the opportunity to make an elaborate scheme of treatises, written by prominent scientists. In an undated manuscript, probably written in preparation for the prospectus, he listed the following plan:[54]

1. Arithmétique et algèbre (M. Mailly)
2. Géométrie et arpentage (M. le cap. Liagre)
3. Probabilités et arithmétique sociale (M. Quetelet)
4. Astronomie (M. Quetelet)
5. Géodésie et construction des cartes géographiques (M. le colonel Nerenburger)
6. Physique (MM. Plateau et Quetelet)
7. Physique du globe et météorologie (M. Houzeau)
8. Chimie (M. Stas)
9. Chimie appliquée (M. Jacquemyns)
10. Mécanique (M. Lefrançois)
11. Technologie (M. Guillery)

[50] A. Quetelet, *Astronomie* (Brussels, 1849), p. 5.

[51] This series is sometimes put under the heading *Bibliothèque populaire*, although it does not appear to have been formally organized.

[52] ARB, *Correspondance A. Quetelet*, 1380, letter from Jamar, 11 February 1846 and draft answer written on it.

[53] Quetelet is most strict on this and complains every time he has not seen a draft before printing (ARB, *Correspondance A. Quetelet*, 1380, letter from Jamar, 22 June 1850) or whenever he has to 'accept' a treatise or an author he has not chosen or agreed upon himself. An example is the discussion of the volume on Public Hygiene, written by an author chosen by Jamar. Quetelet insisted that this work should be included in the series devoted to Education (ARB, *Correspondance A. Quetelet*, 1380, minutes of a letter to Jamar, 15 October 1852 and answer from Jamar, 27 November 1852).

[54] ARB, *Correspondance A. Quetelet*, 2784.

For the series on natural history and medicine, which he also directed, he continued with:

12. Géologie (M. Domalius) [*sic*]
13. Minéralogie (M. Dumont)
14. Botanique (M. Kickx)
15. Zoologie (M. Selys longchamps) [*sic*]
16. Ethnographie, histoire de l'homme et des races humaines (M. Domalius) [*sic*]
17. Anatomie (M. Schwann)
18. Physiologie (M. Gluge)
19. Médecine (M. Vleminckx)
20. Hygiène (M. Mersmann) [*sic*]
21. Chirurgie (M. Burggraeve)
22. Anatomie comparée (M. Cantraine)

An alternative treatise, '23. Chimie appliquée (M. Mareska)', was crossed out, and the number of '24. Art militaire (M. Brialmont)' was changed to 12 (?).

Quetelet's original scheme was later modified, but most of the proposed books were indeed published. Surprisingly, none of the proposed chemistry books were published. Jean-Servais Stas (1813–91) never handed in his manuscript. Edouard Jacquemyns (1806–74), who was retired and engaged in industrial activities, suggested his successor Foppens as a suitable candidate, since he was asked by the government to draw the plans for a collection of 'popular technological textbooks'.[55] Daniel-Joseph-Benoît Mareska (1803–58), who was approached to write a book on applied chemistry, suggested his Liège colleague Joseph Chandelon (1814–85) as the most suitable candidate, but offered to contribute instead a volume on physiological and pathological chemistry.[56] The technological treatise by [H.?] Guillery, which contained some applied chemistry, was transferred to the series on industry (not directed by Quetelet). In the end, no book on chemistry was included.[57] This resulted in the following list of published books:

---

[55] ARB, *Correspondance A. Quetelet*, 1378, letter from Jacquemyns, 21 July 1849.

[56] ARB, *Correspondance A. Quetelet*, 1380, letter from Marseka, 20 August 1849.

[57] Quite a few authors either withdrew their engagement or refused the invitation, such as Edmond de Sélys-Longchamps (1813–1900), Jacques-Olivier De Mersseman (1805–53), Jean-François Vlaeminckx (1800–76), Guillaume Nerenburger (1804–69), André Dumont (1809–57) and François-Joseph Cantraine (1801–63). Quetelet was able to redistribute subjects among his collaborators, and found some new authors: for botany he enlisted Antoine-Frédéric Spring (1814–72); for zoology he turned to Joseph Désiré Hannon, who had already contributed to the *Bibliothèque nationale*; medicine was dealt with by Salomon-Louis Laurillard-Fallot (1783–1875), an author suggested by Jamar; comparative anatomy by Pierre-Joseph Van Beneden (1809–94).

1. Edouard Mailly, *Principes de la science du calcul, arithmétique et algèbre*
2. Jean-Baptiste Liagre, *Éléments de géométrie et de topographie*
3. Jean-Baptiste Liagre, *Traité élémentaire de topographie*
4. Adolphe Quetelet, *Théorie des probabilités*
5. Adolphe Quetelet, *Astronomie*
6. Joseph Plateau, *Physique*
7. Jean-Charles Houzeau, *Physique du globe et météorologie*
8. Jean-Charles Houzeau, *Règles de climatologie ou Exposé sommaire des notions que la science possède sur le cours des saisons*
9. Nicolas-Louis Edouard Lefrançois, *Notions de mécanique générale*
10. Alexis Brialmont, *Précis d'art militaire*
11. Joseph Désiré Hannon, *Traité élémentaire de zoologie*
12. Antoine Spring, *Botanique*
13. Jean-Baptiste-Julien d'Omalius d'Halloy, *Minéralogie*
14. Jean-Baptiste-Julien d'Omalius d'Halloy, *Géologie*
15. Jean-Baptiste-Julien d'Omalius d'Halloy, *Des races humaines ou Eléments d'ethnographie*
16. Louis Fallot, *Aperçu de la médecine dans ses rapports avec les maladies internes*
17. Theodor Schwann, *Anatomie du corps humain*
18. Gottlieb Gluge, *Physiologie*
19. Adolphe Burggraeve, *Chirurgie*
20. Alexandre-Auguste Sovet, *Hygiène privée et publique*
21. Pierre-Joseph Van Beneden, *Anatomie comparée*

The books were not great masterpieces, although they appear to have been widely disseminated in public libraries. The authors enlisted by Quetelet were mostly born between 1800 and 1815, making them in their late forties at the time of writing. They were all well-established scientists, many of them university professors and members of the Royal Academy. The most famous contributor was the geologist Jean Baptiste d'Omalius d'Halloy (178885), who contributed three titles – geology (see Figure 4.3), mineralogy and ethnography – while the anatomist Theodor Schwann (1810–82) was highly regarded as the pioneer of cell theory.

Clearly, Quetelet's approach to organizing the *Encyclopédie populaire* was very different from the strategy used by Jamar in the *Bibliothèque nationale*. Quetelet aimed not to make the *Encyclopédie* a starting opportunity for some young scholars to address a national audience, but more a celebration of Belgian science at its best. And this elite science, Quetelet was quite convinced, was to be found in his personal network. The mathematicians on the list were all former students of his, or junior collaborators at the observatory: Joseph Plateau (1810–83), Nicolas-Edouard Mailly (1810–91), Jean-Baptiste-Joseph Liagre (1815–91), Jean-Charles Houzeau de Lehaie (1820–88). This group was indeed to form a core group in many of his projects.[58] Mailly would later write Quetelet's obituary, Liagre would speak at his funeral and Houzeau would be one of his successors at the observatory. For the other sciences he used his contacts at the Academy or his colleagues at the Military School.

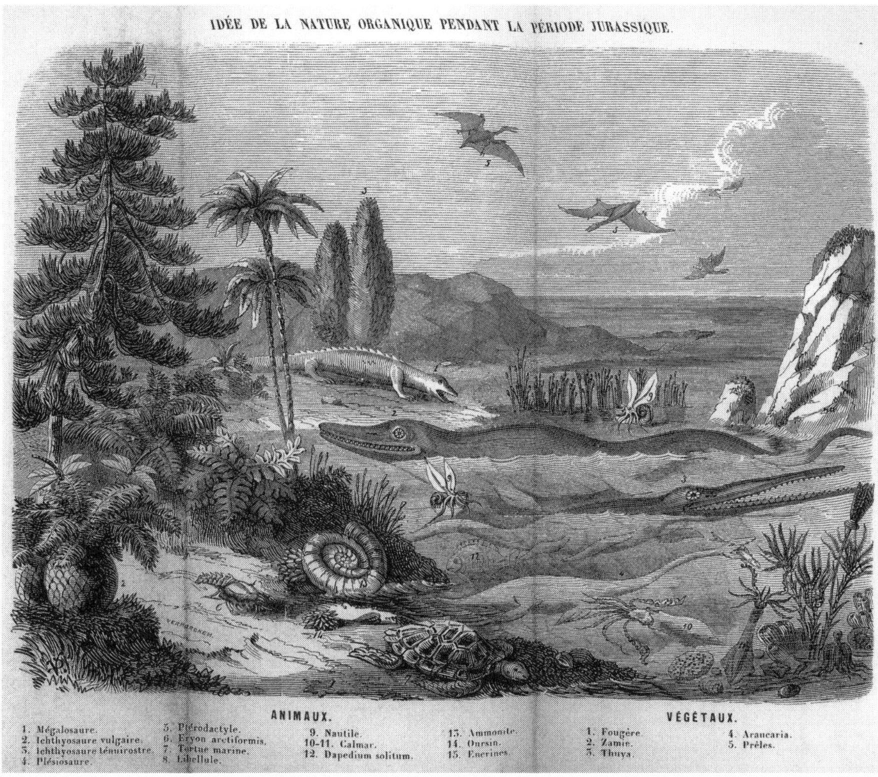

Figure 4.3 A sketch of organic nature during the Jurassic period (d'Omalius d'Halloy, *Géologie*). The picture is adapted from a very similar one accompanying John Phillips's popular articles on geology in *Penny Magazine* (1833).

In his many writings, Quetelet's nationalism was pronounced and undisputed, but at the same time he never presented science as a reflection of purely local interests. Science was universal, and a nation should be proud of the international recognition of its scientists. Quetelet therefore did not urge his authors to put a 'Belgian' veil over the science in their books. Some books contained scant nationalist references, such as engravings representing Adrianus Spieghelius or

58 Quetelet in particular organized networks of meteorological observations and evening sessions in his observatory to spot meteors.

Simon Stevin. Burggraeve included a dedication to Vesalius at the front of his volume on surgery. But other engravings pictured Laplace or Euler just as well. The short historical notes in the volumes by Liagre and Mailly contained no references to Belgium, but merely related the internal development of the respective sciences. The image which resulted was that Belgium had taken its place as a fully developed scientific nation. The few references to national history highlighted the continuity of scientific life in Belgium, but the emphasis was on contemporary science within an international, universal perspective. This is even more striking when compared to the series devoted to industry and commerce, or to agriculture for instance – not under the direction of Quetelet. In his book on technology H. Guillery insists on Belgian resources and ingenuity and focuses on the industries and crafts that are useful for the nation.[59] Jules Kindt in his treatise on the linen industry provides a portrait of one J.-B. Kruck as a frontispiece, 'a citizen deceased a few years ago in obscurity, who is nevertheless the inventor of the mechanical spinning of linen'.[60]

A constant preoccupation in Quetelet's popularization efforts is the promotion of the institutions to which he belonged. His writings present a centralized picture of the scientific community, focusing on the Royal Academy, the Royal Observatory and the Royal Military Academy. Quetelet's omission of chemistry in the science series is probably not completely innocent. Quetelet was promoting his own views and contributions to science by giving much space to his favourite topics: probability theory, 'physique du globe' and meteorology. Although chemistry was an important research domain in Belgian science, it did not fit in well with Quetelet's preferred notion of science in which mathematics played a crucial role. It is indeed fair to say that Belgian science in Quetelet's view coincided with his own scientific research programmes and with the (many) official and central institutions to which he belonged. The *Encyclopédie* and its aura of nationalist elite science legitimated the physical-mathematical notion of science which was dear to Quetelet.

Quetelet's approach to popularization was to show off a fully mature scientific community. Quetelet had indeed managed to enlist an impressive number of good scientists, and had only written two of the 21 books. This was quite different from the section on technology (with only five books written by two authors) or agriculture (with nine books by three authors). The rhetoric of cooperation

---

[59] H. Guillery, Technologie (Brussels, n.d.). The author is mentioned as 'H. Guillery, professeur à l'université de Bruxelles'. At the time of publication Hippolyte Guillery was still a medical student at Brussels University, only to be appointed a professor in 1866. So it is more probable that the treatise was written by the engineer Hippolyte Guillery (1793–1849), who was not, however, a professor at Brussels University. See L. Alvin, 'Guillery (Hippolyte)', *Biographie Nationale*, VIII (1884–85), cols 535–46. A third candidate may be Charles-Etienne Guillery (1791–1861), former colleague of Quetelet at the *Musée* and the Military School, and also a professor at Brussels University and a specialist in chemical technology.

[60] J. Kindt, *Industrie linière. Fabrication des fils et toiles de lin* (Brussels, n.d.), p. 8.

and professional harmony between the elite scientists was often emphasized by Quetelet as secretary of the Academy, and the publication of the *Encyclopédie* was a perfect vindication of his point. Controversy over the demarcation of the different topics – for example between Selys de Longchamps and Hannon, or between Van Beneden and Hannon – was always solved at the Academy, and not with the publisher, Jamar, or even the director of the series.[61] But Quetelet did not apparently hold authority over all his colleagues. Stas, who was supposed to write the book on general chemistry, never gave a page or even an outline, which made Jamar pretty desperate and had other authors postponing their contributions.[62]

The selection of authors throws light on the co-productive opportunities for the emergent scientific community. As most of the authors were selected from the Academy or other Brussels institutions, the impression was firmly suggested that the capital of the nation was at the same time the pinnacle of good science in the country. Anything outside of these circles was immediately discarded by not being represented. When, in 1854, several regional societies proposed the formation of a national federation of societies which would be on the same level as the Brussels Academy, the plan was immediately blown away by the Brussels Academy.[63] As a country could only have one capital, so the scientific community could only have one central and legitimate Academy.

Conclusion

Alexandre Jamar's two publishing efforts represent very different approaches to the popularization and nationalization of science. In the *Bibliothèque nationale*, Belgium was the central concept to which different scholarly fields were applied. Science could only participate when defined around a Belgian topic: local botany and natural history. The *Encyclopédie populaire* also sought to serve the national cause, but its main aim was the education of the masses, who were in danger of being seduced by socialist propaganda. Its goal was to elevate the people to a higher level of knowledge, and science was an ideal means to achieve that goal. Scientists endorsed this new moral role of educating the people, creating for themselves the aura of cultural authority.

In the *Bibliothèque* only junior and relatively unknown scientists would participate. The scope of the series did not put enough emphasis on science to make it an ideal starting point for an academic career. For amateur scientists, such as Jules Deby, this was not a problem. In fact, playing to local audiences was

---

[61] ARB, *Correspondance A. Quetelet*, 1380, letter from Jamar, 13.12.1849 and 16.3.1853.

[62] ARB, *Correspondance A. Quetelet*, 1380, letter from Jamar, 10.7.1852.

[63] J.-B. David, 'Projets pour l'organisation de bibliothèques circulantes et pour la création de sociétés provinciales de littérature et de sciences', *Bulletin de l'Académie Royale de Belgique*, 21 (1854): 238–47.

exactly his ambition as a scientist. The *Encyclopédie* scientific series was of a more professionalized character, and Adolphe Quetelet used them very deliberately to promote the elite status of science (and of the scientists connected to the central institutions he was presiding over). By bringing a large but centralized circle of top scientists into focus he did contribute to feelings of national pride, but only as he was augmenting the prestige of science itself. Belgium had its own scientists; the honour of the country was to honour its scientists.

Quetelet's vindication of Belgian science in the *Encyclopédie populaire* reinforced the image of an elite scientific establishment popularizing science to a lay audience of middle-class citizens. By refusing to put science in a Belgian perspective as the *Bibliothèque nationale* had attempted to do, he emphasized the high standards of science, the quality of the authors writing on scientific topics and the cultural distance between professional science and lay audiences. Whereas Deby had referred to local contributions and the knowledge of non-academic observers, Quetelet and his pupils stressed the formalist, abstract nature of science, adding that they were aware that these books did not do justice to the real complexities of science. The audiences were thus instructed not to think that they had a manual at hand from which to learn the first steps of science. The scientists held the power to instruct the people, and by insisting on the elementary level adopted in their books, they also made it clear that science was a profession in itself, whose practice was in fact reserved to a small elite group, thereby reinforcing their own position.

Paradoxically, by presenting Belgian science as a local representation of an international movement, Quetelet's efforts also emphasized, perhaps without his being aware of it, the peripheral nature of Belgian science. As Belgian contributions were not particularly put in the spotlight, the impression was reinforced that the real progress of science was taking place elsewhere. The Belgian achievement was to assume its position in the world of science, which by implication was outside Belgium and indeed a much larger and global phenomenon. The connecting hinge between local and global science was to be Quetelet's Academy, which thus gained importance and prestige. In this sense, the popularization of Belgian science, as pursued by Quetelet, promoted the local interests of his small, centralized scientific community at the expense of their international ambitions.

Chapter 5
Circumventing the 'Elusive Quarries' of Popular Science: The Communication and Appropriation of Ganot's Physics in Nineteenth-Century Britain

Josep Simon[1]

Adolphe Ganot (1804–87) published two physics textbooks in Paris that went into a large number of successive editions in France and other countries through translation into Spanish, English and other languages. During the second half of the nineteenth century, these made a decisive contribution to physics and its teaching on an international scale.[2] Ganot's motivations as an author and publisher were driven by his position as a teacher preparing mainly medical students for the *baccalauréat-ès-sciences* examination and his earlier involvement in courses for artisans organized by the Association Philotechnique.

His books were often characterized as 'popular', and Ganot certainly had an educational and social agenda as an author and teacher. But their actual readership was particularly diverse, including elite practitioners of science, booksellers, students, teachers, instrument makers, journal reviewers and compilers, theological writers and leisure readers in different countries. Their characterizations of the books were also particularly diverse and encompassed the realms of the popular, leisure and entertainment, pedagogy and research.

When confronted with this case study, one struggles to find clear answers from a general survey of secondary literature on 'science popularization'. Despite the numerous contributions to this field in recent decades, it is easy to feel that it currently suffers from a certain ontological, epistemological and methodological disorientation.[3]

---

[1] I am grateful to Jon Topham for fruitful discussions that helped clarify my insights and improve my conception of this paper.

[2] Josep Simon, *Adolphe Ganot (1804–1887) and his textbooks of physics* (Oxford, 2004), unpublished MSc thesis, and 'La famille Baillière et l'introduction du *Traité de Physique* de Ganot en Angleterre', in Danielle Gourevitch and Jean-François Vincent (eds), *J.-B. Baillière et fils, éditeurs de médecine* (Paris, 2006).

[3] See Chapter 1 by Jon Topham in this volume.

Peter Burke defined the study of popular culture as an 'elusive quarry', acknowledging the double difficulty in having to deal with past cultures that are different to our own and using indirect means and sources.[4] As argued by Jon Topham in this volume, 'science in popular culture' and 'science popularization' are intersecting subjects, which could be made part of a wider context of scholarship by conceptualising 'popularization' through a historical framework centred on 'communication'. The history of popular culture, education theory, the history of reading, the history of technology, and science communication studies are already helping to foster a move from the context of knowledge production to its context of use, and conceptualising communication as an interactive and multidirectional phenomenon.

In this sense, Roger Chartier has remarked that the pretension of defining popular culture by the distribution of objects or texts in society is misleading, as there are in fact no objects that are 'popular' *per se*. It is the way of using these products that really characterizes what is popular. Thus, we should focus on studying the ways in which these are appropriated – the different ways in which culture is actively used and given meanings by individuals.[5] In relation to texts, these are often different from the ones assigned to them by actors such as the author, the printer and the bookseller;[6] hence the relevance of studying not only the social topography of the practice of science, but also the communicative and appropriative processes shaping it.

'Appropriation' is also a useful concept for tackling the study of the international communication of knowledge,[7] a topic whose relevance has been emphasized by James Secord's historiographical proposal of bringing 'communication' to the centre of our analysis and narratives.[8] In recent decades, the major efforts pushing the historiography of science in an international direction have often fallen under the labels of studies on 'science and empire' and on 'centre and periphery'.[9] Interestingly, the models of communication traditionally used in these research areas on a macroscopic level are afflicted by the same problems as those

---

4 Peter Burke, *Popular Culture in Early Modern Europe* (London, 1978).

5 Roger Chartier, 'Culture as Appropriation: Popular Cultural Uses in Early Modern France', in S.L. Kaplan (ed.), *Understanding Popular Culture: Europe from the Middle Ages to the Nineteenth Century* (Berlin, 1984), pp. 229–53, on p. 233.

6 Jonathan R. Topham, 'Scientific Publishing and the Reading of Science in Nineteenth-Century Britain: A Historiographical Survey and Guide to Sources', *Studies in History and Philosophy of Science*, 4/31 (2000): 559–612.

7 K. Gavroglu, M. Patiniotis, F. Papanelopoulou, A. Simoes, A. Carneiro, M.P. Diogo, J.R. Bertomeu Sánchez, A. García Belmar and A. Nieto-Galan, 'Science and Technology in the European Periphery. Some historiographical reflections', *History of Science*, 46/2 (2008): 153–75.

8 For more details see Chapter 1 by Jon Topham in this volume. James A. Secord, 'Knowledge in Transit', *Isis*, 95/4 (2004): 654–72.

9 See Gavroglu, 'Science and Technology in the European Periphery'.

traditionally used – on a microscopic level – in the study of popular culture and science popularization. In this sense, Secord's proposal can certainly be useful for establishing links between the practices of science in local contexts, and those performed across national and international boundaries.

In this chapter I intend to problematize 'science popularization' through a case study analysing the 'communication' of Ganot's textbooks from France to Britain, and their 'appropriation' in the latter. Before proceeding with my case study, I will introduce my approach through my reading and appropriation of the work on the structure of society and scientific practice by Edward Shils and Ludwik Fleck, respectively, and my assessment of the centre/periphery divide as represented by the work of Joseph Ben David. I will argue that the central role that Shils and Fleck give to communication helps to clarify the processes involved in science popularization, and can be used to bring together the study of knowledge circulation in local, national and international contexts.

Focusing on the Processes: Centre and Periphery vs Communication and Appropriation

As I argue in this section, by abandoning a centre/periphery model for a focus on communication and appropriation, the study of Adolphe Ganot's textbooks makes it possible to cross the boundaries that in the history of science traditionally divide the popular, education and research, and the local, the national and the international.

In Europe and the Americas, the nineteenth century was characterized by the rise of secondary education and the introduction – on a systematic basis – of science in the secondary school and university curriculum. Concomitantly, this period was critical for the rise of nation states.[10] Furthermore, it was also the time of the constitution of distinct scientific disciplines such as physics.[11]

In this context, scientists and education administrators observed their foreign peers from their national standpoints. The rhetorical use of the virtues of the other was often used as a political strategy to gain social and professional status, and the organization of education was considered to be a critical matter in ensuring the formation of a professional community and its associate disciplinary identity.[12]

---

[10] A. Anderson, 'The Idea of the Secondary School in Nineteenth-Century Europe', *Paedagogica Historica*, 40/1 and 2 (2004): 93–106; A. Green, *Education and State Formation. The Rise of Education Systems in England, France and the USA* (Basingstoke and London, 1990); E. Roldán Vera, *The British Book Trade and Spanish American Independence: Education and Knowledge Transmission in Transcontinental Perspective* (Aldershot, 2003).

[11] Iwan Rhys Morus, *When Physics Became King* (Chicago, 2005).

[12] Robert Fox and George Weisz, 'Introduction: The institutional basis of French science in the nineteenth-century', in Fox and Weisz (eds), *The Organization of Science*

By the mid-nineteenth century, reference to the serious consequences of the low status of scientific practice in England, in contrast with the virtues of the organization of scientific research and teaching in France and Germany, had a permanent position in rhetorical discourses of British scientists and educators.[13] Contemporarily, French books and periodicals had a wide readership in Britain.[14] In subjects such as physics, English schools and university colleges, teachers and students were highly dependent on French textbooks such as those by Adolphe Ganot. These books were the fruit of a mature system of scientific education developed over the previous half century. By contrast, scientific education in England only started to be instituted on a solid and general basis from the 1860s.[15]

The question of the relative scientific status of France, Britain and Germany during the nineteenth century has traditionally been tackled through the work of Joseph Ben-David.[16] His comparative study of scientific production in the aforementioned countries and the USA contributed to promoting in history of science the use of a centre/periphery model introducing biases in the national and international study of science in a historical perspective. In focusing especially on structural aspects and not giving analytical relevance to how knowledge circulates across national borders, the dynamics of Ben-David's model are rather static. Furthermore, his centre/periphery model has favoured the use of unconceptualized national boundaries and the contraction of the international dimension of our discipline.

On the other hand, in local contexts, Ben-David's use of Kuhn's concepts of 'paradigm' and 'revolution' led him to restrict the production of science to an elite

and Technology in France, 1800–1914 (Cambridge, 1980); Agustí Nieto-Galan, 'The Images of Science in Modern Spain', in K. Gavroglu (ed.), *The Sciences in the European Periphery During the Enlightenment* (Dordrecht, Boston, London, 1999); Graeme Gooday, 'Lies, Damned Lies and Declinism: Lyon Playfair, the Paris 1867 Exhibition and the Contested Rhetorics of Scientific Education and Industrial Performance', in I. Inkster et al. (eds), *The Golden Age. Essays in British Social and Economical History, 1850–1870* (Aldershot, 2000), pp. 105–20; Rudolf Stichweh, *Zur Entstehung des modernen Systems wissenschaftlicher Disziplinen: Physik in Deutschland* (Frankfurt, 1992).

13 Gooday, 'Lies Damned Lies and Declinism'; Matthew Arnold, 'Schools and Universities on the Continent', in R.H. Super (ed.), *The Complete Prose Works of Matthew Arnold* (Ann Arbor, MI, 1964), vol. IV.

14 Josep Simon, 'The Baillières: The Franco-British Book Trade and the Transit of Knowledge', in Robert Fox and Bernard Joly (eds), *Franco-British Interactions in Science Since the Seventeenth Century* (Paris, 2008).

15 D.P. Newton, 'A French influence on nineteenth- and twentieth-century physics teaching in English secondary schools', *History of Education*, 12/3 (1983): 191–201; Robert Fox, 'The Context and Practices of Oxford Physics, 1839–77', in Robert Fox and Graeme Gooday (eds), *Physics in Oxford, 1839–1939. Laboratories, Learning and College Life* (Oxford, 2005), pp. 24–79.

16 Joseph Ben-David, *The Scientist's Role in Society: A Comparative Study* (Englewood Cliffs, 1971), pp. 1–15, 186–7; Fox and Weisz, 'Introduction'.

occupying the esoteric circle of scientific practice, and to limit its communication with other parts of society to extraordinary occasions. Hence, quantifiable research was the driving criterion in his scientific international ranking, and other practices such as education and popularization that could be alternative indicators of the relevance of science in culture and society were dismissed.

Ben-David tried to find a political compromise in the ranking arising from his analysis. Thus, the succession of France, Germany and the USA as centres could be complemented by assigning to Britain a regular second influential position and a brief emergence to the centre in the transit from French to German hegemony. As I have suggested, a somewhat different picture emerges if we focus on education, popularization and international communication. Defining historically the 'centre' and the 'periphery' is – like deciding what is the 'popular' – not a simple matter. It depends on historical perceptions of the self and the other held by scientific practitioners, on national and international historiographies, and on the forms of transnational circulation of science. Conceptualizing this historical problem involves the use of an analytical framework based on communication and appropriation.

In this sense, the work of Edward Shils and Ludwik Fleck offers interesting tools to conceptualize 'communication'. As the editor of *Minerva*, Edward Shils decisively contributed to the promotion of the centre/periphery model in the history of science through the publication of works such as that by Ben-David.[17] However, his own work focused on a centre/periphery model applied to society in a more local perspective.

Shils characterized society as structured by a centre and a periphery, not located in spatial or geographical terms but in relation to a system of symbols, values and beliefs. The centre is structured by the actions of the elites, who are beheld with authority and are responsible for propounding, pursuing and affirming the 'central value system' in society. The latter is the central zone of society not only because it is espoused by the elites, but also because of its intimate affinity to what society holds to be sacred.[18] In fact, the 'central value system' is configured through consensus. Consensus is possible both because and when individuals being part of society can successfully interact through communication, which happens in turn when a common value system is shared.[19]

---

[17] I am indebted to Roy MacLeod, who drew my attention to the relevance of Shils in this context through the presentation of his research in two seminars at Manchester and Leeds in 2006.

[18] Edward Shils, 'Center and Periphery', in *Center and Periphery: Essays in Macrosociology* (Chicago and London, 1975), pp. 3–16.

[19] Edward Shils, 'Society: the Idea and its Sources', in *Center and Periphery*, pp. 17–33.

Shils's macro social model shared many characteristics with the social model of scientific practice proposed by Ludwik Fleck, whose work he had read.[20] Despite the different aims and scale of their approaches, their models have a similar geometry, structuring society in concentric areas occupied by elites and ordinary citizens, respectively. They also have similar dynamics, driven by communication across these areas, and they acknowledge that there is no absolutely privileged direction in this communication.

Fleck's conceptualization of scientific practice through 'thought collectives' configuring and constrained by a 'thought style', and their subdivision into small 'esoteric circles' and large 'exoteric circles', stressed, like Shils's centre/periphery model, the transformative and multidirectional role of communication. Furthermore, it suggested that communication was a fundamental driving agent that shapes science both in local contexts and across national boundaries.[21] Thus, the boundaries that have traditionally been used to separate the elite and the laity and the local and national and the international can be challenged through a more accurate conceptualization based on the study of communication and appropriation.

In addition, Fleck offered a useful characterization of the carriers of scientific knowledge and agents of scientific communication through the definition of three major genres. According to him, a 'thought style' is represented by 'vademecum science', as the carrier of common expert knowledge and the tool binding a 'thought collective'. It is opposed to 'journal science' in that it is comprehensive and consensus-based. It differs from 'popular science' in that it is critical and organized. However, the character of the vademecum is also determined by the fact that every communicative action – including those leading to its configuration – makes knowledge more exoteric.[22] Thus, communication always transforms knowledge, and it acts towards the constitution of 'thought styles' based on social and intellectual consensus. 'Appropriation' is performed through communication, made possible and constrained by a 'central value system', or a 'thought style' and, at the same time, contributing to its configuration.

In the following sections I consider the appropriation and communication of Adolphe Ganot's textbooks and its movement between esoteric and exoteric circles in France and Britain. I will first briefly present the books and their French readers, and will discuss Ganot's own appropriations. Then, I will consider the communication of Ganot's physics in nineteenth-century Britain through six case studies analysing their appropriation by various readerships, including the publisher, the translator, the reviewer, the science teacher, the theological reader, the researcher and the instrument maker, respectively. I will expose the complexity

20 Thomas S. Kuhn, 'Foreword', in Ludwik Fleck, *Genesis and Development of a Scientific Fact* (Chicago, 1979), pp. vii–xi.

21 On Fleck see also Jon Topham's Chapter 1 in this volume. Fleck, *Genesis and Development*, pp. 39, 41, 51, 98–9, 105–7, 109, 112–13.

22 Ibid., pp. 114, 124.

of defining distinct genres and labelling of scientific writing and publishing by showing that Ganot's physics can be viewed not only as 'textbook science', but also as 'popular science' and 'vademecum science'. And I will argue that the wide diversity of processes of communication and appropriation to which Ganot's textbooks were central contributed to giving them the quality of representing a thought style or a central value system defined by Franco-British communication.

The Genesis and Readers of Ganot's *Physique*

Adolphe Ganot was the author of two physics textbooks, the *Traité élémentaire de physique expérimentale et appliquée* (1851) and the *Cours de physique expérimentale à l'usage des personnes étrangères aux connaissances mathématiques* (1859). From 1850, he taught physics and directed a private school in Paris, preparing for the *baccalauréat-ès-sciences* examination controlling access to scientific, medical and engineering studies in France.[23]

As a private secondary school teacher, he was not part of the French scientific elite. However, the pedagogical practices developed in schools like his had an essential role in the configuration of French scientific education, a fact admitted by the elite.[24] Ganot's textbooks were the result of the communication and appropriation processes involving, in the context of the Parisian *Quartier Latin*, an exoteric circle composed of private teachers, printers, booksellers, journal compilers and instrument makers, and an esoteric circle formed by the French scientific elite linked to the state university administration. The pedagogical practices created by the former were subsequently appropriated by the latter, and, in parallel, Ganot's textbooks expanded their context of use by travelling from the exoteric circle of private education to the esoteric circle of state secondary education.[25] Subsequently, as we will see, they were successfully communicated to and appropriated into a larger geographical context of an international character.

In the preface to the *Traité*, Ganot stated that he had designed his book as the complement to his physics cabinet and lectures. The book's subtitle targeted students preparing for the *baccalauréat-ès-sciences*, the final year examination of the School of Medicine, and the entrance examinations for the École Polytechnique and the

[23] Bruno Belhoste, Claudette Balpe and Thierry Laporte (eds), *Les sciences dans l'enseignement secondaire français. Textes officiels* (Paris, 1995).

[24] Jean Baptiste Dumas, 'Rapport sur l'état actuel de l'enseignement scientifique dans les collèges, les écoles intermédiaires et les écoles primaires', ibid., pp. 207–23.

[25] Anon., 'Ganot et Maneuvrier: Traité Élémentaire de Physique (22e édition)', *Revue Générale des Sciences Pures et Appliquée*, 1/15 (January 1904): 746; Simon, *Adolphe Ganot*.

École de Saint-Cyr.[26] A contemporary advertisement[27] added that some parts could also appeal to readers interested in physics for other purposes. From its second edition, a more general public was targeted, including all French secondary and higher education establishments.[28] Designing potential audiences and targeting the widest set of readers was a usual task in the work of nineteenth-century publishers.[29] The *Traité* had large print runs and numbers of editions,[30] only comparable to certain pedagogical treatises, and only surpassed by almanacs, periodicals, trade catalogues or successful novels of the time.[31] Most of Ganot's school customers were students of medicine and pharmacy; only a few of them were science and engineering students.[32] However, in France the book also had readers who were not centrally defined in its subtitle, belonging both to 'esoteric' and 'exoteric' circles: practitioners of science becoming part of the elite of the Académie des Sciences such as Léon Foucault, journal editors such as the ex-Jesuit François-Napoléon-Marie Moigno, instrument designers such as Zénobe Gramme and military engineers such as Arthur-Constantin Krebs. As I will show, Ganot combined his experience as a teacher and 'journal science' reader to produce books that readers confronted not only as textbooks, but also as 'popular science' and as science vademecums.

In writing his second textbook, the *Cours*, he appropriated his first book, the *Traité*.[33] Ganot took special care to explicitly state that the *Cours* was a different book, as his interest was to widen his publishing market by targeting new audiences. However, he certainly re-read and appropriated the *Traité* in order to prepare the *Cours*.

The *Cours*'s subtitle addressed the 'gens du monde', female students involved at different levels in the higher ranks of the French primary education system, and in general, readers who were not familiar with mathematics. The first term – originally coined to designate the leisure aristocratic class – was subsequently

---

26 Adolphe Ganot, *Traité Élémentaire de Physique Expérimentale et Appliquée et de Météorologie* (Paris, 1851).

27 Inserted in a copy of its first edition.

28 Ganot, *Traité* (Paris, 1853).

29 Bernardette Bensaude-Vincent, Antonio García-Belmar and José Ramón Bertomeu-Sánchez, *L'émergence d'une science des manuels: les livres de chimie en France (1789–1852)* (Paris, 2003), pp. 123–8, and James Secord, *Victorian Sensation* (Chicago, 2000), p. 53.

30 The 18th edition of the *Traité* (1880) had a print run of 20,000 copies (Archive of the House of Longman, *Atkinson File*). By then Ganot claimed to have produced 204,000 copies of the book since 1851. Ganot, *Traité* (Paris, 1880).

31 Fréderic Barbier, 'Les marchés étrangers de la librairie française', in H.J. Martin (ed.), *Histoire de l'édition française* (Paris, 1985), pp. 269–81.

32 Adolphe Ganot, *A messieurs les membres du jury de l'expropriation pour cause d'utilité publique* ... (Paris, 1856).

33 This idea was developed by Secord in relation to successive editions of a book. However, here I will concentrate on Ganot's appropriation of his first book in preparing the *Cours*. See Secord, *Victorian Sensation*, pp. 78, 385–93.

appropriated to designate those who were not constrained by a formal education course, and who could read a book or attend a lecture for leisure purposes.[34]

Ganot argued that because physics was one of the most interesting and current subjects, it should also be one of the most 'popular' and better known to the public. Unfortunately – according to him – this was not the case; not only among the laity, but also among the 'literary class'. The reasons were the expenditure required to acquire scientific instruments, and the common use in treatises of algebraic language, which constrained the communication of physics to most of the readers. This had created the misleading impression that it could only be studied by an exclusive set of people.[35] Indeed, mathematics, closely linked to the training of the French elite at the École Polytechnique, possessed the highest social prestige among the sciences. Only during the second half of the century did experimental physics progressively challenge this intellectual and social hierarchy.[36]

Thus, Ganot's *Cours* intended to make physics accessible to everyone by reducing the presence of mathematical formulae, unnecessary – in his opinion – to explain many physical phenomena, and by introducing a good number of high-quality illustrations representing experiments at the moment of their performance (being able to substitute them when instruments were not available)[37] – see Figures 5.1 and 5.2.

As for the *Traité*, the *Cours* widened in the course of its life the public targeted by its subtitle, including students of the secondary school classical curriculum.[38] Ganot's second book also had large print runs and numbers of editions, although both inferior to those of the *Traité*.[39]

According to Ganot, the *Cours* was a completely different book from the *Traité*, especially because of its illustrations and its composition.[40] Most illustrations were new, and they showed instruments being manipulated by human figures, often in domestic and social environments (see Figure 5.1). In contrast, in the *Traité*, only hands emerging from the white of the page appeared (see Figure 5.2), although some experiments were also located outside the classroom.

---

[34] See Émile Littré, *Dictionnaire de la langue française* (Paris, 1863–69), and the online lexicographic corpus FRANTEXT <http://www.frantext.fr/> (accessed 21 October 2007).

[35] Advertisement inserted in Adolphe Ganot, *Cours de physique purement expérimentale, à l'usage des personnes étrangères aux connaissances mathématiques* (Paris, 1859).

[36] Claudette Fournier-Balpe, *Histoire de l'enseignement de la physique dans l'enseignement secondaire en France au XIXe siècle* (Paris, 1994).

[37] Advertisement in Ganot, *Traité* (Paris, 1859).

[38] Ganot, *Cours* (Paris, 1872).

[39] In 1881, the eighth edition of the *Cours*, the last prepared by Ganot, had a print run of 13,000 copies. According to him, since 1859, 64,500 copies of the book had been produced. Ganot, *Traité* (Paris, 1880).

[40] Ganot, *Cours* (Paris, 1859).

524 ÉLECTRICITÉ.

d'une bobine à deux fils : l'un gros, n'ayant que 3 à 4 mètres de longueur; l'autre fin, enroulé sur le premier et d'une longueur de 500 à 600 mètres (fig. 372). Ces fils sont non-seulement recouverts de soie avec soin, mais d'un vernis à la gomme laque destiné

Fig. 372. — Effets physiologiques des courants d'induction.

à les isoler entre eux. Deux bornes de cuivre placées à gauche, sur la planchette qui porte la bobine, sont en communication avec les pôles d'un seul élément de Bunsen. De la borne la plus à gauche part une lame de cuivre qui se rend à une petite roue dentée, mue par un mouvement d'horlogerie, et communiquant avec l'un des bouts du gros fil de la bobine. L'autre bout du même fil se prolonge sur la planchette jusqu'à la seconde borne. Enfin, sur la droite, sont deux autres bornes auxquelles aboutissent les deux extrémités du fil fin qui recouvre le premier; et des mêmes bornes partent deux autres fils qui représentent les prolongements du fil induit. Pour avoir des commotions, on termine ces deux derniers fils par des cylindres de cuivre qu'on prend dans les mains, comme

Figure 5.1 Normal type size. The performance of the demonstration of induction by a magnet in the *Cours*: 'Men, the Body and the Senses'. Ganot, *Cours* (1859), p. 524.

INDUCTION. 691

Cette expérience montre que l'électricité des machines électriques peut, aussi bien que celle de la pile, donner naissance à des courants d'induction.

L'appareil de M. Matteucci sert aussi à démontrer la production des courants induits par l'influence des courants voltaïques. Pour cela, on fait passer le courant d'une pile dans le fil inducteur C, et, en même temps, on fait communiquer les deux fils *h* et *i* avec un galvanomètre. Or, on observe, au moment où le courant inducteur commence ou finit, les mêmes phénomènes qu'avec la bobine décrite ci-dessus (fig. 520), et l'aiguille du multiplicateur est d'autant plus déviée que les deux plateaux A et B sont plus voisins.

698. **Induction par les aimants.** — On a vu que l'influence

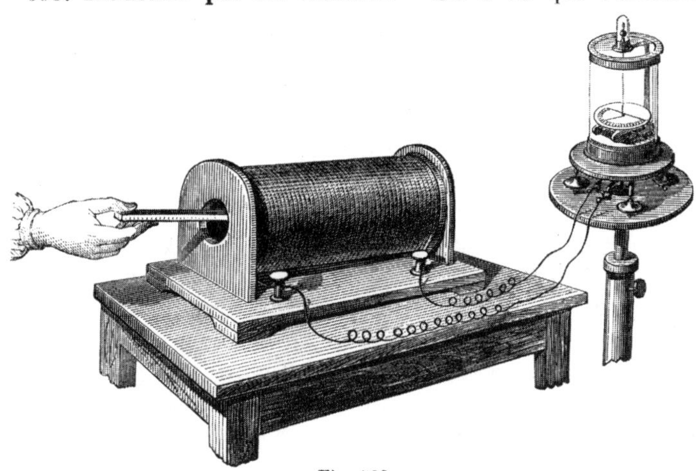

Fig. 522.

d'un courant aimante un barreau d'acier (685) ; réciproquement, un aimant peut faire naître, dans les circuits métalliques, des courants d'induction. M. Faraday l'a démontré au moyen d'une bobine à un seul fil de 200 à 300 mètres de longueur. Les deux extrémités du fil étant mises en communication avec un galvanomètre, comme le montre la figure 522, on introduit brusquement dans la bobine, qui est creuse, un fort barreau aimanté, et on observe alors les phénomènes suivants :

1° Au moment où on introduit le barreau, le galvanomètre indique, dans le fil, un courant induit instantané, inverse de celui qui existe autour du barreau, en assimilant celui-ci à un solénoïde, comme on l'a fait dans la théorie d'Ampère (684).

Figure 5.2 Normal type size. The performance of the demonstration of induction by a magnet in the *Traité*: 'The Lonely Hand, the Galvanometer and the Sight'. Ganot, *Traité* (1859), p. 691.

However, the *Cours* had fewer illustrations than the *Traité*, both absolutely and proportionally. Concerning the composition, the *Cours* used more exemplification, exclamation marks, historical anecdotes and analogies with 'ordinary language'. Instrument and experiment descriptions, the enunciation of laws and definitions, and historical introductions were shortened by limiting the number of elements introduced. Algebraic and numerical applications and the section of problems present in the *Traité* disappeared in the *Cours*. Some of the books, chapters and articles composing the former were eliminated or amalgamated in the latter.

In spite of this, both had the same basic structure and contents. In general, their illustrations described the same experiments, and one could move from the *Traité* to the *Cours* simply by adding a body to the hands of the anonymous manipulator appearing in the former, and a location background. The use of a special typographic system and its pedagogical aims were also coincident in the two books.

By 1859, Ganot had acquired experience as author and publisher in the production, marketing and use of eight editions of the *Traité*. The *Cours* was thus designed to target a different market, whose public grew due to educative reforms expanding the teaching of science in schools. These reforms were propounded by the French scientific and political elite. However, their configuration was the result of the communicative interaction between 'esoteric' and 'exoteric' circles. Ganot's redesign of his first textbook was a result of this interaction, and of his own appropriation of a work which had already played a major role in the configuration of a 'thought style', in the context of French physics education. As a result, a new book was produced, contributing in turn to reshaping this configuration.

Ganot's textbooks were translated into most European languages. Their study thus offers the opportunity to assess the processes of the international communication and appropriation of science in nineteenth-century Europe. The agents in these processes shaped and translated the books into new contexts and meanings, contributing to the configuration of a 'thought style' of an international character. The next section analyses these processes in a case study of their development in Britain, one of the countries in which the communication and appropriation of Ganot's textbooks had a longer and wider extension.

The Communication and Appropriation of Ganot's Physics in Nineteenth-Century Britain

The communication and appropriation of Ganot's physics in Britain was first developed through the agency of international booksellers and international students. The *Traité* was published in English by Hippolyte Baillière, whose family arguably built the most important nineteenth-century international network in the scientific and medical book trade. Their work contributed to defining the communication and appropriation of French science into Britain and vice versa, thus playing an important role in the configuration of a Franco-British 'thought style'.[41] Their role was shared by international students. Educative trips to France and Germany were then a common route for British medical and chemistry students.[42] Back in Britain, many of them engaged in the translation of continental books and articles for scientific journals.[43] The appropriation of foreign works involved the design of new publics, constrained by different cultural and social settings, including different disciplines and school cultures.

Edmund Atkinson (1831–1900), the English translator of Ganot's textbooks, fits into this pattern. Trained as a chemist in England, he subsequently studied in Germany and France.[44] Back in England, he contributed reviews of foreign works to the *Philosophical Magazine*, published by William Francis (who had also studied in Germany). In this context, Baillière commissioned the *Traité*'s translation from Atkinson, through recommendation of his peer, Francis.[45] His translation of Ganot's *Traité* coincided with the start of a long career as a physics teacher, translator and scientific editorial adviser.[46] At the time, only a few physics textbooks designed for high and secondary level education were available in Britain,[47] and the articulation of secondary school education on a national scale, through the organization of science examination systems, was incipient.[48] Atkinson was not a member of the English scientific elite, so he did not have direct input in the design and administration of scientific research and education in England. Initially, he conducted and published research papers as a chemist, but, subsequently, he did not publish research into physics in any specialized journal.

As indicated by James Secord, a book is not only a compilation of pages, but the sum of a wide array of representations including printing and publishing practices, reviews, advertisements, excerpts, prefaces, translations and readers' perceptions and responses.[49] This section will assess some of these representations through the study of the appropriation and communication of Ganot's books in England, and its agents.

---

[41] Simon, 'The Baillières'.

[42] Adrian Desmond, *The Politics of Evolution. Morphology, and Reform in Radical London* (Chicago, 1989); Jean-Claude Caron, *Générations romantiques: les étudiants de Paris et le Quartier latin, 1814–1851* (Paris, 1991).

[43] William H. Brock and Arthur J. Meadows, *The Lamp of Learning: Two Centuries of Publishing at Taylor & Francis* (Bristol, 1998).

[44] [George Carey Foster and Hugo Müller], 'Obituary Notices', *Journal of the Chemical Society, Transactions*, 79 (1901): 888–9.

[45] Brock and Meadows, *The Lamp of Learning*, p. 197.

[46] Archive of the House of Longman, *Atkinson File*.

[47] Newton, 'A French influence'.

[48] Roy MacLeod (ed.), *Days of Judgement: Science Examinations and the Organization of Knowledge in Late Victorian England* (Driffield, 1982) and Graham Birley, *Barbarians at Play: A Case Study of Examinations in the Physical Sciences in Nineteenth Century Britain* (Wolverhampton, 1996).

[49] Secord, *Victorian Sensation*, p. 111.

The Publisher

The appropriation of the *Traité* by Baillière had important consequences for the size, format, text layout and intended audiences of the book. Hippolyte Baillière knew that Ganot's *Traité* was a successful book among medical and science students in Paris, and in 1858 he was already selling it in England.[50] Through his education and training in Paris, and his professional experience in London, he had a good knowledge of different 'thought styles' defining practices and knowledge in the book trade, science, medicine, education and reading in two different countries. For this reason, his communicative and appropriative actions played an important role in the configuration of a Franco-British 'thought style' in relation to these fields.

The first edition of the English translation was published between 1861 and 1863 as the *Elementary Treatise on Physics, Experimental and Applied*.[51] In keeping with Baillière's standard commercial practices, the *Treatise* was issued in England in monthly parts.[52] In France, the book had originally been published in only two parts, probably related to Ganot's course's length and financial strategy. Moreover, the size of the book was increased.

The *Traité* contained a typographic system using a small type size and asterisks to address different kinds of readers, and to introduce recent work that did not have the necessary consensus to be included in the main text yet (see Figure 5.3 in contrast with Figure 5.2). This system was completely eliminated in the English edition – apparently English readers did not like it[53] – having important consequences on the configuration of the book as a new product. Sections in small size type were in most cases introduced to the main text, thus acquiring an authoritative stamp that was lacking in the original.

Baillière addressed his edition to 'Colleges and Schools', a subtitle preserved in all subsequent editions. Thus, it was not deemed necessary to target a particular public preparing for specific examinations. Ganot's identification as a 'professeur' – a term used in France for any teacher in secondary and higher education (comprehended by the idea of 'université') – was transformed by Baillière into 'professor' – strictly used in Britain only to designate university positions, thus enhancing the author's authority in England. The book was used in Britain at many schools and colleges during this period, and was the standard in the preparation of a wide range of science examinations.[54]

---

50 Hippolyte Baillière, *H. Baillière's Catalogue of Recent Foreign Books on Chemistry, Electricity, Physics, Meteorology, &c., &c.* (New York, 1858).

51 The fourth edition (1870) of the book had a print run of 5,500 copies, increased in 1879 to around 7,000 in its ninth edition, which sold 4,000 copies in the first ten months. Archive of the House of Longman, *Atkinson File*.

52 A practice usually identified as characteristic of novels and cheap literature.

53 Anon., 'Fernet's Elementary Physics', *Nature*, 3 (November 1870): 23–4.

54 Newton, 'A French influence'; Fox, 'The Context and Practices of Oxford Physics, 1839–77'.

322 CALORIQUE.

14 degrés à 32, pesant 208 kilogrammes, toujours abstraction faite de la dilatation, ils absorbent une quantité de chaleur égale à 208 (32 — 14) ou 3744 unités; on a donc

$$540\,p + 68\,p = 3744;\ \text{d'où}\ p = 6^{\text{k}}\!,\!1{,}158.$$

VI. Dans un premier vase, on a de l'eau à 11 degrés; dans un second, de l'eau à 91 ; combien doit-on prendre de kilogrammes dans chacun d'eux pour former un bain de 250 kilogrammes à 31 degrés?

Soient x et y les nombres de kilo. à prendre respectivement dans chaque vase, on a d'abord $x + y = 250$ [1]. On obtient une deuxième équation en x et en y, en observant que x kilo. à 11 degrés contiennent 11 x unités de chaleur, et que y kilo. à 91 degrés en contiennent un nombre représenté par 91 y. D'ailleurs, les 250 kilo. de mélange, à 31 degrés, renferment 250 × 31, ou 7750 unités; on a donc l'équation $11\,x + 91\,y = 7750$ [2].

Les équations [1] et [2] étant résolues, on trouve $x = 187$ kil.,5, et $y = 62$ kil.,5.

* THÉORIE DYNAMIQUE DE LA CHALEUR.

344. **Equivalent mécanique de la chaleur.** — Partant de l'idée que le développement de la chaleur est dû à un mouvement vibratoire des molécules, soumis aux lois ordinaires de la mécanique, plusieurs géomètres et physiciens travaillent depuis quelques années au développement d'une théorie nouvelle qu'ils désignent sous le nom de *Théorie dynamique de la chaleur*, et dans laquelle ils se proposent non-seulement de faire voir qu'une quantité de chaleur donnée peut se transformer en travail mécanique (381), et réciproquement; mais de calculer le travail mécanique que peut produire une quantité de chaleur déterminée, ou quelle quantité de chaleur peut développer un certain travail mécanique. On sait, en effet, que la chaleur peut produire un travail mécanique, comme il arrive dans l'expansion des vapeurs et dans la dilatation des gaz; et que réciproquement on peut développer de la chaleur par une action mécanique, telle que la percussion, la pression ou le frottement (383 et 384). La théorie dynamique de la chaleur est pleine d'actualité, et mérite de fixer l'attention des physiciens et des mécaniciens, car elle peut apporter des améliorations de la plus haute importance aux machines à vapeur et aux machines à air chaud.

Montgolfier paraît être le premier physicien qui ait avancé qu'il y a identité de nature entre le calorique et le mouvement, en ce sens non-seulement que la chaleur est une cause de mouvement, et le mouvement une cause de chaleur; mais en ce sens encore que la chaleur et le mouvement sont deux formes différentes, deux effets d'une seule et même cause; en un mot, que la chaleur peut se convertir en mouvement et le mouvement en chaleur.

Se basant sur ces considérations théoriques, Montgolfier inventa, en 1800, une machine qu'il appela *pyro-bélier*, au moyen de laquelle, pensait-il, le travail journalier d'un cheval-vapeur (381) ne devait plus demander qu'une dépense de quelques centimes. Le principe du pyro-bélier consistait à dilater par la chaleur une certaine quantité d'air, toujours la même, enfermée en vase clos; à faire servir cette augmentation de volume et d'élasticité à soulever une colonne d'eau; puis à restituer à cette même masse d'air, la chaleur dépensée par la dilatation et *convertie en effet mécanique*, pour lui restituer la force élastique perdue, et ainsi de suite.

M. Séguin aîné, neveu de Mongolfier, a présenté à l'Institut (le 3 janvier 1855) un mémoire dans lequel il décrit une nouvelle machine à vapeur fondée sur les idées théoriques ci-dessus.

Figure 5.3 The mechanical equivalent of heat. Small type size and asterisk marking knowledge outside of formal education curricula and – according to Ganot – lacking consensus. Atkinson's translation eliminated the asterisk, used normal type size and reshaped this section by a more detailed account, privileging German and English results against French. Ganot, *Traité* (1859), p. 322.

The Translator

Edmund Atkinson appropriated Ganot's books in several ways. First, together with the publisher, he contributed to the design of their intended audience. Second, he reshaped the contents and form of the text in significant ways, according to his pedagogical practice as a teacher, and to the evolving context of scientific education in England. Atkinson's translations were performed while he was teaching at Cheltenham College, and subsequently at the Royal Military College, Sandhurst. This process was constrained by the internal mechanics of the books, vademecums in the sense coined by Fleck. Third, he appropriated Ganot's section of problems as an author. Finally, he appropriated the *Treatise* (and thus the *Traité*) for his translation of the *Cours*.

In the preface to the *Treatise*, Atkinson claimed that he had previously used the *Traité* as a teacher. He valued the book highly for its large number of editions and translations, its 'clearness and conciseness' and 'methodical arrangement', and the quality of its illustrations. However, because of its close link to the 'French systems of instruction', he thought it necessary to make 'alterations and additions' to meet the needs of the English student. In its first edition, the translation was often literal and, in general, did not significantly supplement the *Traité*'s contents. However, it is characterized by more synthetic sentences, more reduced historical introductions, different examples, more algebraic formulae (still simple ones), a more quantitative and mathematical approach, the recalculation for London of observational data given for Paris, and a stronger anti-realist approach in relation to the physical agents and theories of electricity. In subsequent editions, Atkinson introduced new contents and significantly changed some sections.

The first parts to be completely reshaped were the introductory chapters on mechanics, which Ganot had limited to a minimum due to the greater independence in relation to physics of this subject in France, in contrast with Britain.[55] In addition, Atkinson introduced new contents related to recent research conducted in Britain, and results published in English or German. Sometimes additional illustrations were added, often referring to local instrument makers. New articles were typically introduced at the end of chapters, keeping the general structure of the book intact. The *Traité* displayed results produced in France, but also in England, Germany and other countries. However, Ganot's knowledge of foreign research was based on its appropriation by French journal writers and translators.[56] Following Fleck's distinction of 'journal' and 'vademecum' science, Ganot's book was not a mere aggregation of journal articles, but a closed organized system – a vademecum. The

[55] Maurice Crosland and Crosbie Smith, 'The Transmission of Physics from France to Britain: 1800–1840', *Historical Studies in the Physical Sciences*, 9 (1978): 1–61.

[56] Ganot cited Moigno's *Cosmos,* Germer Baillière's *Revue des cours scientifiques* and the more 'elitist' *Annales de chimie et de physique* and *Journal de physique*, John Tyndall's *La Chaleur*, Rudolf Clausius' *Théorie mécanique de la chaleur* and Angelo Secchi's *L'unité des forces physiques*.

tight internal mechanics of the structure of the *Traité*, and perhaps the possibility of saving labour on the basis of Ganot's regular new editions,[57] are factors configuring Atkinson's appropriation. Hence, he respected the general structure and most of the contents of the *Traité*'s successive editions, and, in general, only introduced additions at the end of chapters. Thus, for instance, Atkinson never completely reconfigured the *Treatise* in terms of the driving concept of energy conservation, promoted in Britain from the 1860s. But he introduced an article on this subject at the end of the first book of the *Treatise* as early as in 1868.

The close links between translation and authorship are especially noticeable for the case of Atkinson's appropriation of the *Traité*'s problems. The book's appendix of examination questions and problems was a characteristic feature of his origin in the context of French private preparatory schools, subsequently generalized in French state schools and, concomitantly, in most French physics textbooks. Although there were some previous books on physics problems, Ganot arguably pioneered the configuration of physics textbooks as closed systems organizing knowledge in ways mapping all the sides of pedagogical practice at schools including student interrogation and examination. Atkinson included this section from the *Treatise*'s seventh edition (1875), arguing that teachers and other users of the book had told him of the need for this. He found many of Ganot's problems devoid of interest in having only algebraic or geometrical solutions. Hence, he added new problems focusing on the use of physical principles, based on his teaching experience and that of his colleagues.[58] In addition, the appendix was published separately in 1876 with Atkinson identified as the author.

After Baillière's death, Atkinson successfully negotiated the handing over of the *Treatise* to Longman, the most important publisher of this period in London.[59] In 1868, he proposed the translation of Ganot's *Cours*.[60] Its first edition was published in 1872 as the *Natural Philosophy for General Readers and Young Persons*.[61] Atkinson intended to address it to students at a more elementary level of instruction. It was therefore a priority to eliminate the *Treatise*'s mathematical formulae. Considering it difficult to produce a coherent work by expurgating the

---

57 Ganot sent – at least on certain occasions – a copy of his book to its English publisher, in order to help the work of the translator, and apparently he regularly sent engraving stereotypes for each new edition. Archive of the House of Longman, *Atkinson File*.

58 'Preface' in Adolphe Ganot, *Elementary Treatise on Physics Experimental and Applied* (London, 1875).

59 Topham, 'Scientific Publishing', p. 584.

60 Archive of the House of Longman, *Atkinson File*.

61 Its second edition (1875) had a print run of 5,000 copies and sold around 2,600 copies in the first seven months. The print run of the third edition (1878) was increased to 7,000 copies and the fourth (1881) and fifth (1884) attained 9,000 and 10,000 copies respectively (Archive of the House of Longman, *Atkinson File*).

Treatise,[62] he thought it better to translate Ganot's *Cours*, which already had an extensive circulation in France. His translation introduced modifications aimed at targeting students in the upper classes of boys' and girls' schools, and candidates for the University of London's entrance examination. It was also considered suitable for the general reader who wished to acquire knowledge of the main physical phenomena and laws in 'familiar language'.[63] In his opinion, there were many English books that targeted this audience but most of them were not sufficiently suitable for teaching purposes, or were too expensive.[64]

Atkinson's translation of the *Cours* shared the textual and conceptual characteristics of his translation of the *Traité*. Again, the asterisks system (see Figure 5.3) was eliminated. Furthermore, some articles were eliminated, and articles and illustrations from the *Traité* and the *Treatise*, not included in Ganot's *Cours*, were introduced. In addition, from the third edition (1878) Atkinson provided an appendix of questions, which was never included in the *Cours*. Thus, he not only appropriated the *Cours* for his *Natural Philosophy*, but also the *Traité* (through his previous appropriation of the *Treatise*) in the making of his second physics textbook.

The Reviewer

The subsequent editions of the *Treatise* and *Natural Philosophy* were reviewed in a wide range of English periodicals that addressed different readerships in Victorian England. The reviewers addressed a wide range of audiences (from working-class to middle- and upper-class readers) and considered the books suitable for such diverse readerships.[65] The reviewer of Ganot's books was a special kind of reader

---

62 As previously pointed out, Fleck's concept of 'vademecum' helps to understand the fact that the *Traité* was a closed organized system.

63 Note here the analogy with the qualification of *Vestiges of Natural Creation* as a book displaying 'familiar knowledge'. Secord, *Victorian Sensation*, p. 100.

64 'Preface' in Adolphe Ganot, *Natural Philosophy for General Readers and Young Persons* (London, 1872).

65 Periodicals such as the *Athenaeum* (weekly, of liberal orientation, one of the most influential and independent Victorian literary journals) and the *Academy* (linked to the Oxford reform movement and appealing to an educated middle- to upper-class readership); journals addressed at middle- to upper-class readers with a general interest in science, such as the quarterly *Popular Science Review* (one of the best-known periodicals of its time in this spectrum), the widely circulated *Nature* (a weekly incorporating science 'popularization' among its aims)- and the *Quarterly Journal of Science* (addressing middle- to upper-class science amateur readers); popular weekly illustrated journals such as the agricultural *Land and Water*, and cheap mass-circulating science weeklies addressed at working-class readers such as the *English Mechanic*; scientific and medical specialized journals such as the weekly *Telegraphic Journal*, the institutional weekly *British Medical Journal* and the leading independent weekly *Lancet*. See *Waterloo Directory of English Newspapers and Periodicals, 1800–1900, Series 1 and 2.*

– a paid reader – and thus his appropriation of the books, when not anonymous, involved an exercise of social and professional legitimization.[66]

The books were commonly praised for their print layout and the large number and quality of illustrations – considered a distinctly French feature[67] – for the large number of subsequent French and English editions; for their structure and contents;[68] and for the usefulness of their questions section, particularly for self-taught students.[69] In the 1870s, the *Treatise* was already considered a standard physics textbook.[70] Despite the publisher's and translator's intended differentiation between the two books and their readerships, different reviewers and readers accorded to them common aspects.

Natural Philosophy was explicitly addressed at candidates for the London Matriculation examination.[71] Nonetheless, both textbooks had a questions section, and in fact the *Treatise* was also used for this purpose by some students.[72] Following Atkinson, it was often stressed that *Natural Philosophy* was not an abridgement

66 Gowan Dawson, Richard Noakes and Jonathan Topham, 'Introduction', in Geoffrey Cantor et al. (eds) *Science in the Nineteenth-Century Periodical: Reading the Magazine of Nature* (Cambridge, 2004), pp. 1–34.

67 Anon., 'Elementary Treatise on Natural Philosophy', *Nature*, 3 (31 August 1871): 343–4; Anon., 'Natural Philosophy for General Readers and Young Persons', *Quarterly Journal of Science*, 9 (July 1872): 379 and Anon., 'Ganot's Physics', *Popular Science Review*, 13 (1874): 415.

68 Anon., 'Ganot's Elementary Treatise on Physics', *The Lancet*, 81/2068 (1863): 444; Anon., 'Elementary Treatise on Physics Experimental and Applied, for the Use of Colleges and Schools', *The Lancet*, 114/2925 (1879): 429 and George F. Rodwell, 'Ganot's Physics', *Nature*, 5/8 (February 1872): 285–7.

69 Anon., 'Elementary Treatise of Physics', *Quarterly Journal of Science*, (1876): 110.

70 Anon., 'Ganot's Elementary Treatise on Physics Experimental and Applied', *British Medical Journal*, 6 (April 1872): 370; Anon., 'Ganot's Popular Natural Philosophy', *Popular Science Review*, 11 (1872): 297–8; Anon., 'Natural Philosophy for General Readers and Young Persons', *Quarterly Journal of Science*, 9 (July 1872): 379; Anon., 'An Elementary Treatise on Physics, experimental and applied', *The Lancet*, 110/2828 (1877): 691–2 and Anon., 'Natural Philosophy for General Readers and Young Persons', *The Lancet*, 111/2848 (1878): 466.

71 Anon., 'Natural Philosophy for General Readers and Young Persons', *Quarterly Journal of Science* (1872): 379 and Anon., 'Natural Philosophy for General Readers and Young Persons', *The Lancet*, 111/2848 (1878): 466.

72 This is the case of Oliver Lodge, who subsequently became a well-known physicist. It might be, however, that Lodge referred to *Natural Philosophy* when talking of 'Ganot's Physics'. As discussed later there was a certain confusion in the familiar designation applied to the two books. Oliver J. Lodge, *Past Years: An Autobiography* (London, 1931), pp. 79–80.

of the *Treatise*.[73] The former was often qualified by reviewers, publisher and translator as 'more popular' than the latter, but sometimes the same designation was used to refer to both.[74] The relative status of the two books in relation to what was 'popular' was not clearly defined. Instead, they served both functions linked to formal education, self-instruction and popularization, and different readers assigned different functions to them according to their own expectations and perceptions.

For Ganot, the *Cours* was 'popular' because it was 'accessible to everyone'. For different English reviewers, the 'popular' aspect of the *Treatise* was linked to the subject matter, the type of descriptions, the easy reading and the 'sensation'.[75] In parallel, they considered it significant to stress that the books were accurate and covered a wide subject matter. Thus, both the form and the contents of the books are aspects linking them with the social and the cultural, and their qualities seemed to appeal to a wide range of audiences both in 'esoteric' and in 'exoteric' circles. Furthermore, as reflected by reviewers, the continued editions and the test of time provided them with authority. All these factors made Ganot's textbooks 'standard' works.[76] In fact, as indicated by Fleck, the 'standard' is one of the qualities of the 'popular'.[77] Their enduring role in the processes of communication and appropriation between 'exoteric' and 'esoteric' circles, within and between two different countries, are congruent with their important contribution to the definition of a Franco-British 'thought style' constraining physics, education organization, pedagogy, printing and the book trade in different ways.

The reviews of Ganot's books were usually positive. Nevertheless, some included criticisms, in general related to the relative presence of English actors in them[78] and to the different conceptual structure of French and British physics, in particular on the use of the concept of energy conservation, promoted by William

[73] J. Stuart, 'Ganot's Natural Philosophy', *Nature*, 5 (29 February 1872): 339; although some reviewers did not hesitate to state exactly the opposite. Anon., 'Natural Philosophy for General Readers and Young People', *Quarterly Journal of Science*, Oct., 12 (1875): 528.

[74] Anon., 'Ganot's Physics', *Popular Science Review* (1874): 415.

[75] Anon., 'Ganot's Physics', *Popular Science Review*, new series, 3 (1879): 406 and Anon., 'Ganot's Elementary Treatise on Physics, Experimental and Applied', *British Medical Journal* 6 April (1872): 370. For the meanings of 'sensation' in nineteenth-century England see Secord, *Victorian Sensation*, pp. 11–17.

[76] Fyfe has discussed on categories 'classic' or 'standard works'. Olesko uses the concept of 'canon', but restricts her discussion to the context of physics laboratory instruction. Aileen Fyfe, 'Publishing and the classics: Paley's *Natural theology* and the nineteenth-century scientific canon', *Studies in History and Philosophy of Science*, 33 (2002): 729–51 and Kathryn Olesko, 'The Foundations of a Canon: Kohlrausch's *Practical Physics*', in D. Kaiser (ed.), *Pedagogy and the Practice of Science: Historical and Contemporary Perspectives* (Cambridge, MA, 2005), pp. 323–55.

[77] Fleck, *Genesis and Development*, p. 112.

[78] R. Hunt, 'Elementary Treatise on Physics Experimental and Applied', *The Athenaeum*, February 21 (1843): 260; Anon., 'Electricity for Students', *Popular Science*

Thomson and Peter G. Tait's *Treatise on Natural Philosophy*.[79] These concerns are discussed in the next section, in relation to the appropriation of Ganot's textbooks by a particular kind of reviewer: the science teacher.

The Science Teacher

Science teachers, like Edmund Atkinson, were major readers of Ganot's textbooks. Scientific education in England developed rapidly from the 1860s through the consolidation of national systems of examinations contributing to the strengthening and configuration of the science school and university college curriculum. School science examinations were held by the old universities (Oxford and Cambridge), the new liberal university (London), the Department of Science and Art, and the Society of Arts. Private schools played an important role in the science movement, and from the 1840s foundations such as Cheltenham and Marlborough College were important agents in the constitution of a science curriculum.[80]

George F. Rodwell was a science master at Marlborough[81] in 1872 when he wrote a long review in *Nature* on the *Treatise*.[82] Like Atkinson, he had conducted research as a chemist before focusing on physics teaching. In 1873, he published a small textbook on natural philosophy based on his lectures at Guy's Hospital.[83]

Rodwell regretted the scarce presence of the key concepts of 'Thomsonian Physics'. Thomson and Tait had established in Britain a new set of canonical concepts in mechanics involving the reconfiguration of kinematics, statics and dynamics, and the substitution of 'force' by 'energy' as the fundamental entity in nature. The conservation of the energy principle was an essential part of this conceptual structure.[84]

In spite of this, Rodwell considered that Ganot's book included many new relevant articles on instruments and experiments. Four instrument illustrations – extracted from the *Treatise* – were appropriated for reviewing purposes, and Rodwell intended to show the reader how familiar he was with their main characteristics and manipulation.

Review, 6 (1867): 69–70; Anon., 'Natural Philosophy', *Popular Science Review*, 10 (1871): 72 and Rodwell, 'Ganot's Physics', 285–7.

79 Crosbie Smith and M. Norton Wise, *Energy and Empire: A Biographical Study of Lord Kelvin* (Cambridge, 1989), pp. 348–9.

80 Trevor Hearl, 'Military Examinations and the Teaching of Science, 1857–1870', in MacLeod (ed.), *Days of Judgement*, pp. 109–49.

81 John Foster Kirk, *A Supplement to Allibone's Critical Dictionary of English Literature and English and American Authors* (Philadelphia, 1891).

82 Rodwell, 'Ganot's Physics'.

83 George F. Rodwell, *Notes of a Course of Nineteen Lectures on Natural Philosophy: delivered at Guy's Hospital during the session 1872–3* (London, 1873).

84 Smith and Wise, *Energy and Empire*, p. 349.

Rodwell's book followed a similar plan to that of Ganot's *Treatise*, but was shorter. However, it assigned an important role to the concept of energy where the *Treatise* gave more relevance to 'force' and 'work'. In addition, it encouraged students to make their own instruments by dispensing with unnecessarily expensive instrument collections [like those displayed by Ganot].

In 1884, Silvanus P. Thompson (1851–1916) – a university lecturer in physics – cited Ganot's book in a general review of elementary physics textbooks, assigning – like Rodwell – a major role to the principle of energy conservation. While considering both the *Treatise* and *Natural Philosophy* to be among the best English elementary physics books, he thought that, despite Atkinson's efforts, they were still affected by 'Ganot's academically conservative way of treating physical problems'. Instead, he considered that Balfour Stewart's *Elementary Lessons in Physics* – the major promoter of the 'doctrine of energy' in the range of elementary textbooks – was far superior.[85] A decade earlier, Thompson had probably used Ganot's textbooks to prepare the University of London BSc examinations, and, subsequently, he purchased one of his books just before becoming a physics lecturer at Bristol.[86]

The conservation of energy was also an important factor on the agenda of another kind of science reader, one who considered that science and theology were congruent systems explaining the mechanisms of the world. This reader's appropriation and communication of Ganot's textbooks is discussed in the next section.

The Theological Reader

Ganot's books played a standard role in the cultural and social strategies of theologians of different kinds, vindicating – through their communicative interaction with the 'esoteric' circle of nineteenth-century British science – their role in the configuration of the 'thought style' of science. This interaction is analysed in this section through the analysis of the use of Ganot's work by Henry Drummond and Helena Blavatsky.

Henry Drummond (1851–97) studied science and theology in Edinburgh, where he was a lecturer in natural science and theology professor at the Free Church College. He travelled extensively, giving lectures to students and the working classes which were subsequently compiled in successful books. Drummond was considered a heterodox theologian and a 'cultivated amateur man of science', and renowned as a skilful orator. His work was driven by the idea that nature and the spiritual world were operated by the same principles.[87]

---

85 Silvanus Phillips Thompson, 'Daniell's Physics', *Nature*, 30 (15 May 1884): 49–51.

86 Jane Smeal Thompson and Helen G. Thompson, *Silvanus Phillips Thompson, D.Sc., LL.D., F.R.S. His Life and Letters* (London, 1920), p. 3.

87 D.W. Bebbington, 'Drummond, Henry (1851–1897)', *Oxford Dictionary of National Biography* (Oxford, 2004) <http://www.oxforddnb.com/> (accessed December 2007). All

Drummond relied on the status of Ganot's 'Physics' in 'The Third Kingdom', one of his lecture papers,[88] in which he analysed the relations between the 'Spiritual', the 'Inorganic' and the 'Organic' by drawing on analogies with scientific concepts and theories such as 'energy', 'force' and 'evolution'. The 'Spirit' was represented as a type of 'Energy' operating through physical phenomena analogous to those in nature. The difficulty of observing 'Spiritual forces' was compared to that of observing physical forces.[89] In this context, Drummond introduced a colourful parable: an extraterrestrial reader of Ganot's 'Physics' and Grove's 'The Correlation of Physical Forces', landing in Labrador, would be puzzled to learn that even if many physical forces were available there, they were ignored and had never been used. Therefore, thinking that spiritual energies were not real was as inconclusive as thinking that physical forces were not real where they were not used.

Ganot's 'Physics' was considered by Drummond to be both an authoritative and familiar physics work for the audiences and readers he addressed – students, but also workers – that he could use to argue for the correlation of all forces, including the spiritual. Like Atkinson, and in contrast to Rodwell and Thompson, Drummond did not have any problem with combining Ganot's physical explanations through forces with Thomson and Tait's focus on energy and its conservation.

Like Drummond, Helena Petrovna Blavatsky (1831–91) used science to support her theological work, and considered that theology and science shared a common ground: scientific laws were insufficient to understand the universe, and needed to be completed with other laws of nature that were only accessible to occult wisdom. However, she was explicitly confrontational against elite scientific writers such as T.H. Huxley and John Tyndall on this matter

Like Drummond, Blavatsky travelled extensively, publicizing Theosophy – her theological system – in public lectures and private meetings. In 1888, she published *The Secret Doctrine*, a complex work covering a large number of topics, including cosmic evolution, the origins of the universe, the history of humankind and reincarnation.[90] There, she provided a frame to defend Theosophy against the criticism of scientists, citing a wide range of works, including 'standard books on Science' such as Ganot's 'Physics'. Blavatsky argued that basic concepts in science such as 'ether', 'energy' and 'force' were as 'mysterious' as those of Theosophy, and their agents even less well defined than those of her system. After discussing 'ether' and 'energy', she cited Ganot's 'Physics' definition of 'molecular energy or

---

references to the *DNB* in this chapter were accessed electronically on the aforementioned date.

88 Contained in a posthumous compilation of his papers. Henry Drummond, *The New Evangelism and Other Papers* (London, 1899).

89 Drummond used the example of gravity, that he considered to be a force for a long time unnoticed in the history of mankind. Ibid., pp. 150–5 (I have used the American edition simultaneously published in New York by Dodd, Mead and Company).

90 R. Davenport-Hines, 'Blavatsky, Helena Petrovna (1831–1891)', *Oxford Dictionary of National Biography*.

forces' to argue that, paradoxically, the same scientists that criticized Theosophy completely ignored the nature and agent of these forces.[91]

Thus, for Blavatsky, like for Drummond, Rodwell's and Thompson's discussion on the need to adopt 'Thomsonian physics' privileging 'energy' and its conservation over 'force' was not relevant. Instead, Ganot's *Treatise* represented a standard work in physics, well known by the lecture audiences, which could be used to argue for the congruence of science and theology. Furthermore, it could be used to attack the authority of the scientific elite on science.[92] Thus, communication between individuals such as Drummond and Blavatsky, belonging to the 'exoteric circle' of science, and the British scientific elite was not only intellectual, but also of a strongly social character.

The social character of the 'thought community' constrained by Ganot's textbooks was also defined by the practices of scientific instrument design and its actors. In the next section, I analyse the appropriation of Ganot's books by British instrument makers and scientists involved in instrument design.

The Researcher and the Instrument Maker

The communicative power and close relation to modern developments in science of Ganot's textbooks' illustrations – highlighted by all their reviewers – played a major role in the appropriation of his work by different actors engaged in instrument design. In this context Ganot's *Treatise* was a science vademecum, that is, a reference work representing the 'thought style' constraining a 'thought community'. This view was not only shared by members of the British scientific elite such as William Thomson, but also by individuals cohabitating with Thomson in the social context of engineering and instrument design, such as Sebastian Ferranti and the heads of the firms Horne & Thornthwaite and Elliott Brothers.

William Thomson used Ganot's *Treatise* as a 'vademecum' in the context of his daily research practice. In 1863, he wrote to Thomas Andrews in reply to his demand for technical details about Holmes's electric light apparatus for lighthouses. Andrews was then a professor of chemistry at Queen's College, Belfast, and maintained a close professional and personal relationship with Thomson and Tait.[93] Thomson stated that he could not send him exact details, but could alternatively provide descriptions of Nollet's French apparatus for which he had previously

---

[91] Helena Petrovna Blavatsky, *The Secret Doctrine: The Synthesis of Science, Religion, and Philosophy* (London, 1888).

[92] In fact, the British scientific élite opinions on the relations between science and theology were not at all homogeneous. Scientific materialism was not only rejected by Drummond and Blavatsky, but also by scientists of the status of Peter G. Tait and Balfour Stewart. P.J. Hartog, 'Stewart, Balfour (1828–1887)', rev. Graeme Gooday, *Oxford Dictionary of National Biography*.

[93] Arthur Harden, 'Andrews, Thomas (1813–1885)', rev. Anita McConnell, *Oxford Dictionary of National Biography* and Smith and Wise, *Energy and Empire*, pp. 352–5.

taken notes. Furthermore, he remarked that Ganot gave an illustration of one of these apparatus, although of different proportions.

Indeed, Ganot's *Traité* and the first edition of the *Treatise* included four illustrations and a detailed description of this apparatus, built in Paris. Thomson used Ganot's book as a vademecum which made it possible to fish for data, in particular on French instrumentation. Furthermore, he specifically cited Ganot's illustration, rather than its written description, showing that Ganot's illustrations were important for physicists designing their own instruments in the laboratory or the workshop.

This was also the case for several London instrument makers who inserted at the end of Ganot's *Treatise* lists of instruments displayed in the book and available in their shops.[94] Hence, instrument makers such as Horne & Thornthwaite and Elliott Brothers appropriated the *Treatise* for their commercial purposes. The selection of instruments portrayed in the *Treatise* contributed to defining their trade catalogues and thus, to a certain degree, their production strategies. This production was diversified by targeting different cultural and social communities, including the scientific elite, schools and amateur scientists. This diversity is also represented in Ganot's *Treatise* that, in turn, contributed to defining it.

The illustrations and text of Ganot's *Treatise* were also important for the research by the engineer Sebastian Ferranti.[95] He first encountered the book in his school days, as an introductory 'standard work' to physics,[96] then probably in his (unconcluded) engineering studies at University College, London, and finally, both as a 'standard work' and a 'vademecum' in the course of research leading to his successful design of easy-reading electrical meters.

According to Ferranti, he started his research by reading the articles on the action of a current upon another in Ganot's *Treatise*.[97] The latter included an illustrated discussion of Ampère's laws applied to this phenomenon in several cases,[98] and some of Ferranti's diagrams and explanations are simplified versions of Ganot's illustrations.[99] Thus, he used Ganot's work as a 'vademecum' in which the standard knowledge defining the 'thought style' of his 'thought collective' could be checked. But, in addition, having confronted the book on his own since his childhood, his use of it to fish for core ideas that could be developed into

---

94 Ganot, *Treatise* (1866) and (1868); copies available at the Bodleian Library, Oxford.

95 I thank Graeme Gooday for guiding me towards Ferranti's work.

96 John F. Wilson, 'Ferranti, Sebastian Ziani de (1864–1930)', *Oxford Dictionary of National Biography*.

97 Gertrude Ziani de Ferranti and Richard Ince, *The Life and Letters of Sebastian Ziani de Ferranti* (London, 1934), p. 239 and Sebastian Ziani de Ferranti, 'On the Ferranti Electricity Meter and its Evolution', *Transactions of the Royal Scottish Society of Arts*, 14 (1896): 52–3.

98 Ganot, *Treatise* (London, 1863), pp. 648–57.

99 Especially figures 499, 500, 507 and 508. Ibid., pp. 652 and 654.

successful engineering designs conforms to Fleck's analysis of popular knowledge as 'the major portion of every person's knowledge' that often 'set the standard for the content of expert knowledge'.[100]

Conclusion

Ganot's books were originally intended principally for an audience enrolled in formal education in a particular economy of teaching. However, both books had complex and varied real readerships. One of the intended aims of one of Ganot's books was to 'popularize'. However, both books were considered to be 'popular' by different readers. The *Traité* and the *Cours*, and their English translations, were linked by a complex relation of appropriations performed by their author, their translator and in general their readers.

Readers appropriated the books in different and contradictory ways. For different readers they represented school manuals, aids for the preparation of examinations, introductions to science, sources of self-instruction, popular science books, 'sensational' readings, symbols of French physics, sources supporting the fight against scientific materialism, science vademecums, and instrument trade catalogues, respectively. It is their ways of reading and using, of appropriating and communicating the books that defines their status.

The perception of the books varied among readers but also changed over time. The multiplicity of readers and responses to the books contributed to their long survival and rise to the status of 'standard'. The books were an essential tool in the multilateral communication between different 'esoteric' and 'exoteric' circles in France and Britain, and the diversity of their readers tells us how diverse their appropriations and communications were. Such diverse, international and enduring processes determined the contribution of Ganot's textbooks to an international 'thought style' or a cultural centre within a 'central value system' by becoming science vademecums, constraining a large 'thought collective' comprehending all their readers located both in 'esoteric' and 'exoteric' circles.

In this process, the distinctive French features of Ganot's books were appropriated into the British context by adapting them to the distinctive features of British physics, printing and publishing, pedagogy, reading practices and instrument making. In turn, the original characteristics of Ganot's books helped shape the British context in all these fields and secured for them a central position in British culture and society.

[100] Fleck, *Genesis and Development*, p. 112. In addition, Ferranti's historical narrative of the development of his meter might have also had a pedagogical and academic intention in starting by introducing a standard work known to any reader.

Chapter 6
The Circulation of Energy: Thermodynamics, National Culture and Social Progress in Spain, 1868–1890

Stefan Pohl-Valero[1]

Studying the construction of the science of energy in the Victorian context, Crosbie Smith has argued that terms such as force, energy and thermodynamics should be employed as categories of historical actors that were constructed within specific local contexts and in relation to particular audiences.[2] Furthermore, several studies have shown that the meanings of thermodynamics underwent multiple transformations and were used in many discourses. In the Victorian context, for instance, William Thomson and his allies used the laws of thermodynamics to characterize the visible creation of the universe, stressing that the dissipation of energy (and the thermal death of the universe as its consequence) was the 'strongest weapon in the armoury against anti-Christian materialists and naturalists'.[3] Later, in the hands of such scientific popularizers and social thinkers as Belfour Stewart, Peter Tait and Thomas Henry Huxley, thermodynamic arguments were explicitly used to support social discourses. The economic system of capitalism was portrayed as a thermodynamic system where the dissipation of energy (in the form of heat) was used to characterize – and criticize – a communist society destined for thermal death.[4]

---

[1] This chapter is based on part of my D.Phil thesis, Stefan Pohl-Valero, 'La circulación de la energía: Una historia cultural de la termodinámica en la España de la segunda mitad del siglo XIX' (Universitat Autònoma de Barcelona, 2007). I would like to thank the research group 'Los públicos de la ciencia' at the CEHIC, Universitat Autònoma de Barcelona, for their generous support, and especially Agustí Nieto-Galan, who supervised my thesis. An early version of this chapter was presented to the History of Science Workshop, HPS, University of Cambridge, in February 2006, where I benefited from the comments of the workshop participants.

[2] Crosbie Smith, *The Science of Energy: A Cultural History of Energy Physics in Victorian Britain* (Chicago, 1998), Introduction. See also, Crosbie Smith, 'Force, Energy, and Thermodynamics', in Mary Jo Nye (ed.), *The Cambridge History of Science: Volume 5. The Modern Physical and Mathematical Sciences* (Cambridge, 2003), pp. 289–310, p. 290.

[3] Smith, 'Force, Energy, and Thermodynamics', p. 302.

[4] Greg Myers, 'Nineteenth-Century Popularizations of Thermodynamics and the Rhetoric of Social Prophecy', in Patrick Brantlinger (ed.), *Energy and Entropy: Science and Culture in Victorian Britain* (Bloomington, 1989), pp. 307–38, pp. 327–9 and 331–4.

In general, science historians have drawn attention to the cultural resources involved in the emergence of thermodynamics. Political economy, industrial culture and social and theological perceptions and metaphors influenced the thinking of the natural philosophers that formulated the laws of thermodynamics.[5] Moreover, cultural historians and literary critics have explored the impact of the new image of nature, articulated by thermodynamics, for a modern conception of society and culture.[6] The metaphor of the human body as a thermal machine, governed by the laws of thermodynamics, helped to shape social thought across the ideological spectrum.[7] Throughout Europe, thermodynamics was appropriated by liberal, conservative and communist intellectuals, and its laws informed such fields as physiology, sociology and economics, reflecting the wide circulation of these laws within the European public sphere over the second half of the nineteenth century.[8] As Greg Myers has noted, thermodynamics, like Darwinism, 'has been intertwined with social thought, influenced by it and influencing it since the earliest formulations'.[9]

This article analyses the manner in which the laws of thermodynamics, laden with natural, moral and social meanings, were appropriated by a specific group of Spanish scientists and social reformers. During the 1870s, Doctor of Physics Enrique Serrano Fatigati elaborated an educational programme for physics that was based on an evolutionary and energetic interpretation of nature. Serrano's efforts to construct a unified science of nature, merging astronomy, thermodynamics and biology, were portrayed as an example of national-original scientific culture at the service of social reform. Serrano and his circle formed part of a group of school and university teachers who were in disagreement with the scientific and political establishment of the Bourbon Restoration. In 1876, some of this group founded a private institution in Madrid that aimed to offer an educational system that was an alternative to the official one, in clear response to state and Church control

[5] Norton Wise, 'Work and Waste: Political Economy and Natural Philosophy in Nineteenth Century Britain (I)–(III)', *History of Science*, 27 (1989): 263–301, 391–449, and 28 (1990): 221–61; Crosbie Smith and Norton Wise, *Energy and Empire. A Biographical Study of Lord Kelvin* (Cambridge, 1989); Crosbie Smith, *The Science of Energy*.

[6] Greg Myers, 'Nineteenth-Century Popularizations of Thermodynamics'; Anson Rabinbach, *The Human Motor: Energy, Fatigue, and the Origins of Modernity* (Berkeley, 1992); Bruce Clarke, *Energy Forms: Allegory and Science in the Era of Classical Thermodynamics* (Ann Arbor, MI, 2001).

[7] Rabinbach, *The Human Motor*, pp. 2–5.

[8] See, for instance, Kurt Bayertz, 'Naturwissenschaft und Sozialismus: Tendenzen der Naturwissenschafts-Rezeption in der deutschen Arbeiterbewegung des 19. Jahrhunderts', *Social Studies of Science*, 13 (1983): 355–94; Phillip Mirowski, M*ore Heat than Light: Economics as Social Physics: Physics as Nature's Economics* (Cambridge, 1989); Álvaro Girón-Sierra, *En la mesa con Darwin: evolución y revolución en el movimiento libertario en España (1869–1914)* (Madrid, 2005).

[9] Myers, 'Nineteenth-Century Popularizations of Thermodynamics', p. 307.

over education. Serrano's energetics programme, containing scientific and moral lessons, was taught at this institution. Later, in the 1880s, the energetic conception of nature was used by Serrano's advocates – especially the chemists Laureano Calderón and José Rodriguez Mourelo – to characterize society and the way it should be organized.

In particular this chapter will consider the interplay between public meanings of natural science and the social and political agendas that shaped this process of appropriating scientific theories. Special attention will be given to the way in which a specific meaning of entropy – among other theories – articulated a discourse of natural and social progress. I will also be examining the role of science communication, which includes articles in cultural journals, textbooks, public lectures and school education, and the hegemonic hostility against materialism in which such appropriation practices took place. A further social elaboration of energy concepts will be analysed at the end of the chapter, calling to attention the rescue of thermodynamic metaphors in order to *naturalize* – and therefore legitimize – differences between social classes.

The Thermodynamic Evolution of the Universe

During the 1870s, the physics and chemistry school teacher Enrique Serrano Fatigati (1845–1918) published several articles in cultural journals describing an organic and unified conception of nature.[10] Serrano was a liberal intellectual who fought for the freedom of education and for a democratic spirit in Spanish society of the time. Serrano's scientific interests included physics, chemistry, biology, botany, geology and public health. Citing the ideas of Friedrich von Schelling (1775–1854) as his inspiration,[11] Serrano elaborated a monistic worldview that unified the organic and non-organic. He also proposed the unification of the

[10] The main works in which Serrano developed his system are: Enrique Serrano Fatigati and Salvador Calderón Arana, *Estudio de filosofía natural: total organización de la materia* (Madrid, 1870); Enrique Serrano Fatigati, 'El progreso de la materia', *Revista de la Universidad de Madrid*, 2 (1873a): 69–98; Enrique Serrano Fatigati, 'Las actividades de la naturaleza', *Revista de la Universidad de Madrid*, 2 (1873b): 631–650; Enrique Serrano Fatigati, 'La evolución en la naturaleza', *Revista de la Universidad de Madrid*, 3 (1874): 292–317, 496–512; Enrique Serrano Fatigati, 'Apuntes para un programa de física', *Revista de la Universidad de Madrid*, 5 (1875): 60–87, 361–404; Enrique Serrano Fatigati, 'Una lección de física general', *Revista de la Universidad de Madrid*, 6 (1876): 71–97, 167–80; Enrique Serrano Fatigati, 'Concepto de la física fisiológica', *Revista Europea*, 9 (1877): 434–9, 463–7; Enrique Serrano Fatigati, *Física biológica: estudios sobre la célula* (Madrid, 1877).

[11] Serrano explicitly mentioned Schelling's *Von der Weltseele* (1798). See Serrano, 'Las actividades de la naturaleza', p. 641. In this work, Schelling regarded nature as an organism and revived the ancient Stoic doctrine of a single soul pervading all of nature. His general vision of nature could be described as vitalistic monism. Frederick Beiser, 'German Romanticism' in *Routledge Encyclopedia of Philosophy CD-ROM*, 1998.

material and spiritual worlds through the idea of natural energy. Everything in the universe, according to Serrano, had an active soul that could be characterized as natural energy. Thus, following the laws of thermodynamics, the whole of nature was in constant transformation and progressive development.

Using Laplace's nebular hypothesis, Serrano transferred the idea of the organization of celestial matter to the realm of the organic world. As Serrano observed, although Laplace had aimed to prove the stability of the solar system, his nebular hypothesis – combined with modern physics – was the best proof of the constant development of nature.[12] Influenced by the Lamarckian mechanism of the inheritance of acquired characteristics, Serrano explained the variability of living forms as a consequence of external influences. Based on this theory, he was able to relate the progressive transformation of matter (exemplified by this nebular hypothesis) with the transformation of living organisms.

Serrano attempted to show the evolution of the universe as a continuous chain of natural phenomena. Through the ongoing transformation of cosmic matter, the Earth and the other planets were formed, followed by plants, animals and humans. There was no essential difference between organic and inorganic nature, life differing from the inorganic only in its higher organization. According to Serrano, the whole of nature was in constant evolution, but this evolution was teleological. Unifying the organic and inorganic worlds was fundamental for Serrano's system, emphasizing the organic character of all matter. For Serrano, this organic character meant that all matter had a soul – an internal, active force conscious of itself – which directed its transformation. As Serrano remarked, some philosophers called this active principle 'God'. Serrano preferred to call it 'natural energy or activity'.[13] For him, then, the two laws of thermodynamics were the fundamental laws governing the universe. The first law demonstrated the unity of nature and its constant activity and transformation. In his effort to present nature as a single living unity, Serrano used the words 'activity' and 'energy' indiscriminately, retaining an older meaning of energy – one related to the concept of life and will. Once it was accepted that all forces present in the universe were different manifestations of the activity of nature, the transformation of life and the transformation of natural forces became strictly analogous.

The second law of thermodynamics was that governing the direction of this transformation. Referring to William Thomson's 1852 article 'On a universal tendency in Nature to the dissipation of mechanical energy',[14] Serrano discussed and rejected the thermal death of the universe as the consequence of nature's tendency for thermal dissipation.[15] His arguments were based on the infinity of the universe and on a dynamic impossibility. Given a future state of thermal

---

12 Serrano, 'La evolución de la naturaleza', p. 294.

13 Serrano, 'El progreso de la materia', p. 97.

14 William Thomson, 'On a Universal Tendency in Nature to the Dissipation of Mechanical Energy', *Philosophical Magazine*, 4 (1852): 304–6.

15 Serrano, 'El progreso de la materia', p. 75.

equilibrium, Serrano argued, an initial state of pure potentiality would be expected, something that he found impossible. The important aspect of energy dissipation, he stressed, was its directional character, not its imposition of a beginning and an end. These two temporal points were situated in the infinite, like a line that increases asymptotically. In short, the whole universe was an organic being that, according to the laws of thermodynamics, was in constant transformation following an irreversible line of progression:

> The species also transform themselves, fulfilling the phenomenon in the same way as occurs with energy, which is a transformation that passes through several progressive and regressive states, heading, as a whole, in one, single direction.[16]

This progression was material and moral. Serrano emphasized that, just as cosmic matter and living organisms evolved into more complex organizations, so too the moral world became better. The progress of individuals involved the search for what is right, for beauty and for truth. Furthermore, the thermodynamic evolution of the universe demonstrated, to Serrano, that life existed on other planets, and – more fully still – that there were higher forms of life (material and moral) on older planets.[17] The high status of thermodynamics and astronomy were the pillars on which Serrano based his evolutionary worldview. Even more important than palaeontological and geological evidence,[18] the new science of thermodynamics was used by Serrano as the principal and most authoritative scientific demonstration of nature's progressive evolution. The social relevance for promoting this progressive worldview was highlighted in a review of Serrano's work. The idea that the universe and nature represented a place of change, of continuous progression in fact, was portrayed as the best example of the erroneousness of conceiving the institutions and society as something stationary.[19] Nevertheless, if the progression of nature should persuasively imply the need for changes in society, a proper exposition of this worldview was required.

---

[16] Serrano, 'La evolución en la naturaleza', p. 502: 'Las especies se transforman también, cumpliéndose aquí el fenómeno de la misma manera que en la energía se realiza: esto es, pasando por diversos estados progresivos y regresivos, dirigidos en su conjunto en una sola y misma dirección.'

[17] See, for example, Serrano, 'El progreso de la materia', p. 92.

[18] The palaeontological and geological arguments used by Serrano in his evolutionary worldview are analysed in Francisco Pelayo, 'Repercussions of Evolutionism in the Spanish Natural History Society', in Thomas Glick, Miguel Angel Puig Samper and Rosaura Ruiz (eds), *The Reception of Darwinism in the Iberian World* (Dordrecht, 2001), pp. 95–110, 105–6.

[19] Alfredo Calderón, 'Una idea. Estudios de filosofía natural. Total organización de la materia', *Revista de España*, 29 (1872): 187–98, p. 193.

Materialism and the Public Circulation of Science

The evolutionary worldview set out by Serrano was especially concerned with emphasizing that his system was not materialistic. During the 1870s, many of the local proponents of new scientific theories shared this concern, which in turn reflected the hegemonic idea that materialism was the principal cause of moral degradation and social disorder. In the early 1870s, conservative intellectuals were fairly skilled in disseminating the idea that the public discourses of the likes of Ludwig Büchner, John Tyndall or Ernst Haeckel were all the same version of atheistic materialism. This scientific materialism was portrayed as the principal intellectual support for communism and anarchism. In order to analyse the process of scientific appropriation of Serrano's thermodynamic evolution, it is important to take into account the cultural and political characteristics shaping the public circulation of science in the Spanish context of the 1870s.

In 1868 a bourgeois revolution took place in Spain. *La Gloriosa*, as the revolution was called, reflected a long process that sought the establishment of a modern, liberal nation state. Although the monarchy returned to power only six years after the democratic revolution, the period between 1868 and 1874 was crucial for the construction of Spain as a modern state. After the uprising, the press and literature experienced a period of freedom in which it was possible, for instance, to import and publish all manner of books. With the return of the monarchy, the press was again censored, but book publishing remained unfettered.[20] Additionally, the individual liberties proclaimed by the Constitution of 1869 fostered the meeting of private individuals in different institutions. These liberties generated not only the proliferation of clubs and other cultural institutions where the bourgeoisie got together to discuss political and cultural subjects, but also the emergence of working-class cooperatives and resistance societies.[21]

As a consequence, a broad and heterogeneous public (with contrasting political and social values) began to gain wider access to diverse books and articles that used scientific theories to support ideological postures. As in the rest of Europe, science articulated several discourses that sought the progressive secularization of society and the state. It was no surprise that Büchner's famous book *Kraft und Stoff* (1855) was published just after the revolution, in 1868, by an anarchist publisher in Madrid.[22] The book presented a materialistic interpretation of nature. Stressing that matter and force were inseparable concepts supporting the idea of

---

[20] For a general survey of the history of printing and reading in Spain, see Victor Infantes, François Lopez and Jean-François Botrel (eds), *Historia de la edición y la lectura en España: 1472–1914* (Madrid, 2003).

[21] For the political and cultural relevance of the so-called 'sexenio democrático', see, Rafael Serrano García (ed.), *España, 1868–1874: nuevos enfoques sobre el sexenio democrático* (Valladolid, 2002).

[22] Ludwig Büchner, *Fuerza y materia. Estudios populares de historia y filosofía naturales* (Madrid, 1868).

an eternal and cyclical universe, Büchner rejected the existence or necessity of God.[23] Similarly, the *History of the Conflict between Religion and Science* (1874) by William Draper, published in Spain only two years after its original publication, strongly criticized Catholicism. The book put forward the view that science and Catholicism were traditional enemies. The prologue of the Spanish translation was written by one of the former presidents of the first Spanish Republic, Nicolás Salmerón (1837–1908).[24] Political instrumentalization of science was reflected in a great number of scientific texts. For instance, in the prologue to Büchner's *Aus Natur und Wissenschaft*, the anarchist publishers of the 1873 Spanish version claimed that the popularization and democratization of science was an important instrument for the revolution and democratization of Spanish society.[25] Besides Büchner's and Draper's books, Spanish society saw and discussed other popular texts written by such scientists as John Tyndall and Ernst Haeckel that used energy conservation as one of the principle scientific arguments to question traditional religious values.[26]

Following the translation of *Kraft und Stoff*, the conservative elites, despite condemning the freedom of the press, resorted massively to them in order to attack materialism and spread a public image of harmony between religion and science. The conservative reaction to scientific materialism was adapted to the new liberal social order in which public opinion was essential. Several books and articles were published rejecting the above mentioned writers' ideas, and Catholic scientific journals were founded to counterbalance views of a more liberal kind.[27] In the case of Draper's book, the public debate was encouraged by a special prize awarded by the Royal Academy of Moral and Political Sciences for the best essay on the harmony

---

23 For Büchner's materialism, see Frederick Gregory, *Scientific Materialism in Nineteenth-Century Germany* (Dordrecht, 1977).

24 John William Draper, *Historia de los conflictos entre la religión y la ciencia* (Madrid, 1876).

25 Ludwig Büchner, *Ciencia y naturaleza: ensayos de filosofía y de ciencia natural* (Malaga, 1873), pp. v–vi.

26 Significantly enough, the famous Belfast address given by Tyndall before the annual meeting of the British Association for the Advancement of Science in 1874 was translated and published in a Spanish journal that same year. John Tyndall, 'La evolución histórica de las ideas científicas', *Revista Europea*, 2 (1874): 469–76, 500–13. For thermodynamics arguments used by Haeckel, see for instance Ernst Haeckel, 'Teoría evolutiva del universo y de la tierra', *Revista Europea*, 12 (1878): 392–9.

27 Conservative and Catholic journals such as *La Defensa de la Sociedad* or *La Ciencia Cristiana* were founded in the 1870s, paying special attention to the materialistic implications of scientific theories. Texts against Büchner's ideas were numerous. See, for instance, Zeferino González, *El positivismo materialista* (Madrid, 1872); Francisco Caminero, 'El doctor Büchner o el catecismo de los materialistas', *La Defensa de la Sociedad*, 4 (1873–74): 439–51; Antonio María Fabié, *Examen del materialismo moderno* (Madrid, 1875).

between the Catholic religion and science.[28] All these texts and articles emphasized that 'real science' could not reject the idea of God, and stressed the social dangers that an eternal and mechanically determined universe could produce.

Scientific materialism was portrayed not only as an ideology against the Catholic Church, but also as the main source of moral degradation and therefore the intellectual fermentation of social disorder. When the first Spanish congress of socialist workers was held in Barcelona in 1870 and anarchist movements started to emerge, the principal cause of the workers' unrest debated in the public sphere was materialism. The Paris Commune of 1871 is one of the clearest international examples of social disorder, related by the contemporary press to materialism represented by people like Büchner. Despite different political orientations, many of the Spanish elites helped to portray scientific materialism as a common social threat that had to be fought. Many political and cultural journals discussed the possible social disorders that the inappropriate presentation of modern science might give rise to. Science was presented as a key element for the progress of civilization; however, the absolute centrality of moral progress was also emphasized. The idea that real and proper progress was the combination of the material and the moral, based on Catholic principles, was repeated over and over again. This kind of progress was, as one article literally observed, the only one that could prevent the unrest and agitation of the working classes.[29]

The general perception that scientific materialism was a common threat to Spanish society gained wide circulation, and its hegemonic status was a result of the expansion and consolidation of the press and literature during the 1870s. As Cecilio Alonso has noted, the developments of print technology and the expansion of publishing houses, among others, were factors that transcended the field of political interests and converted the press and popular books into universal instruments of cultural communication.[30] Nevertheless, the communication of science clearly reveals ideological concerns with materialism. If the anarchists were using science for revolution, the popularization of science by the intellectual leaders in Spanish society was also used – in this case, to spread traditional moral lessons. For example, Emilio Huelin,[31] an engineer and important popularizer

[28] For the reception of and the debate over Draper's book in Spain, see A. Mayordomo, 'La recepción en España de la obra de J.W. Draper: una perspectiva del conflicto entre la religión y la ciencia', *Historia de la Educación*, 4 (1988): 145–58.

[29] Francisco Pareja de Alarcón, 'Estudios filosófico-religiosos', *La Defensa de la Sociedad*, 5 (1874): 628–85, p. 674.

[30] Cecilio Alonso, 'El auge de la prensa periódica', in Víctor Infantes, François Lopez and Jean-François Botrel (eds), *Historia de la edición y la lectura en España: 1472–1914* (Madrid, 2003), pp. 559–67, p. 559.

[31] Huelin has been portrayed as one of the most important Spanish scientific popularizers of the 1870s. See Leoncio López-Ocón, 'El fomento de la educación y de la ciencia en la sociedad española del Sexenio democrático', *Boletín Institución Libre de Enseñanza*, 28–9 (1997): 127–48, pp. 137–40.

who wrote articles on science in different journals, used the public exposition of science to condemn materialism. In many of his articles, he stressed the importance of maintaining the idea of God when studying modern science. In a text dated 1873, in which he explained the history, technical characteristics and industrial applications of petroleum, he started by referring to how this fuel was used by the communists when trying to burn down Paris during the Commune. The tragic events of the Paris Commune, Huelin stressed, had been the result of scientific materialism and its moral degradation represented by persons such as Büchner or Haeckel.[32]

Under such a context of intellectual hostility towards scientific materialism, it was no surprise that the exposition of Serrano's energetic evolution reflected the shared concern of many Spanish intellectuals regarding the best way of speaking about science in various forums. In an increasingly heterogeneous public sphere, the communication of science became more complex and dangerous: the images of science produced could be used for purposes other than those originally planned, or suffer public rejection. And thermodynamics and evolutionism were at the very centre of this cultural and social public negotiation.

In the Ateneo of Madrid, one of Spain's most important and influential cultural institutions of the time, many intellectual leaders used to meet to discuss political and cultural affairs.[33] The central issue chosen for debate in 1875 reflected the moral and social concerns regarding a thermodynamic evolutionary interpretation of nature. That year, the guiding question for the debate was: 'should the life of organized beings be considered the transformation of universal energy?'[34] The president of the institution, conservative politician and university professor José Moreno Nieto (1825–82), referring to Ernst Haeckel's works, summarized its evolutionary thoughts in the following terms: the universe can be

---

[32] Emilio Huelin, 'Petróleo cual agente de bárbaros incendios y de otras distintas maneras considerado', *Revista de España*, 31 (1873): 5–25, pp. 5–9. Other widely published scientific popularizers, such as Louis Figuier, condemned materialism as well. In one of Figuier's texts published in Spain, he stressed the importance of scientifically demonstrating the Christian idea of the immortality of the soul in order to avoid social disorder. Significantly enough, Figuier portrayed the Paris Commune as the practical consequence of scientific materialism. Louis Figuier, *Después de la muerte; ó la vida futura según la ciencia* (Barcelona, 1873), p. xiii. For details of Figuier as a scientific popularizer, see Catherine, Benedic, 'Le Mondes des vulgarisateurs', in Bruno Béguet (ed.), *La Science pour tous. Sur la vulgarisation scientifique en France de 1850 à 1914* (Paris, 1990), pp. 30–49.

[33] For the history and cultural relevance of the Ateneo, see José Luis Abellán, *El Ateneo de Madrid: historia, política, cultura, teosofía* (Madrid, 2006); Inman Fox, 'Spain as Castle: The Making of a National Culture', in Penny Schine Gold and Benjamin C. Sax (eds), *Cultural Visions: Essays in the History of Culture* (Amsterdam, 2000), pp. 197–222.

[34] Rafael M. De Labra, *El Ateneo de Madrid. Sus orígenes, desenvolvimiento, representación y porvenir* (Madrid, 1878), p. 169: '¿Puede y debe considerarse la vida de los seres organizados como transformación de la fuerza universal?'

conceived as initially inert cosmic matter that through some random forces began organizing itself, generating the stars, the planets and life. All these processes can be characterized as the vibration of matter, without any final aim. The whole mechanism behind this transformation is the struggle for life. This doctrine, stresses Moreno, could only generate social chaos and the ruin of society.[35]

Even more popular than Darwin,[36] Haeckel and his monistic worldview were portrayed as one of the best examples of the materialistic spirit of the time; and as we have seen, his name, along with that of Ludwig Büchner, appeared in many articles that discussed moral degradation in Spanish society. Although the public exposition of Haeckel's system developed from a mechanistic to a pantheistic interpretation of nature,[37] his evolutionary philosophy of monism was especially criticized for its mechanistic and non-teleological form, and for its effort to reduce all vital and spiritual processes to physicochemical laws. Criticizing Haeckel's evolutionary thoughts, one article railed 'thanks to this theory we know now that mechanical causes are enough to explain the development of individuals, and there is no need to invoke direct causes that work in harmony with an ultimate purpose. [...] Blessed mechanics, from now on, they will be used to explain everything.'[38]

The exposition of Serrano's system made a great effort to avoid such a random interpretation of evolution. For Serrano – who was one of the participants in the Ateneo debate of 1875 – the second law of thermodynamics was the best proof that the universe evolves following a distinct path of progress. Although Serrano used some of Haeckel's experiments to confirm his idea of the universe as an organic being in constant evolution, he claimed to reject 'the erroneous and materialist conclusions of this German sage'.[39] The problem with Haeckel's system, according to Serrano, was his emphasis on mechanizing the concept of the organic. Serrano's

---

35 José Moreno Nieto, *Discurso pronunciado por el Ilmo. Señor D. José Moreno Nieto el día 3 de noviembre de 1876 en el Ateneo Científico y Literario de Madrid con motivo de la apertura de sus cátedras* (Madrid, 1876), pp. 28–9.

36 See Thomas F. Glick, *Darwin en España* (Madrid, 1982); Francisco Pelayo, *Ciencia y creencia en España durante el siglo XIX: la Paleontología en el debate sobre el darwinismo* (Madrid, 1999).

37 Paul Weindling has argued that the metamorphosis of Haeckel's description from a mechanistic-minded anti-clerical view in the 1860s to a Darwinian pantheist view in the 1890s has to be understood in relation to the German culture of the time and to public opinion. Paul Weindling, 'Ernst Haeckel, Darwinismus, and the Secularization of Nature', in James R. Moore (ed.), *History, Humanity and Evolution: Essays for John C. Greene* (Cambridge, 1989), pp. 311–27, p. 313.

38 Marcellino Venturoli, 'Haeckel, la teoría evolutiva y su teoría antropogénica', *La Ciencia Cristiana*, 7 (1878): 31–40, p. 39: 'Gracias a esta teoría hemos podido comprender que para explicar el desarrollo del individuo, bastan las causas mecánicas, y que es enteramente ocioso invocar causas directas obrando en armonía con un fin, es decir, las antiguas causas finales. [...] dichosa mecánica que va a ser desde ahora suficiente para todo.'

39 Serrano, 'La evolución de la naturaleza', pp. 300–301.

energetic monism precisely avoided the mechanization of nature and allowed the teleological character of its transformation. These were Serrano's main conceptual strategies to secure a wide public acceptance of his system.

Transpositions of Scientific Theories

In the second half of the nineteenth century, energy conservation and the nebular hypothesis were commonly used as general arguments to support progressive evolutionism.[40] Haeckel and Büchner stressed the importance of these theories in their worldviews.[41] Similarly, many of the Spanish promoters of evolutionism (which in general subscribed to the idea of progressive development) emphasized that evolution was an extension of the general laws of physics. For instance, Rafael García Álvarez (1828–94),[42] professor of natural history, began his book on evolutionism by explaining the nebular hypothesis and the conservation of energy. Concluding his work, he characterized evolution in the following terms:

> The transformist theory, which has been so well developed by Darwin, and which we have just outlined, is a stage of the general theory of evolution, which in turn is a particular case of the great cosmic hypothesis, the transformation and conservation of force.[43]

It was therefore not surprising that Serrano elaborated his teleological system based precisely on astronomy and thermodynamics. In particular, the nebular hypothesis was a major tool in the articulation of a general science of progress, and was closely

---

40 For an example of concept transpositions in the construction of a science of progress in early Victorian Britain, see Simon Schaffer, 'The nebular hypothesis and the science of progress', in James R. Moore (ed.), *History, Humanity and Evolution: Essays for John C. Greene* (Cambridge, 1989), pp. 131–64. See also, James Secord, *Victorian Sensation: The Extraordinary Publication, Reception, and Secret Authorship of 'Vestiges of the Natural History of Creation'* (Chicago, 2001).

41 See, for example, Büchner, *Fuerza y materia. Estudios populares de historia y filosofía naturales*, p. 251; Ernst Haeckel, 'Teoría evolutiva del universo y de la tierra', *Revista Europea*, 12 (1878): 392–9, 431–6, p. 395.

42 García Álvarez has been portrayed as one of the leading Spanish naturalists that accepted and supported Darwinism. Pelayo, 'Repercussions of Evolutionism in the Spanish Natural History Society', p. 109.

43 Rafael García Álvarez, *Estudio sobre el Trasformismo* (Granada, 1883), p. 110: 'La teoría transformista, cuyos principios fundamentales, tan admirablemente desenvueltos por Darwin, acabamos de exponer, es una fase como dejamos dicho, de la más general, la teoría de la evolución, que a su vez puede considerarse como un caso particular de la gran hipótesis cósmica, la trasformación y persistencia o conservación de la fuerza.'

related to the idea of life on other planets.[44] These ideas were widely circulated in the 1870s. In particular, popular books by Camille Flammarion captured public interest by setting out the theories of the origin of the universe and the scientific possibility of life on other planets. Flammarion's *Plurality of Inhabited Worlds*, in which the author stressed the idea of moral and material progress, was published in Spanish at least seven times in the 1870s.

The perception of material and moral progress related to astronomy was reflected in literature, and particularly in science fiction. For instance, the author of a novel that appeared in a cultural journal in 1870 incorporated the nebular hypothesis and the plurality of inhabited worlds in his narrative. The work narrated a cosmic voyage made by a young Spaniard and a German professor. In the novel, the nebular hypothesis was extensively explained and much astronomical information was given. After several adventures, the stellar voyagers arrived on Saturn, which was inhabited by a superior society. Over many pages, the author describes the outstanding intellectual, moral and social development of Saturn's society.[45]

Particularly interesting in Serrano's exposition were the local characteristics that allowed him to use the second law of thermodynamics as the law for all progress. First of all it is important to mention that, in the Spanish context, Thomson's cosmic interpretation of energy dissipation was not used in the 1870s to fight anti-Christian materialism.[46] The local promoters of thermodynamics resorted to quite different strategies in order to present a 'non-controversial' image of the new science. One of the most important popularizers of thermodynamics in Spain, the engineer and liberal politician José Echegaray (1832–1916), used this science as a key legitimating element in the institutionalization of mathematical physics in Spanish universities.[47] Echegaray and his allies – particularly the

---

44 See Schaffer 'The nebular hypothesis and the science of progress'. For a general survey of the idea of extraterrestrial life, see Michael J. Crowe, *The Extraterrestrial Life Debate, 1750–1900: The Idea of a Plurality of Worlds from Kant to Lowell* (Cambridge, 1986).

45 The novel, published originally in 1870 in the journal *Revista de España*, has been reproduced in Tirso Aguimana de Veca, 'Una temporada en el más bello de los planetas', in Nil Santiáñez-Tió(ed.), *De la luna a Mecanópolis. Antología de la ciencia ficción española (1832–1913)* (Barcelona, 1995), pp. 57–82.

46 It was only in the 1880s when Spanish conservative intellectuals started to popularize the thermal death of the universe in order to scientifically prove that it had a beginning and an end. Some of these authors gave a special mention to the popular book by natural philosophers Balfour Stewart and Peter Tait, *The Unseen Universe* (London, 1875). See, for example, Marc Antonie Duilhé de Saint-Projet, *Apología científica de la fe cristiana* (Valencia, 1886); Manuel Polo Peyrolón, 'Materialismo', in *Diccionario de ciencias eclesiásticas*, vol. VII (Valencia, 1888), pp. 134–9; Francisco Íñiguez e Íñiguez, *Aplicación del análisis matemático a las demás ciencias* (Madrid, 1887).

47 See Stefan Pohl-Valero, 'La termodinámica como elemento legitimador de la física teórica y aplicada en la España de la segunda mitad del siglo XIX', *Quaderns d'Història de l'Enginyeria*, 7 (2006): 73–114.

engineers Gumersindo Vicuña (1840–90) and Francisco Rojas (1832–1909) – went to great lengths to define the epistemological and ontological characteristics of modern physics. In the 1870s, these three engineers, who were the first holders of the chair of mathematical physics at the University of Madrid, published several books and journal articles on physics and especially on thermodynamics, becoming local authorities in these subjects.[48] It was stressed by these engineers that the characteristics of modern physics were in perfect harmony with religion. An important aspect of this argument was based on the ontological interpretation of energy. Inert matter (ponderable and ethereal) and its motion were the basic elements in nature that produce all physical phenomena. Reducing the concept of energy to matter and motion allowed them to argue that an external and creating force would have put matter into motion at the beginning of the universe.[49] This argument was especially directed against Büchner's idea that the concepts of matter and force were inseparable, which in turn was the central argument supporting the idea that the universe was eternal and self-regulating. The thermal death of the universe as a consequence of energy dissipation was barely mentioned by these engineers in the 1870s. Rojas alone outlined the idea in his book on the history of thermodynamics, but rejected the use of thermodynamics for such prophecies. As he observed, the application of the laws of thermodynamics to the whole universe was beyond human understanding.[50]

Consequently, the cosmic meaning of energy dissipation was highly malleable and adaptable to different discourses. In the Spanish context of the 1870s, the second law did not imply an inconvenient issue for evolutionary-progressive thought, as was the case, for example, with Herbert Spencer.[51] Among the Spanish

---

48 See especially, José Echegaray, *Tratado elemental de termodinámica* (Madrid, 1868); José Echegaray, *Teorías modernas de la física. Unidad de las fuerzas materiales* (2nd edn, Madrid, 1873); Gumersindo Vicuña, *Teoría y cálculo de las máquinas de vapor y gas con arreglo a la termodinámica* (Madrid, 1872); Gumersindo Vicuña, *Elementos de física al alcance de todo el mundo* (Madrid, 1874); Gumersindo Vicuña, *Manual de física popular* (Madrid, 1878); Francisco Rojas, *Termodinámica. Su historia, sus aplicaciones y su importancia* (Barcelona, 1876); Francisco Rojas, *El problema físico y el problema químico se resolverán en el mecánico. Memoria presentada ante la Real Academia de Ciencias de Barcelona* (Barcelona, 1877).

49 Echegaray, *Teorías modernas de la física*; Rojas; *El problema físico y el problema químico se resolverán en el mecánico*.

50 Rojas, *Termodinámica. Su historia, sus aplicaciones y su importancia*, p. 142.

51 Clarke, *Energy Forms*, p. 67. For the British context, Stephen Brush has indicated the contradictions between the second law of thermodynamics and evolutionary worldview: 'The second law became a cornerstone of the evolutionary worldview as understood by physical scientists in the late nineteenth century. The "Thermal death" concept already shows that it implies decay rather than progress, and thus seems to contradict the optimistic flavour of evolutionary theories in biology and the social sciences.' Stephen G. Brush, 'The Nebular Hypothesis and the Evolutionary Worldview', *History of Science*, 25 (1987): 245–78, p. 263.

promoters of evolutionary worldviews, Serrano was one of the few people, if not the only one, who was trained in physics and also a specialist in thermodynamics. The public meaning of different scientific theories and his scientific interests allowed him to elaborate a teleological evolutionary system that combined the nebular hypothesis, thermodynamics and organic evolution, and one in which energy dissipation was used as the natural law establishing the entire progress of the universe.

Energetics: A National Science

In 1875, Serrano published an article entitled 'Points in a Programme of Physics'.[52] This text, which was translated and published one year later in the *Philosophical Magazine* by the British chemist Edmund Mills,[53] was an outline of how physics should be taught. The first part of the article was an epistemological discussion of physics in which he highlighted the importance of philosophical investigations so as to avoid continuous confusion by an inextricable labyrinth of facts. The necessary starting point for physics, Serrano stressed, was the recourse to first principles. This was not only important for the development of physics, but was also necessary in every branch of knowledge, especially ethics. With natural energy as the first principle, it was then unnecessary to question the nature of the particular phenomena and thus the mechanization of nature. Instead of using this kind of hypothesis, it would be better to express the phenomena in terms of energy. In the same year and journal that featured Serrano's English article, Mills also published an article 'On the First Principles of Chemistry' in which he asserted the living character of all matter, the inadequacy of atomic theory and how his view of nature, starting with the first principles, was significant for social values. As he sets out in his text, 'in the long run, no part of ourself is dissociated from another part; and the principles which underline chemical work this hour, become in the next the springs of moral action.'[54] This article was translated by Serrano and published in a Spanish journal a year later.[55]

After his 1875 article, Serrano started publishing a series of texts designed to form a textbook for the new style of physics; this science, called *energetics* by Serrano, would also include the branches of chemistry and biology. The first section was dedicated to a systematic study of energy and to understanding nature as an

---

[52] Serrano, 'Apuntes para un programa de física'.

[53] Enrique Serrano Fatigati, 'Points in a Programme of Physics', *Philosophical Magazine*, 1 (1876): 455–61.

[54] Edmund Mills, 'On the First Principles of Chemistry', *Philosophical Magazine*, 1 (1876): 1–16, p. 15.

[55] Edmund Mills, 'Los principios fundamentales de la química', *La instrucción pública*, 1 (1877): 22–34.

organic unit. Following W.J.M. Rankine's 'Outlines of the science of energetics',[56] Serrano showed how to deduce mechanics from the principles of energy.[57] As he pointed out, the other sections of the book should include a detailed description of the way in which the energy was transformed in all branches of nature (including planets, vegetables and animals). The exposition of energy dissipation was termed by Serrano as the 'teleological' aspect of the study.[58]

These publications reflected Serrano's effort to elaborate a new and alternative education programme for physics that offered an organic, unified and evolutionary concept of nature, and that was based on the science of natural energy. Such a pedagogical programme differed dramatically from the official one established in Spanish schools and universities. Serrano's programme included both alternative epistemological and ontological approximations for the study of nature, and was designed to offer natural and social lessons. Moreover, he supported his idealistic approximation of nature with experimental work. Some of his results for new determinations of the mechanical equivalent of heat and for the influence of external conditions on the transformation of microorganisms were published in Swiss, English and French journals.[59]

The difference between the official physics based on mechanical hypothesis and Serrano's organic energetics was evident. In fact, Serrano did have certain institutional problems with his approximation to physics. In 1872 in Madrid, he presented his doctoral thesis on physics. His work, which analysed the best methods for calculating the specific heats of solids and liquids, was rejected by the viva jury. In this work Serrano did not mention the nature of heat, saying only that it was a kind of energy. Serrano presented a second thesis in 1874 and finally obtained his doctorate. The title of this second work was much more in keeping with the model of physics that the likes of Vicuña, Rojas and Echegaray were trying to establish in Spanish universities: 'Heat is a form of motion'.[60] Nevertheless, and

---

56 W.J.M. Rankine, 'Outlines of the science of energetics', *Edinburgh New Philosophical Journal*, 2 (1855): 120–41.

57 Serrano, 'Una lección de física general'. As Crosbie Smith has argued, Rankine's article aimed to consolidate the 'seemingly complex formulations of 'energy' and 'thermodynamics' into concise 'sciences' on firm and unchallengeable axiomatic foundations.' Crosbie Smith, *The Science of Energy*, p. 164.

58 Serrano, 'Concepto de la física fisiológica', p. 466.

59 Enrique Serrano Fatigati, 'On a new determination of the mechanical equivalent of heat', *Philosophical Magazine*, 47 (1874): 155–6; Enrique Serrano Fatigati, 'On Saccharomyces cerevisiae', *Annals and Magazine of Natural History*, 18 (1876): 187–9; Enrique Serrano Fatigati, 'Note sur les bactéries et les générations espontanées', *Archives des Sciences Physiques et Naturelles*, 59 (1877): 383–4; Enrique Serrano Fatigati, 'Influence des diverses couleours sur le développement et la respiration des infusoires', *Comtes Rendus de l'Académie des Sciences*, 89 (1879): 959–60.

60 Serrano's thesis is registered in his academic record at the University of Madrid. Archivo Histórico Nacional, Universidades, 6190, exp. 27.

despite his experiments on thermodynamics and a certain international recognition for this work, Serrano was never recognized by Echegaray and his colleagues as a local authority on thermodynamics.

The need for an alternative space to teach Serrano's style of physics, and science in general, became evident in 1875. In this year, at the very outset of the Bourbon Restoration, all state teaching staff received instructions to follow Catholic dogma in their work. Many (Serrano among them) protested, and some were stripped of their posts, including Serrano's friends, chemist Laureano Calderón (1847–94) and geologist Augusto González Linares (1845–1904).[61] As a consequence, some teachers and intellectuals founded a private educational institution in Madrid where the freedom and secularization of scientific education was highlighted. Under the leadership of Giner de los Rios (1839–1915) and Gumersindo Azcárate (1840–1917), the Institución Libre de Enseñanza (ILE) was founded in 1876. The institution was planned as a free university for the education of a liberal elite destined to reform and modernize Spain. Although the ILE failed in creating a university, it focused on primary and secondary education, becoming an important place for the history of Spanish culture. The ILE brought together an intellectual community that represented liberal values such as secularization, freedom of education and a democratic spirit. Laureano Calderón and Augusto González Linares were among the founders, and Serrano was one of the institution's shareholders.[62] Teaching of the new science of energetics began at the ILE using the works of Serrano and Mills.[63]

Serrano's energetics not only provided a pedagogical alternative to official scientific education, but was also presented as a national product, a native and original science that recognized that the ontological base of physics and its epistemological approach played a very important social function. These ideas were emphasized in an article that appeared in the journal *Revista Europea* in 1879, written by one of the ILE's teachers.

The article, later published in book format, commented on the works of Serrano and González Linares and was titled 'The new natural philosophy in Spain'.[64] Alfredo Calderón (1850–1907), author and brother of Laureano Calderón, opened the text by criticizing the tendency in science to separate living forms from inert matter, stressing how this tendency was present in public education. The best way

---

[61] M. Ruiz de Quevedo, *Cuestión universitaria: Documentos coleccionados* (Madrid, 1876), p. vi.

[62] For the history of ILE, see Antonio Jiménez García, *El Krausismo y la Institución Libre de Enseñanza* (Madrid, 1986). For a list of the founders and shareholders of the ILE, see Antonio Jiménez-Landi, *La Institución Libre de Enseñanza y su ambiente* (Madrid, 1973), pp. 703–20.

[63] *Boletín del Instituto Libre de Enseñanza*, 1 (1877), pp. 3–4.

[64] Alfredo Calderón, 'Movimiento novísmo de la filosofía natural en España', *Revista Europea*, 13 (1879): 449–54, 481–7, 545–51, 587–93, 680–89, 712–20, 737–44, 785–91, 814–18.

to avoid this separation was to start with an organic conception of nature. For Alfredo Calderón, Serrano's system in the field of physics best exemplified this approach.[65] According to Calderón, Serrano's works introduced a completely new way of considering the main problems in physical sciences. Thanks to this new approach, Calderón stressed, the science of natural energy now included living beings in its study, and no longer treated them as an exception to their laws. In that sense, Serrano was presented by Calderón as the Spanish scientist who was constructing a new and native form of evolutionary biological physics; a science that was original and suitable to the Spanish social context.

The spirit of this new science was highlighted by Calderón as the proper way to avoid the positivistic and materialistic spirit of the time.[66] According to Calderón, the science developed by Serrano merged 'tradition and progress'; that is, it returned to the first principles but at the same time took into account the results of the experimental sciences. These characteristics were presented as the principal aspects for the development of an original and national science. Furthermore, Serrano's energetics was presented as a necessary educational tool for the moral and intellectual 'regeneration' of Spain.[67] Calderón was attempting to define a national scientific culture that was original and that contained the liberal social values represented by an educational space such as the ILE.[68] Serrano's energetics, with its organic, teleological and progressive character, was a perfect cultural product for imagining a national identity in support of the political and social ideals that certain liberal intellectuals were trying to promote for a modern Spanish nation state.

The Social Machine

The efforts of Serrano and his circle to promote energetics as the fundamental science for the study of nature reflected an increasing perception that thermodynamics could resolve the deep social tensions that were characteristic of a capitalistic society in the middle of an industrializing process. In the early 1880s Serrano's advocates expressed their belief that energetics was the ideal science for the study of society. During this decade, and especially at the Ateneo of Madrid, the chemists Laureano Calderón and José Rodriguez Mourelo (1857–1932) defended energy as the foundational concept in nature, and rejected the traditional concept of matter. As with Serrano, matter was explained as the way energy transformations were perceived, and nature was explained as an organic being following a path of

65 Ibid., p. 685.

66 Ibid., p. 718.

67 Ibid., p. 485.

68 Inman Fox has analysed the ILE as one of the Spanish institutions that provided a cultural space for the construction of a national identity at the service of liberal political ideals. E. Inman Fox, *La invención de España: nacionalismo liberal e identidad nacional* (Madrid, 1997); Inman Fox 'Spain as a Castle: The Making of a National Culture'.

progress established by the second law of thermodynamics. Both Calderón and Rodríguez Mourelo emphasized that energetics was the starting point for the study of human beings and society.[69] Moreover, it was claimed that this science could inform such fields as physiology, sociology and psychology.[70]

In the early 1880s, the central government started a series of social reforms aimed at improving the conditions of the working class. Although the government founded a national commission of social reforms in 1883, the *Comisión de Reformas Sociales* (CRS), the commission was not able to produce tangible results and the debate it originated focused on the material and moral health of the workers.[71] Nevertheless, for Spanish intellectual elite, natural science in general – and energetics in particular – represented an important element in their reformist projects. The Ateneo of Madrid was one of the main spaces in which these projects were discussed. If in 1875 the central question for debate was, as mentioned above, the general and even metaphysic issue 'Should the life of organized beings be considered the transformation of universal energy?', the question for debate in 1883 reflected an interest that applied more to society: 'Are the laws of the struggle for survival in the individual, and the principle of energy conservation in the social organism, enough for the establishment of modern sociology?'[72]

For the likes of Calderón and Rodriguez Mourelo, the answer was evident. As Calderón stated in 1884, pedagogy and sociology should study individuals as complex machines, where the distribution and characteristics of energy of each person would inform on their responsibilities and position in society.[73] But besides the possibilities that energetics seemed to offer for the study of human beings, the very idea of perceiving society as a thermal machine was being consolidated in the mentality of the intellectual leaders of Bourbon Restoration Spain. As the economic

---

69 José Rodríguez Mourelo, *Concepto actual del cosmos* (Madrid, 1880); José Rodríguez Mourelo, 'Concepto de la energía', in *Curso de ciencias naturales. Conferencias* (Madrid, 1883), pp. 49–85; Laureano Calderón, 'Concepto de la materia', in *Curso de ciencias naturales. Conferencias* (Madrid, 1883), pp. 21–48; Laureano Calderón, 'Discurso de apertura del curso 1884. Sección de Ciencias Exactas, Físicas y Naturales', in *Discursos leídos en el Ateneo Científico, Literario y Artístico de Madrid con motivo de la apertura del curso de 1884* (Madrid, 1884), pp. 99–136.

70 Rodríguez Mourelo, 'Concepto de la energía', pp. 56–7; Calderón, 'Discurso de apertura del curso 1884. Sección de Ciencias Exactas, Físicas y Naturales', p. 131.

71 Antonio Buj Buj, 'La cuestión urbana en los informes de la comisión de reformas sociales', in Horacio Capel, José M. López Piñero and José Pardo (eds), *Ciencia e ideología en la Ciudad (II). I Coloquio Interdepartamental. Valencia, 1991* (Valencia, 1994), pp. 73–86. See also Juan Ignacio Palacio Morena, *La institucionalización de la reforma social en España, 1883–1924: la Comisión y el Instituto de Reformas Sociales* (Madrid, 1988).

72 Quoted in Antonio Jiménez García., *El Krausopositivismo de Urbano González Serrano* (Badajoz, 1996), p. 148: '¿Son suficientes la ley de la lucha por la existencia en el individuo, y el principio de conservación de la energía en el organismo social para constituir la Sociología moderna?'

73 Laureano Calderón. 'Discurso de apertura del curso 1884', p. 131.

debates about free trade reflected, optimizing the energy available in nature was becoming one of the new values of the liberal society of the Restoration.[74]

The metaphor of society as a thermal machine was also reflected in the public debates about the so-called 'worker question'. Similarly to social Darwinism, the analogy between the natural order and the social order was used to demonstrate that social differences obeyed a natural law. In 1891 Laureano Calderón was able to explain in detail how energetics should inform the organization of society. That year, Calderón was one of the participants at the Ateneo of Madrid discussing the role of natural sciences in the solution of the 'social question'. Calderón portrayed society as a thermal machine governed by the laws of thermodynamics. He stressed that, just like a thermal machine, society should follow certain natural laws in order to be able to work and progress. The second law of thermodynamics demonstrated that there must be a difference in temperature for the machine to be able to produce work. And for the social machine, this temperature gradient was represented by the social differences between the working class and the bourgeoisie. Calderón was defending a liberal nation state, which he characterized as being established by democratic principles, following the idea of laissez-faire individualism. For him, the intervention of the state or religion in balancing society and in restricting individual initiatives would only lead to the shutting down of the social machine.[75]

Conclusion

The appropriation of scientific ideas exposed in this chapter was intended to characterize this practice as a social and natural knowledge-producing process, avoiding the idea of a bipolar framework of analysis where science is produced in a generating centre and then received as a well-established product in a passive periphery. During the 1870s, Enrique Serrano developed a detailed evolutionary worldview based on thermodynamics. One of its main features was its teleological and progressive character. Serrano used the second law of thermodynamics as the law governing the progressive development of nature and society. He championed an organic and unitary concept of nature in which the science of energetics would be its general science. The intention of this new science was to provide an alternative educational programme to the official one, and offer natural and social lessons.

In the late 1870s, Serrano's energetics was portrayed as an original national science that was suited to Spanish social circumstances. It is worth noting that the

---

[74] See, for example, José Echegaray, 'Aplicación de las fuerzas naturales a la industria y al comercio', in *Conferencias del curso de 1879 a 1880. Círculo de la Unión Mercantil* (Madrid, 1881), pp. 105–25; Pedro Estasen, *La Protección y el libre cambio: consideraciones generales sobre la organización económica de las nacionalidades y la libertad de comercio* (Barcelona, 1880).

[75] Laureano Calderón, *La cuestión social y las ciencias naturales: discurso pronunciado en el Ateneo de Madrid en 1890–1891* (Madrid, 1891).

image of a national science that Alfredo Calderón was constructing was part of the efforts of many Spanish intellectuals in the search for national identities and values.[76] The Spain of the second half of the nineteenth century sought to construct a modern liberal nation state, imagining a community with different sources of national identities, merging old and new traditions, and, as with the rest of Europe, science became a discursive indicator of modernity and progress. It is also important to observe that one of the favourite debates in the Spanish press was the scientific position of Spain in comparison with other European countries, and the existence or not of a proper and 'native' scientific culture. In other words, science was presented in the public sphere as an important issue for national identity and pride. But the political and social disorders after the 1868 liberal revolution had left a taste for moderation. Although many liberal intellectuals continued their efforts to limit the Church's control over society, for example in the freedom and secularization of education, they also recognized the importance of some traditional values as indispensable elements of social stability and control. Science in general and the new physical concepts and evolutionary theories in particular were at the very centre of this tension.

Scientific education and the popularization of science were perceived by the Spanish elites as a necessary aspect of a modern nation state. But their adequate presentation and interpretation was also important. The case study presented here seeks to portray this concern. Once Serrano's evolutionary worldview had been, on one hand, legitimized by the science of energy, and on the other, separated from materialism given its teleological and non-mechanical character, the path was open for its public acceptance, and therefore for its use in the service of social education and reform.

Furthermore, this process of scientific appropriation reflects the importance of taking into account science as a public and cultural product. Serrano's cosmology was not a monolithic theory received from a science-generating centre, but rather an amalgam of scientific subjects that were 'in the air'. Theories such as the nebular hypothesis, the plurality of worlds or the unity of physical forces (the conservation of energy) were very popular and widely discussed in the public sphere at the time. They were enveloped by controversy but, infused with varying significance, they were also objects of fascination. These subjects were therefore ideal for social and political purposes. In a rhetorical sense, they were common places (ones that use familiar language and images in order to explain new concepts) ready to be adapted to many discourses.[77] One of these was the legitimation of a new liberal and capitalistic society characterized as a thermal machine.

---

[76] For a general survey of the construction of national identities in the Spain of the nineteenth century, see José Álvarez Junco, *Mater Dolorosa* (Madrid, 2001).

[77] For the use of 'common places' in the popularization of thermodynamics in the Victorian context, see Greg Myers, 'Nineteenth-Century Popularizations of Thermodynamics and the Rhetoric of Social Prophecy'.

Chapter 7
Electric Adventures and Natural Wonders: Exhibitions, Museums and Scientific Gardens in Nineteenth-Century Denmark

Rikke Schmidt Kjærgaard[1]

The Power of Place

The number of public places for science increased from the mid-nineteenth century. New media directed towards the upper classes, the growing middle class and even the new working class now presented a world of exotic adventure and strange knowledge.[2] Moreover a plethora of opportunities to experience science first hand was offered, not only in text and images, but also in settings giving the audience the illusion of being in the rainforest watching the wildlife or going back millions of years in Earth's history. Learning about science was linked to ideas of public enlightenment, education and entertainment in one package combining armchair reading with scientific exploration in safe environments. It was not enough to satisfy one's scientific curiosity through books and magazines. More was needed and more was offered, as one of the most prolific Danish science popularizers and a naturalist by training, Jens Orten Bøving-Petersen, put it in 1912: 'Knowledge of rocks and minerals is not gained through reading and images. What is needed is looking for yourself – in the collections of the museums and in nature.'[3] The new public places for science increased the potential for exhibitors, inventors and scientists to make scientific knowledge available to a wider audience. However, science in the popularized version was very different from that of the research laboratory. By offering the very distant past and the glorious future and by appealing to people's curiosity, hope and horror, the real innovation of popular science at exhibitions, in museums and in scientific gardens was 'a science for the

---

[1] I am grateful to Peter C. Kjærgaard, Graeme J.N. Gooday and the editors for valuable suggestions and advice.

[2] See Peter Broks, *Media Science Before the Great War* (Basingstoke, 1996), for a detailed study of the emerging media's role in science popularization from Victorian optimism to Edwardian pessimism.

[3] Jens Orten Bøving-Petersen, *Vor Klodes Saga: En skildring af Jordens udvikling og fortidsliv* (Copenhagen, 1912), Introduction.

people' in ways no book or journal could ever match. Instrumental in its success was the power of place.[4]

All segments of Danish society were attracted to the exhibitions, the new scientific museums and gardens presenting the life of plants and animals from all over the world. This chapter argues that this new way of experiencing anything from electric contraptions to bicycling primates first hand was instrumental in forming public opinion about science and technology in nineteenth-century Denmark. Specific communication strategies for popularizing science emerged in the form of special exhibition newspapers, massive media coverage of exhibitions in popular magazines and specific learning strategies within exhibition settings. One of the means of popularizing was the construction of specifically designed narratives. One example is the rhetoric and settings of the so-called 'living exhibitions' found in the Danish Zoological Garden and in Danish exhibitions in general: telling who we are by telling how somebody else is. Another is the case of how official committees carefully staged a narrative to include everybody in World's Fairs by sending working-class members abroad using government money in order to participate in the much publicized events, when in fact these exhibitions were mostly intended for and visited by the middle and upper classes. Furthermore, this study will show that, as science specialization increased by the turn of the century, the divide between a professional interest in communicating scientific knowledge and working exclusively for popularizing it became more apparent.[5]

---

4 The notion of the 'power of place' is borrowed from Sophie Forgan, 'Building the Museum: Knowledge, Conflict, and the Power of Place', *Isis*, 96/4 (2005): 572–85. For other recent studies of the importance of the new physical public spaces for science popularization, such as museums, see the following articles in the special edition on museums and the history of science in *Isis*, 96/4 (2005): Samuel J.M.M. Alberti, 'Objects and the Museum', pp. 559–71; Sally Gregory Kohlstedt, '"Thoughts in Things": Modernity, History, and North American Museums', pp. 586–601; and Jim Bennett, 'Museums and the History of Science: Practitioner's Postscript', pp. 602–8. Furthermore David N. Livingstone argues for the general importance of place for the production of scientific knowledge in *Putting Science in Its Place: Geographies of Scientific Knowledge* (Chicago and London, 2003). Janet Browne adopts a slightly different, but equally important use of the notion 'power of place' in the second volume of the Darwin biography *Charles Darwin: The Power of Place* (London, 2002).

5 Peter Bowler argues that it is a myth that the professionalization of science itself caused a gap between scientists and popularizers; see Bowler, 'Experts and Publishers: Writing Popular Science in Early Twentieth-Century Britain, Writing Popular History of Science Now', *British Journal for the History of Science*, 39 (2006): 159–87. However, as the professionalization of science continued, the gap widened between the scientific content of the specialized journal article and the popular magazine story. According to Bowler we should be careful not to think – as previous historical studies have argued – that professional scientists now gave up the idea of writing for a popular audience.

A New Scientific Experience

The emphasis on public enlightenment and educational ideology during the nineteenth century made the market for popular science grow and turned popular science into a genre of its own, or rather into several specialized genres.[6] We have a fairly good understanding of nineteenth- and early twentieth-century science popularization when it comes to the printed media, with respect to the science popularizers, publishers and readers.[7]

Although the Royal Danish Academy of Sciences and Letters had followed the publication patterns of similar institutions since the mid-eighteenth century, scientific issues had not been communicated systematically in any accessible form to anyone outside the Danish scientific community.[8] A century later this picture had completely changed as publication channels became more accessible with technological advances in print media, and as more public places became available to larger audiences.

Compared to Denmark the market for popular science was in any respect much larger in scale in Britain, France and Germany.[9] In this context the World's Fairs were the most prominent and spectacular, making a sensational start with the 1851

---

[6] See Roger Cooter and Stephen Pumfrey, 'Separate Spheres and Public places: Reflections on the History of Science Popularization and Science in Popular Culture', *History of Science*, 32 (1994): 237–67; Bernadette Bensaude-Vincent, 'In the Name of Science', in J. Krige and D. Pestre (eds), *Science in the Twentieth Century* (Amsterdam, 1997), p. 319–38; Jonathan R. Topham, 'A View from the Industrial Age', *Isis*, 95 (2004): 431–42; James Secord, 'Knowledge in Transit', *Isis*, 95 (2004): 654–72; and Peter Broks, *Understanding Popular Science* (Buckingham, 2006).

[7] See Bowler, 'Experts and Publishers'; Jonathan R. Topham, 'BJHS Special Section: Book History and the Sciences. Introduction', *British Journal for the History of Science*, 33 (2000): 155–8; James Secord, *Victorian Sensation: The Extraordinary Publication, Reception and Secret Authorship of 'Vestiges of the Natural History of Creation'* (Chicago, 2000); Bernard Lightman, '"The Voices of Nature": Popularizing Victorian Science', in Bernard Lightman (ed.), *Victorian Science in Context* (Chicago and London, 1997); the three volumes originating in the *Science in the Nineteenth-Century Periodical Project*: Geoffrey Cantor, Gowan Dawson, Graeme Gooday, Richard Noakes, Sally Shuttleworth and Jonathan R. Topham (eds), *Science in the Nineteenth-Century Periodical: Reading the Magazine of Nature* (Cambridge and New York, 2004); Geoffrey Cantor and Sally Shuttleworth (eds), *Science Serialized: Representations of the Sciences in Nineteenth-Century Periodicals* (Cambridge MA, 2004); Louise Henson, Geoffrey Cantor, Gowan Dawson, Richard Noakes, Sally Shuttleworth and Jonathan R. Topham (eds), *Culture and Science in the Nineteenth-Century Media* (Aldershot, 2004).

[8] Hans Henrik Hjermitslev, Casper Andersen and Peter C. Kjærgaard, 'Populærvidenskab og folkeoplysning', in Peter C. Kjærgaard (ed.), *Lys over landet – Dansk Naturvidenskabs Historie 3, 1850–1920* (Aarhus, 2006), pp. 345–74.

[9] Ibid., p. 346.

Great Exhibition in London.[10] Throughout the century Danes visited World's Fairs in large numbers. Many Danish reports confirm the impression of international and Danish scientific grandeur, and through these reports in newspapers, magazines, journals and booklets this experience was shared with the many people confined to a reading experience.

In the fairytale *The Dryad* from 1868 Hans Christian Andersen described the Fair in Paris in 1867 as a 'great table on Christmas Eve, an Aladdin's-castle of industry and art'.[11] Twenty-five years later the president of the Danish Industrial Association repeated Andersen's metaphor when describing the World's Fair in Chicago in 1893.[12] By the use of specific narratives, in this case the fairytale universe of Aladdin, the authors referred to wealth, sumptuousness and greatness.

Light and electricity were without doubt the largest topics and among the most spectacular and popular themes at exhibitions by the late nineteenth century and the early twentieth century. A Danish traveller recalled the light at the Paris 1889 World's Fair in the following passage:

> Think of the exhibition as a random night. A cannon is fired and as if by magic the inner garden and the surrounding buildings are illuminated. The central dome glares in golden splendour. … The electric light below the arches burns in rows, and if you look under the ceilings of the galleries it will seem as if they are filled with stars. In every panel in the ceiling there is an incandescent lamp. The galleries in the garden are filled with electric flares and in the trees you find small coloured flares. … From the Eiffel Tower you can see to the Trocadero Palace, where the main building is marked by lines of fire. The statues of the large cascade in front of the building sparkle in the light. It's a running stream of fire that tumbles down its stairs. Suddenly a bundle of rays emerge from the top of the Eiffel Tower. They seem rather thick and encourage you to think that they

---

[10] Jeffrey A. Auerbach, *The Great Exhibition of 1851: A Nation on Display* (New Haven, 1999); Robert Brain, *Going to the Fair. Readings in the Culture of Nineteenth-Century Exhibitions* (Cambridge, 1993); Robert W. Rydell, *All the World's a Fair: Visions of Empire at American International Expositions, 1876–1916* (Chicago, 1984). For an emphasis on American World's Fairs in the twentieth century see Robert W. Rydell, *World of Fairs: The Century-of-Progress Expositions* (Chicago, 1993), which also, as an introduction, includes a chapter on American World's Fairs from 1851 to 1915 (pp. 15–37). For a detailed collection on primary sources referring to the World's Fairs see Brain, *Going to the Fair*.

[11] Hans Christian Andersen, *H.C. Andersens samlede skrifter* (Copenhagen, 1880), vol. 15, p. 160. The phrase 'wonder-castle' is commonly used as the English translation for 'Aladdin's Castle'.

[12] Margit Mogensen, *Eventyrets tid. Danmarks deltagelse i Verdensudstillingerne 1851–1900* (Odense, 1993), p. 264.

> can sweep everything aside. They represent the most powerful electric force of light so far and darkness is defeated by their power.[13]

At the World's Fair in Paris in 1878 the director of the Tivoli amusement park in Copenhagen saw the carbon arc light. Ten years later Tivoli was filled with electric lighting at the Nordic Industrial, Agricultural and Art Exhibition in Copenhagen (Figure 7.1). At a time when there was no public electric power this was indeed a marvellous sight for the Danes. The first public Danish electric power plant was built in the provincial town of Køge in 1891, and in 1892 Copenhagen joined in.[14] Inspired by the lights in Tivoli and the spectacular Nordic Industrial, Agricultural and Art Exhibition in 1888 the Danish poet Christian Richardt wrote about the 'progress in the North' and the 'spark of Prometheus' existing among Danish men of science.[15]

Figure 7.1 The main building at the Nordic Industrial, Agricultural and Art Exhibition in Copenhagen in 1888. Handcoloured print.

[13] Adolf Bauer, *Studier paa Verdensudsillingen i Paris 1889* (Copenhagen, 1889), p. 12 (my translation).

[14] Peter C. Kjærgaard, 'Kundskab er magt', in Peter C. Kjærgaard (ed.), *Lys over Landet*, pp. 11–40.

[15] Christian Richardt, *Samlede digte* (Copenhagen, 1895), vol. 3, pp. 101, 104.

During the 1870s more professional, scientific and specialist exhibitions focusing on one theme, often technical or scientific, emerged – in Denmark as in the rest of the western world. The fascination and international attention these specialized exhibitions enjoyed was instrumental in promoting domestic science and technology in the late nineteenth century. The popular Danish magazine *Illustreret Tidende* (*Illustrated Times*) wrote about the Paris 1881 exhibition on electricity:

> The main result of this unique exhibition will without doubt be that electric light from this day on will become the joint property of nations. The occasionally short duration of the use of the electric light is most likely in the past. It will without doubt make itself indispensable in public use and pervade homes and family life. This will be an advantage for the health and beauty of houses and cities; in short, an advantage for common well-being.[16]

Electricity continued to be one of the dominant features symbolizing the future and progress well into the twentieth century. One example is the Danish national exhibition in Aarhus in 1909. This was not a specialized exhibition, but a broad-ranging regional fair presenting the products and aims of national and local industries. For this event the exhibition committee had chosen an 'Electricity Tower' as the landmark for the exhibition. The Electricity Tower was used as a symbol on the exhibition poster; it featured in the many official photographs from the exhibition and was printed on souvenir mugs. Even the award given for the most outstanding object or event was a silver miniature of the Electricity Tower. In the festschrift made for the exhibition electricity again played a major role. The chapter 'Use of electricity in life' presented electrochemical processes, electrical railways, X-rays, medical treatment using electricity, cooking by the use of electricity, vacuum cleaners and wind-powered production of electricity.[17] The 1909 Aarhus exhibition was the first provincial exhibition to stand up to the great exhibitions in Copenhagen. Contemporary sources wrote about Aarhus as a modern provincial town, and alluded to the Hans Christian Andersen story of the ugly duckling turning into a beautiful swan.

The focus of the national exhibitions was mainly on applied science – both in Copenhagen and in Aarhus. Industry and agriculture were richly displayed, and the artefacts were centred on electricity and the chemistry of everyday life. Copenhagen was still the political and financial centre, but around 1900 provincial towns, such as Aarhus, became players in a gradual process towards dissolving the capital's scientific, technological and educational monopoly. National pride went hand in hand with the presentation of science and technology, and regional events such as the Nordic exhibitions or international ones such as the World's Fairs grew into some of the biggest public events of the nineteenth century.

---

[16] *Illustreret Tidende*, 25 September 1881, p. 650 (my translation).

[17] L. Friis, *Festskrift. Landsudstillingen i Aarhus* (Aarhus, 1909).

Exhibitions and Academic Science

The new public places for science gave visitors a chance to encounter science and technology first-hand, and Danes an opportunity to present their ideas and products. At the 1851 exhibition in London a working model of an electromagnetic machine made by the Danish engineer Søren Hjort was displayed, and writing paper decorated with algae and moss, made by the Danish amateur biologist Caroline Rosenberg, was among objects exhibited in Paris in 1867.[18] Although there was plenty of science and technology on show at the exhibitions, it is important to note that it was quite different from academic science. At the exhibitions everything was jumbled together and reported as part of the same package, although working scientists would not necessarily subscribe to the way everything was connected and seen to belong in the same group. Still, this changed from the early World's Fairs in the mid-nineteenth century to the early twentieth century. Initially exhibitions served as an important source for scientists to gather information about new developments as well as for the exchange of knowledge, ideas and inventions, while at the same time serving as parks of pleasure and entertainment for the general public. By the early twentieth century scientists had found other venues to present and learn about inventions and ideas as specialized journals and organizations became more and more professional in targeting information towards their members. Consequently the great exhibitions gradually evolved into more of a market than a suitable platform for promoting original scientific research and popular science.[19]

Inventories from the World's Fairs reveal that the Danish contributions were many and varied. Denmark followed the international trend by exhibiting a great variety of objects, ranging from insects to compasses to artificial limbs.[20] Participation in the World's Fairs increased during the 1850s and 1860s with public interest and professional success, a trend also true for the Danish exhibitors. By the 1880s the number of exhibitors in scientific and technological areas fell,

---

[18] The areas and buildings dedicated to women did not contain much science and technology, and women were rarely represented by inventions or handicraft within technical or scientific areas. On women in Danish science, their admission to university education and career paths in nineteenth-century Denmark see Peter C. Kjærgaard, 'Videnskabens kvinder', in Kjærgaard (ed.), *Lys over Landet*, p. 331–40.

[19] Paul Greenhalgh, *Ephemeral Vistas: The Expositions Universelles, Great Expositions and World's Fairs, 1851–1939* (Manchester, 1988).

[20] *Beretning om Kongeriget Danmarks Deltagelse i Verdensudstillingen i Paris 1867*, published by the Danish Exhibition Committee (Copenhagen, 1868); *Fortegnelse over de ved Verdensudstillingen i Paris 1878 prisbelønnede danske udstillere* (Copenhagen, 1878); *Fortegnelse over de danske Deltagere i Verdensudstillingen i Paris 1900 – Industriafdelingen* (Copenhagen, 1900); *Officiel beretning om Danmarks deltagelse i Verdensudstillingen i Paris 1900*, published by the Danish Exhibition Committee (Copenhagen: 1902), p. 40; and Jim A. Bennett, *Science at the Great Exhibition* (Cambridge: 1983).

mainly because of 'exhibition fatigue' by the end of the 1870s and increasingly better venues for communication among scientists outside the exhibitions.[21]

Academic science and laboratory life rarely formed part of the exhibitions. Nonetheless, the exhibitions gave science a face that people could recognize and identify with. In this way the exhibitions were instrumental in science and technology gradually becoming part of Danish public life and, eventually, the everyday lives of Danes.

The Working-Class as Public

Although the main audience for the exhibitions was found among the growing middle classes, great attention was given during the second half of the nineteenth century to the new working class as public. This was the case at the internationally publicized World's Fairs in London and Paris, and also at regional exhibitions around Denmark. Several communicative strategies were used in educating and enlightening the Danish working class through exhibitions and museum collections. Part of this was clearly symbolic in nature and remained at a rhetorical level. But there were also serious and genuine efforts to raise the educational standards of working-class people. The folk high school movement working from the mid-nineteenth century had been instrumental in educating the sons and daughters of farmers in rural Denmark.[22] In the major cities, notably in Copenhagen, several groups began to take an interest in educating the workers too. A more formalized strategy was employed by the University Extension Movement from the early 1890s, but there were also earlier attempts.[23]

One strategy was to bring working-class members to the World's Fairs using government money. There are several examples of the Danish government and private foundations giving grants to individuals or even larger groups for travelling to exhibitions around the world. Individual recipients were always middle- and upper-class members, professionals or academics. In 1867 laboratory director V. Stein received a travel grant from Det Reiersenske Fond, allowing him to visit the World's Fair in Paris. Among the applicants was also a Dr M. Jeppesen, who instead received a grant from the Agricultural Society. Jeppesen's reason for going was to learn about fertilizers in order to utilize the discovery of phosphate on the Danish island of Bornholm.[24] This was clearly in the interests of the foundation, but it is less evident why a director of a chemical laboratory should be given financial

---

21 Vagn Dybdahl, 'Dansk industri og de store udstillinger 1889–1914', *Erhvervshistorisk årbog*, 12 (1961): 7–26, p. 8.

22 Casper Andersen and Hans Henrik Hjermitslev, 'Videnskab på landet', in Kjærgaard (ed.), *Lys over landet*, pp. 251–60.

23 For the importance of the University Extension Movement in Denmark see Hjermitslev, Andersen and Kjærgaard, 'Populærvidenskab og folkeoplysning'.

24 Mogensen, *Eventyrets tid. Danmarks deltagelse i Verdensudstillingerne 1851–1900.*

support. However, individual applicants all had special interests in the exhibitions, and were often craftsmen, architects, teachers, journalists or academics. It was also a characteristic of the recipients that they lived in Copenhagen. In other words, the funding committees favoured applicants from the capital, proving that personal relationships and professional networks were highly important for getting grants.

People applying for the various grants – and indeed the recipients of them – were well-connected, elite professionals. No individual worker would apply for, or even hope to receive, a grant to travel to a World's Fair. Instead other people – out of ideological, mainly political reasons – would apply for the workers. The first known example in a Danish context was the so-called 'Hekla Expedition', organized and paid for by the government, taking workers to the World's Fair in London in 1862. The Danish destroyer *Hekla* was fitted out to carrying about 200 civilians on the 'workers' journey' (Figure 7.2).[25] Departing from Copenhagen on 19 June 1862, the workers were supposed to stay in London for nine days. Officially the journey was for the working-class, primarily people without the money or skills to go on their own. Still, this was not without exceptions.

Figure 7.2 Workers aboard the *Hekla* travelling to the World's Fair in London, 1862. *Illustreret Tidende*, 3/152 (24 August 1862), print.

[25] Margit Mogensen, 'Hekla og verdensudstillingen 1862', *Siden Saxo*, 3 (1989): 25–32.

On the list of passengers we find a number of prominent people and academics, for example the chairman of the Danish exhibition committee, Professor Christian Gotfried Hummel, and the teacher, politician and newspaper editor Christian Vilhelm Rimested from the workers' union *Arbejderforeningen* (established in 1860), both members of the committee organizing the expedition. As this trip was designed to reflect the government's care for the working class, not surprisingly a number of journalists were also on board to promote the journey in Danish newspapers. This allowed the public in Denmark to follow the travellers all the way to London and back. Curiously, the nature of the expedition was not lost on the journalists. A report in the popular magazine *Illustreret Tidende* diverted the readers by telling the story of how the ship was forced to anchor at the northern Danish province of Skagen because of a heavy storm. This unexpected stay became an anthropological study in itself for the natives of Copenhagen – and was apparently more interesting to read about than going to the London Exhibition.[26] Soon after, a similar expedition was organized by the union for craftsmen, *Haandværkerforeningen* (established in 1840). On this occasion the participants were chosen from skilled workers and industrialists. Several later expeditions to World's Fairs were supported by the Home Office. By including even a few workers for private and government money in a highly publicized way, a narrative of open access to new technologies and non-exclusive privileges was created, generating an image of a science for all.

Museums

The social ideology originating in the 1830s to include the poor and the growing working class as part of a public enlightenment programme proliferated and was also one of the main reasons for making Danish museums public from the mid-nineteenth century.[27] There were a number of reasons for doing this, and also various educational strategies – for instance to raise the moral standards of workers. By observing the upper classes visiting the museums the lower social classes were expected eventually to acquire a more civilized behaviour. This way, it was argued, the educational level of the working class would improve and the

---

26 *Illustreret Tidende*, 24 August 1862, p. 382.

27 Although Bedini argues that the early royal and private collections from the seventeenth and eighteenth centuries were open to the public, this needs to be defined further. The collections were only open to a particular kind of public: friends of the king and of private collectors, prominent upper-class people, scientists and travellers and paying visitors on a few occasions; see Silvio A. Bedini, 'The Evolution of Science Museums', *Technology and Culture*, 6 (1965): 1–29. This limited access is confirmed by Kenneth Hudson, *A Social History of Museums* (London, 1975), especially in the first chapter, entitled 'Entry as a privilege', pp. 8–30.

intellectual and moral state of the entire population would increase.[28] In the late 1880s the idea was described in the following way: 'a museum and a library with unrestricted access is as necessary for the mental and moral health of the citizens as good hygiene, water supplies and street lights are for our physical health and well being'.[29] For this strategy to succeed, many argued it was vital that admission to Danish museums was free, as had happened around Europe during the nineteenth century.

Danish public museums have a long and complicated history.[30] The Royal Natural History Museum was founded in 1805 and was one of the first museums in Denmark with a partly public agenda. The intention was that collections of colourful insects, beautiful shaped shells, exotic birds and other natural objects, from Greenland to Brazil, should be connected to induce an image of public use and benefit. However, the collection did not live up to the ambitious plans. Around 1820 the European birds were almost the only animals on public display. Neither did the public access programme work particularly well. A visit to the museum had to be announced at least a day in advance and it was the preserve of the upper class only to visit the public part of the museum.

The primary aim of the first museums, exhibitions and zoological gardens was to impress visitors by means of 'splendour, venerable character or ferocious looks'.[31] This idea goes back to menageries, private collections and the Royal 'Kunstkammer'. During the late nineteenth century museums often displayed coins, ancient relics and ethnological materials alongside natural history specimens of all

---

28 On education through industrial museums in nineteenth-century Britain, see R.G.W. Anderson, 'What is Technology? Education through Museums in the Mid-Nineteenth Century', *British Journal of the History of Science*, 25 (1992): 169–84. With an 'open to the public' policy in the 1820s, English industrial museums were well versed in communicating to the public. Discussing George Wilson, director of the Industrial Museum of Scotland from 1855 to 1859, Anderson describes how glassworkers visited the museum to hear a lecture on glassmaking by Wilson: ibid., pp. 178–9. Furthermore, in 1857 the educational collections of the South Kensington Museum in London opened to the public. Also see Brigitte Schroeder-Gudehus (ed.) *Industrial Society & its Museums* (Paris, 1993).

29 Tony Bennett, *The Birth of the Museum: History, Theory and Politics* (London, 1995), p. 18. The quote originates from Thomas Greenwood, *Museums and Art Galleries* (London, 1888). On the importance of social hygiene in a Danish context, see Jens Erik Kristensen and Lars-Henrik Schmidt, *Lys, luft og renlighed: den moderne socialhygiejnes fødsel* (Copenhagen, 2004).

30 For a history of early Danish collections and museums in the eighteenth and nineteenth centuries, see Mogens Bencard, 'Museerne og verdensordenen. Kunstkammerets opståen og grundidé', *Nordisk Museologi*, 1 (1993): 3–16; Helge Kragh (ed.), *Fra Middelalderlærdom til Den Nye Videnskab – Dansk Naturvidenskabs Historie 1, 1000–1730* (Aarhus, 2006), pp. 170–8; and Mogens Bencard, 'Adam Wilhelm Hauch og museerne', in Mogens Bencard (ed.), *Krydsfelt – Ånd og Natur i Guldalderen* (Copenhagen, 2000), pp. 192–201.

31 Bedini, 'The Evolution of Science Museums', p. 26.

kinds to increase popular appeal, and hence to attract as many visitors as possible – which also helped justify the museums' educational role.[32]

The late nineteenth century experienced an unprecedented explosion in the creation and expansion of natural history museums around the world. In 1910 the *Encyclopaedia Britannica* had registered 2,000 scientific museums in existence.[33] The new Zoological Museum in Copenhagen opened to the public 2 November 1870. Comprising collections from the Royal Natural Museum, the zoological collection from the University of Copenhagen and donations from private collections and expeditions, the new museum was to be more than a *Kunstkammer* or a collection for study and educational purposes. The aim was openness and enlightenment. As with most of the European museums it sought a perfect combination of education and amusement for a new audience.[34] Until the late 1890s there was no division between a public and a scientific collection, though. Everything was on display and there were often lots of examples of the same specimen.[35]

> All new museums are divided into a repository and a collection for the public. The repository, accessible only to scientists and students, contains that which is not fit for the public collection … while this, on the other hand, only contains samples of the various animal groups beautifully displayed, often organized in 'biological groups' to mimic their natural environments. Such a separation was not known 50 years ago. Our museum [the Zoological Museum in Copenhagen] was not originally designed according to this plan. Almost everything was on display.[36]

During the early days of the Zoological Museum the richness of deep-sea life was not yet known and variation among species was not appreciated. At the beginning of the nineteenth century Danish museum exhibitions were a matter of quantity. But, as pointed out by Sheets-Pyenson, quantity without quality was meaningless.[37] By the mid-twentieth century, though, it was more about quality. These examples give us an understanding of the difficulties connected to representations within museums in the second half of the nineteenth century. Classification of specimens

---

32 Susan Sheets-Pyenson, 'Cathedrals of Science: The Development of Colonial Natural History Museums during the late Nineteenth Century', *History of Science*, 25 (1987): 279–300, p. 281.

33 Ibid.

34 Ibid.

35 K. Stephensen, 'Zoologisk Museums Historie', *Naturens Verden* (1921): 31–46, 67–79, 123–38.

36 Ibid., pp. 137–8 (my translation).

37 Sheets-Pyenson, 'Cathedrals of Science'. Sheets-Pyenson looks at this matter in relation to colonial museums, but the statement is also true for European museums in mid- and late nineteenth-century Denmark.

and objects developed gradually from an emphasis on curiosity, magical powers and precious content – as in the rest of Europe – to illustrating the works of nature and evidence of human skills.[38] In addition to the Zoological Museum a new Botanical Museum opened in 1878, and in 1893 a new Mineralogical Museum completed the triad of natural history museums in Copenhagen. All were administrated by the university.[39]

The nineteenth century meant changes in centralism, professionalism and openness for scientific collections. The idea of a public educational experience followed the ideology of enlightenment and public knowledge seen at the World's Fairs and other exhibitions in the 1860s and 1870s.[40] Social ideology was also a key issue in Danish popular science writing during the 1880s and 1890s.[41] By the turn of the century scientific collections were no longer a privilege for specialists or the cultural and financial elite. Ordinary Danes were able to visit museums to explore the geology of Denmark and neighbouring countries, experience plant and animal life from millions of years ago and compare them with contemporary specimens from all over the world. From 1916 the Danish Zoological Museum started staging temporary exhibitions, giving popular lectures and guided tours in the collections – now lit by electric lighting. From the early twentieth century the Danish natural history museums were grounded equally in research and education, leaving room for professional and scientific activities as well as for families' Sunday visits.[42]

Zoological and Botanical Gardens

Zoological and botanical gardens constituted other important new venues for first-hand experiences of science and nature. Prior to the opening of public zoological gardens in Europe, menageries served as one of the only possibilities for ordinary people to see exotic animals.[43]

---

38 Bedini, 'The Evolution of Science Museums', p. 25.

39 Torben Wolff (ed.), *Københavns Universitet 1479–1979, "Det matematisk-naturvidenskabelige Fakultet'* (Copenhagen, 1979).

40 Greenhalgh, *Ephemeral Vistas*.

41 Hjermitslev, Andersen and Kjærgaard, 'Populærvidenskab og folkeoplysning', p. 353.

42 Ragnar Spärck, *Zoologisk Museum i København gennem tre aarhundreder* (Copenhagen, 1945).

43 Robert J. Hoage and William A. Deiss (eds), *New Worlds, New Animals: From Menagerie to Zoological Park in the Nineteenth Century* (Baltimore, 1996); Nigel Rothfels, *Savages and Beasts: The Birth of the Modern Zoo* (Baltimore, 2002). The nineteenth-century general relationship between humans and animals in the context of Victorian England is discussed by Harriet Ritvo, *The Animal Estate: The English and Other Creatures in the Victorian Age* (Cambridge, MA, 1987).

The popularity of menageries peaked in the nineteenth century when they transformed from a privilege for the social elite into something for everyone to see. Although there is no complete record of the number of menageries that visited Denmark in the nineteenth century, or of the routes they chose, we know that there were quite a lot. Contemporary sources provide us with useful information about the animals at the travelling menageries and what people thought of them. Some of the menageries had populations of animals comparable to most zoological gardens, and provided people who did not have an easy access to entertainment in the major towns with experiences of 'natural wonders'. In the summer of 1888 Wombwell's Menagerie visited Denmark. The event was captured in a striking passage from a short story by the novelist Johannes V. Jensen, providing a good sense of how it was received by a Danish nineteenth-century rural population. Wombwell's Menagerie – allegedly one of the largest of its kind in the world – was visiting a small village en route between two major Danish towns:

> The menagerie was moving forward as a large train and could apparently house the entire population of a county. It had a herd of adult elephants, a wagon of lions; and in the rest of them were the most extraordinary creatures in the world … The show lasted two hours, that is, people could walk around in the tent for two hours watching the animals. The audience was grateful that particular night in Graabølle village. People were neither formed by any education in natural history, nor lunches in the Zoological Garden. They looked at things from a rather down-to-earth and correct perspective; they went from what they knew to the new, through experience, a road that has never led anyone astray.[44]

In 1859 the Zoological Garden in Copenhagen was founded by the amateur naturalist Niels Kjærbølling.[45] Despite the dislike in academic scientific circles of Kjærbølling's popular approach to science, the educational effect of the Zoological Garden is not to be mistaken. Books provided something towards the scientific education of the public and museums also contributed; but a full

---

44 Johannes V. Jensen, *Himmerlandshistorier* (Copenhagen, 1944), pp. 108, 127 (my translation).

45 For a detailed version of the history of the Zoological Garden in Copenhagen see Julius Schiøtt, *Zoologisk Have og dens Fremtid* (Copenhagen, 1900); Julius Schiøtt, *Zoologisk Have gennem 50 aar* (Copenhagen, 1909); and Jørgensen, *Zoo – en historie om dyr og mennesker gennem 125 år* (Copenhagen, 1984). Prior to the opening in Copenhagen zoological gardens were established all over Europe. The first was in London in 1827, followed by Dublin in 1832, Bristol in 1835, Amsterdam in 1838, Antwerp in 1843 and Berlin in 1844. For a history of zoological gardens and menageries in Europe see Hoage and Deiss (eds), *New Worlds, New Animals*.

knowledge of nature was only obtained when people had experienced how living nature behaved in real life.

Kjærbølling got the idea for a zoological garden in Copenhagen after visiting Berlin's Zoological Garden in 1851 as part of the social programme of an ornithology meeting. Addressing the importance of a zoological garden, Kjærbølling succeeded in persuading the Danish king and the Ministry for National Matters to give him a piece of land in 1859 for the purpose of establishing a national zoological garden in Copenhagen. Kjærbølling had almost no money, but managed to found a rather scanty garden resulting in cages in bad condition and only a few common animals. Five days before the opening in 1859 he desperately advertised for animals in local newspapers.

The Zoological Garden in Copenhagen was not particularly large. There were eagles, owls, chickens, ducks and rabbits, a fox on a chain, a seal in a tub and a turtle in a bucket. In the winter the seal was put down, being too expensive to keep alive during the months when the garden was closed. Over the years many seals suffered the same fate. Wintertime was used for improving the cages and acquiring more animals. After a few years Kjærbølling managed to get a few foreign animals for the garden as well, among them a camel from Moscow, a polar bear from Spitzbergen and a vulture from Messina. However, Kjærbølling was determined to give the Copenhageners a zoological garden to match their city, and by 1865 the animal population had increased – now including exotic animals such as an American wild boar, an emu, several monkeys and a kangaroo. There were no lions, tigers or elephants though, as in most of the other European zoological gardens; but, despite the rather modest range of animals, the garden had become an integral part of Copenhagen cultural life. Records of visitors show that the Zoological Garden was highly popular. Kjærbølling played a central role in popularizing zoology not only behind the scenes, but also directly by walking among the visitors every day telling stories about the animals.[46]

As a direct experience of living nature only the Botanical Garden could compare to the Zoological Garden, and both quickly became popular with the Copenhagen public. Before the 1870s the primary function of the Botanical Garden in Copenhagen was to serve as a medicinal herb garden used for teaching medical and pharmaceutical students, with very limited access to the general public. Intentions to make the Botanical Garden public surfaced in the 1860s and in 1871 decisions about a new location for the garden were made. A commission was set up to find the right location and, as it turned out, the commission also suggested to the University of Copenhagen – which owned the Botanical Garden – that they should build a Zoological Garden at the same time, connected to the Botanical Garden.[47] Neither Kjærbølling nor this particular popular approach to science and nature represented by the Zoological Garden gained much respect

---

46 Jørgensen, *Zoo*.

47 Ragna Heyn Oxenløwe, 'Bygninger, politik og penge' in Kjærgaard (ed.), *Lys over landet*, pp. 61–94.

within the Danish scientific zoological community. The professor of zoology Japetus Steenstrup was one of Kjærbølling's strongest opponents, and at the same time one of the most powerful Danish university scientists in the second half of the nineteenth century. As a member of the 'garden commission' Steenstrup was not entirely without interest in the suggestion of a new Zoological Garden connected to and run by the university. But, as it turned out, the commission in this case had overestimated the capability and willingness of the university to pay for free entertainment for the Copenhagen public. The idea of a public Botanical Garden for educational use was a threat to the existing scientific medical garden and therefore not something to be desired by the university.

Eventually, though, the many negotiations between the garden commission and the university resulted in a new Botanical Garden. The area, presenting numerous exotic plants in central grounds with easy access, opened to the public in 1874. In the decades that followed, animals and plants from all over the world gradually became part of the cultural and scientific lives of the Danes.

Living Exhibitions

Around 1900 the Zoological Garden in Copenhagen appointed a new director. The former secretary of the Danish Industrial Association and editor of several popular science ventures, Julius Schiøtt, was appointed and pushed for changes. Schiøtt received help from Carl Hagenbeck in Hamburg, a German who had specialized in animal trading and had a history of delivering animals to the Zoological Garden in Copenhagen. Upgrading the selection of animals in the garden was just part of the plan. With Schiøtt as director a new concept was introduced – the so-called 'caravans of people'. The Tivoli in Copenhagen had tried its luck with caravans of people from 'exotic cultures' in the 1880s, but for the Zoological Garden the innovative combination of nature and culture breathed life into the idea again in the early twentieth century, with the wildly popular 'exhibition' of, for example, Japanese and Indian caravans from Asia. In 1896 an arena for animal shows was erected. By the turn of the century the shows also included people from foreign cultures.[48]

The caravans had proven to be reliable successes at the World's Fairs. Here the caravans worked as 'living exhibitions' and were mostly collected and brought to Europe from the British and French colonies. One of the most popular features, for example at the Philadelphia exhibition, was the construction of a hospital interior with live acting nurses in a kind of living exhibition.[49] The fascination of technological and scientific inventions was certainly a primary motivation for many visitors to the exhibitions, but for the majority of visitors the marketplace and the appeal of the foreign and mysterious played an even

48 Jørgensen, *Zoo*.

49 Greenhalgh, *Ephemeral Vistas*.

larger role. In Copenhagen, Schiøtt had an eye for people's fascination with adventures from distant parts of the world and the attraction of the unknown. Hence, religious rituals, demonstrations of cooking and presentations of craft were scheduled along with feeding times for the animals. The Zoological Garden had simply acquired a new species for visitors to encounter in its natural habitat – *Homo sapiens*.

Professional and popular interest in anthropology increased in the 1860s and 1870s with new findings and scientific discussions about the origin of humans and cultures around the world. The scientific interest was marked by numerous lectures, publications and meetings, as well as the creation of anthropological societies and specialist journals all over Europe and North America. The professional anthropological wave was followed closely by a popular one. The scientific study of humans was feeding directly into the popular fascination with questions of who we are and where we came from. The many new technologies for making popular science available were instrumental in the creation of popular anthropology.[50]

Important for this success was also the first-hand experience at zoological and botanical gardens as well as at exhibitions. The anthropological studies of both humans and animals were a fascinating attraction for most visitors. Prior to the first living exhibition putting 'primitive cultures' on display in 1867, specific exhibitions for studying living animals and plants had their origins in Paris in 1860, incidentally coinciding with the publication of Charles Darwin's *Origin of Species*. The Jardin Zoologique d'Acclimatation was built especially for the World's Fair with the specific purpose of studying and helping the popularization of botany and zoology. Dedicating a specific building to exhibiting the living world was a success and paved the way for the development of living exhibitions later in the nineteenth century.[51]

The fascination and success of the living exhibitions needs to be seen in the context of European imperialism and relations with foreign peoples and cultures.[52] Colourful reports in the popular press from the deepest jungles of Africa and the heights of the Himalayas became part of weekly armchair entertainment.

[50] For a history of anthropology see Fredrik Barth, Andre Gingrich, Robert Parkin and Sydel Silverman, *One Discipline, Four Ways: British, German, French, and American Anthropology* (Chicago, 2005). See Nélia Dias, "The Visibility of Difference – Nineteenth-Century French Anthropological Collections" in Sharon Macdonald, *The Politics of Display – Museums, Science, Culture* (London, 1998), pp. 36–52, for a detailed history of anthropological collections associated with the Museum of Natural History in Paris and the Anthropological Society of Paris in the nineteenth century. Scientific and popular anthropological accounts of human genealogy in a nineteenth- and early twentieth-century Danish context is discussed by Poul Duedahl, 'Det videnskabelige menneske', in Kjærgaard (ed.), *Lys over landet*, pp. 431–2.

[51] Livingstone, *Putting Science in Its Place*, pp. 57–9.

[52] For a discussion of some of the many aspects of science and colonialism see Roy MacLeod (ed.), *Nature and Empire: Science and the Colonial Enterprise* (Chicago, 2000).

The popularity of discussions and images of race reflected the image of the European rather than people from other parts of the world and mostly reproduced racial stereotypes uncritically.[53] The living exhibitions now provided an opportunity to experience these things firsthand. Their combination of exotic culture and safe confirmation of the racial hierarchy appealed strongly to European audiences, and the number of living exhibitions increased accordingly from 1889 to 1914.[54] The most popular caravans were touring the largest exhibitions and zoological gardens in Europe – including the Zoological Garden in Denmark. They started out as mere entertainment, but gradually became educational examples of 'primitive' human cultures providing an opportunity to experience how modern human beings went through different cultural 'evolutionary stages', not surprisingly arriving at modern European man.

Contrary to the Zoological Museum, the Zoological Garden in Copenhagen became an important venue for the promotion of evolutionary theory and visitors were often reminded of the relationship with the great apes.[55] Orang-utans and chimpanzees were among the major attractions. In 1898 the garden had acquired two orang-utans and two chimpanzees. A colourful poster was made with the words 'Human Apes' printed in bold letters and displaying an image of an orang-utan staring visitors right in the face. But death-rates were high among the primates. One of the orang-utans died the day after it was delivered to the garden; the other died within seven weeks.[56]

By the end of the nineteenth century people were used to the idea of humans being related to other primates. In the Zoological Garden in Copenhagen the chimpanzees were eating with napkins around their necks and a white cloth on the table – if they were not occupied riding bicycles. These animal shows were another kind of living exhibition, showing the similarities between primates and humans. In 1909 the Zoological Garden in Copenhagen had three chimpanzees, a male and two females. The male was called 'Master Link' as a reference to the popular expression 'missing link' and as a direct response to critics of evolutionary

53 See Douglas A. Lorimer, 'Science and the Secularization of Victorian Images of Race' in Bernard Lightman (ed.), *Victorian Science in Context* (Chicago and London, 1997), pp. 212–35.

54 Greenhalgh, *Ephemeral Vistas*.

55 Japetus Steenstrup, who was also opposing the Zoological Garden while promoting his own ideas of a zoological garden managed by the university, was the powerful director of the Zoological Museum and, although on friendly terms with Charles Darwin, he was an ardent opponent of evolution; see Peter C. Kjærgaard and Niels Henrik Gregersen, 'When Darwinism Comes to Denmark: The Early Danish Reception of Charles Darwin's *Origin of Species*', *Ideas in History*, 1 (2006): 151–75.

56 Rikke Schmidt Kjærgaard, 'Videnskabens offentlige rum', in Peter C. Kjærgaard (ed.), *Lys over landet*, pp. 375–98, p. 393.

theory (Figure 7.3).[57] For the director of the Zoological Garden, Julius Schiøtt, there was no doubt. The chimpanzee was indeed a link between primates and people, a connection between animals and humans. As far as Master Link was concerned, he was even a particularly intelligent and gentle specimen.[58]

Figure 7.3 The male chimpanzee Master Link on his bicycle at the Zoological Garden in Copenhagen, around 1900.

57 Although Charles Darwin refrained from explicitly expressing the relationship between humans and apes in the *Origin of Species*, the idea spread quickly in scientific circles and in popular accounts of the origin of man. The findings of human-like remains, like the Neanderthals in 1856, during the second half of the nineteenth century inspired many to search for 'missing links' to document the connection. Among them was Eugène Dubois, who dedicated his life to finding fossil records of human ancestors in order to prove Darwin right; see L.T. Theunissen, *Eugène Dubois and the Ape-Man from Java: The History of the First 'Missing Link' and Its Discoverer* (Dordrecht, 1989).

58 Schiøtt, *Zoologisk Have og dens Fremtid*; Schiøtt, *Zoologisk Have gennem 50 aar*.

On Display

An increase in the number of physical spaces making scientific knowledge visible and available coincided with growing demand for popular literature on science and technology. Although we have a good understanding of how printed popular science production and consumption worked in relation to lectures, meetings and educational settings, we must be careful not to overlook the importance of the new public spaces dedicated to science and the possibilities for first-hand experiences of nature. Science found its natural place in exhibitions and in zoological and botanic gardens. The opening of museums to a larger audience also increased the public space for science.

Even with its limited opening hours and little effort to explain its exhibits, the Zoological Museum in Copenhagen was an immediate success after opening in 1870. The museum was only open to the public two days a week – Wednesday and Sunday – and then just for two hours. Nonetheless, the Copenhagen public used the time well. Going to a museum on a Sunday, walking in the botanic garden or visiting an exhibition became an integral part of life in the capital. Clearly there was also a social component that should not to be overlooked – that going to these events and places was as much a matter of seeing as being seen. As countless photographs and illustrations of people at exhibitions, in gardens and museums show, the new public places for science provided an equal opportunity to put both nature and audience on display. Even with its restrictions the Zoological Museum was visited by 50,000 to 85,000 people a year.

It was no longer an upper-class privilege and by the turn of the century it had become customary for Danes to visit gardens, exhibitions and museums. Putting science on display, demonstrating new technologies and inviting people to experience nature and culture first hand proved its worth. It is difficult, of course, to gauge the depth of the effects that the many new places for science popularization had on public opinion about science. However, taking into account the numbers of visitors to exhibitions and scientific gardens, along with the many reports from exhibitors, curators and visitors themselves, it becomes clear that the new places for science and technology made a difference and received a lot of attention.

Often the attention was framed in a particular narrative to serve particular purposes, for example to promote a certain social ideology as when groups of workers were sent to the World's Fairs. Such events were staged to support an image of an all-inclusive audience, rhetorically making the growing working class count. Specific narratives were also used in a different fashion to sustain public interest in exhibitions over long stretches of time. At the 1909 exhibition in Aarhus the narrative of Louis Blériot's aeroplane served this purpose as visitors waited the entire run of the exhibition for the plane to show up. A special exhibition newspaper followed people's enthusiasm and disappointment over several weeks through daily reports of the non-event. When the exhibition was drawing to an end and Blériot's aeroplane had still not been seen in the air, the newspaper manipulated a picture showing a plane flying over the exhibition area.

After a month of anticipation the aeroplane finally arrived. This was the first time an aeroplane had been seen in Aarhus, so it is no wonder it received massive attention. In the exhibition newspaper the story about the French flying wonder turned into a narrative with daily updates reflecting new technologies, hopes for the future and regional aspirations eventually mixed with distrust and irony. Not all were happy about the vulgarization of academic science. Some argued that it distorted a true understanding and representation of nature. The professor of zoology at the University of Copenhagen, Japetus Steenstrup reacted strongly but in vain against the popular Zoological Garden in Copenhagen, and argued for a university controlled garden instead. But as many professional scientists were engaged in science popularization themselves, the new public spaces for science gained many supporters.

Presenting natural wonders and technological inventions supported the growing interest in understanding and commanding nature. When trying to understand how science popularization worked, at which levels it operated and what it meant for public opinion, analysing the importance of the new public spaces for science is crucial. Exhibitions, museums and scientific gardens were indeed popular attractions and have to be taken into account in a comprehensive study of nineteenth- and early twentieth-century popular science. Exhibitions were also instrumental in inspiring and feeding other public places with ideas, artefacts and inventions. The director of the Tivoli amusement park in Copenhagen, for instance, went to exhibitions around Europe and returned with electric lights, trains and other attractions. This was quite common in Denmark, as everywhere else in Europe and North America.[59] Although specific local contexts defined the circumstances particular to the Danish exhibitions, gardens and museums, the main reception and use followed the general pattern. As elsewhere local, national and regional pride were among the chief ideological vehicles driving the initiatives and innovations of science popularization. The Danish case demonstrates that there is no simple picture of centre and periphery when it comes to filling the physical space for science popularization. Ideas, objects, plants, animals and humans crossed boundaries and were appropriated for various purposes at different times. A small provincial town could serve as a powerful local centre and at the same time be peripheral to the capital, as was the case with Aarhus. In the minds of the Danes there was little recognition of being on the European periphery. Rather the country enjoyed a privileged position as the link between the European continent and the other Scandinavian countries.

[59] Robert W. Rydell, 'Museums and Cultural History: A Review Article', *Comparative Studies in Society and History* 34 (1992): 242–7.

Chapter 8
Genres of Popular Science: Urania and the Scientific Theatre

Gábor Palló

In spring 1897, a performance was held in the Budapest Opera House, the most elegant theatre in town, on the stage where Mozart and Verdi's operas were normally performed. The organizers wanted to see whether theatre could be a suitable medium for popularizing science. According to a contemporary report, the play, entitled *An Excursion to the Moon*, described the Moon as a celestial body, and spoke about lunar and solar eclipses. The audience was large, but was not particularly impressed with the performance. However, in October a group of people decided to establish the Urania Scientific Theatre.[1] Somewhat later, and connected to the theatre, a new society was established under the name of Urania with the aim of spreading scientific knowledge to a wider audience.

With this decision to set up a scientific theatre a new genre of science popularization started in Hungary. The theatre followed other forms of popularization that were already operating successfully. The Hungarian National Museum was set up in 1808 to store and display books, paintings and a collection of minerals. The Museum of Natural Sciences grew out of this collection but remained part of the National Museum until after World War II. Interested people organized various scientific societies, including the Association of Natural Scientists, established in 1841, aimed at informing an educated audience – mostly schoolteachers of natural scientific disciplines – about the latest developments in various scientific subjects. At their meetings, prominent professors delivered lectures on the new results achieved in science all over the world. The oral form was supplemented by printed texts of the lectures published in the journal *Természettudományi Közlöny* (*Journal of Natural Sciences*), launched in 1869, and in books produced by the association's own publishing house. Around the turn of the nineteenth century, representatives of various disciplines (geography, physics, mathematics, chemistry, pharmacy etc.) organized scientific societies and published their journals in Hungarian. In harmony with the received views of the time, these societies considered communication with a lay audience to be a one-way street. The speaker or writer possesses knowledge and transmits it to the ignorant public in a simplified and attractive form. The growing number of societies, journals and books led to an unprecedented boom in science communication in Hungary. This increase was in keeping with the

---

1 'Theatre and Arts', *Országos Hírlap* [National Newspaper] 1/1 (1898): 25.

fervour for modernization spreading throughout the Austro-Hungarian monarchy, particularly in Budapest. Modernization changed the whole social, economic and political structure, mainly in large cities. New kinds of vehicles were seen in the streets, many newspapers appeared advertising all kinds of wonders – including new medical treatments and previously unimaginable new gadgets – and gas lamps provided light in streets and homes.[2] In this atmosphere, some risk-taking entrepreneurs sought to supplement the science popularization genres with a new one – the theatre, a fashionable form of entertainment in Hungary.

Science and Theatre

Urania represented one variant of scientific theatres. Throughout history, it has had several variants, although theatre is not an obvious medium for presenting scientific content. In relation to science, the term 'theatre' has been used both metaphorically and literally. The metaphorical usage is the older one.

Metaphorically, 'theatre' often expresses doubts about something suspicious that might be deceitful or untrue, as opposed to science that claims to tell the truth. Francis Bacon, when analysing the four characteristic fallacies of human thinking, mentioned theatre as one of them. The 'idol of theater', as he called it, creeps 'into men's minds from the various dogmas of philosophies, and from wrong laws of demonstration. These I call Idols of the Theater; because in my judgment all the received systems are but so many stage-plays, representing worlds of their own creation after an unreal and scenic fashion.'[3] In this criticism of human thinking, theatre occurs as a metaphor for any kind of act on stage seen by an audience believing it to be real because of its appearance, or special dramaturgy, but without rightful arguments. In this approach, theories, dogmas and phenomena can be instilled into people who are unable to judge the reality of the performance. Theatre is based on illusion, on appearance, lacking reliable knowledge.

The literal usage of 'theatre' can also be connected with Baconian empiricism, although it denigrated theatre into a medium for communicating unfounded knowledge. Facts played a central role in Baconian science, but whether a singular fact existed or not was difficult to prove. Therefore, facts could only be considered well established if an unbiased audience testified to their reality. For this purpose, experiments should be presented onstage to the audience. This kind of presentation can be considered scientific theatre in a non-metaphoric sense. Some historians, such as Steven Shapin, Simon Schaffer, Jan Golinski and others,

---

2 For the everyday life of contemporary Budapest see John Lukacs, *Budapest 1900: A Historical Portrait of a City and Its Culture* (New York, 1988).

3 Francis Bacon, *The New Organon: Or true directions concerning the interpretation of nature.* 1620. Standard translation by James Spedding, Robert Leslie Ellis and Douglas Denon Heath (Boston, 1863). An Internet version: <http://www.constitution.org/bacon/nov_org.htm> (accessed 12 February, 2005).

have pointed out in various contexts that experimental demonstrations to the lay audience played a decisive role in institutionalizing modern science. In the early Royal Society, science, including chemistry, became a kind of public spectacle.[4] Iwan Morus extended investigations to early nineteenth-century electricity. He showed that Michael Faraday performed experiments for the public as part of exhibitions displaying electrical devices. Morus highlighted the significant role played by the social status of the audience in the reception of Faraday and his rival William Sturgeon's scientific results.[5] In David Livingstone's analysis, public theatrical demonstrations of electric phenomena represented a connection between theatre, the experimental laboratory and museums.[6] In addition, the philosopher Robert Crease analysed recent experimental science in terms of a theatrical performance. Looking for a phenomenological, hermeneutical framework, he used 'theatrical analogy' in the sense that scientific experimentation and theatrical performances share many common features, as both are created, and witnessed, and both represent some characteristic phenomena. Crease discusses science as a cultural practice when he compares experimental with theatrical presentations, representations with recognition.[7]

Another more recent variant of scientific theatre has focused on scientists. Michael Frayn's drama, *Copenhagen*, and several other theatrical plays have attempted to convey views of general interest related to science to a large audience. These plays followed the tradition started decades before Frayn's drama was shown, perhaps with Bertolt Brecht's *The Life of Galileo* or Friedrich Dürrenmatt's *The Physicists*. Unlike experimental demonstrations, these dramas spoke about the moral and political dilemmas of the great scientists who had to face the conservative opinions of their peers or the unwanted consequences of their highly appreciated achievements. The borders between science and other kinds of social and intellectual life seem more blurred in these plays than they apparently are in the received views of science. While sympathizing with Galileo in his fight for truth, the audience is shown the significance of the heliocentric worldview in Galileo's times. The person, the moral, the belief and the conviction appear intertwined on the stage. The audience continues to think about Galileo after the show and without even noticing it; some matters of science become negotiable between the

---

4 Steven Shapin, 'The House of Experiment in Seventeenth-Century England', *Isis*, 79 (1988): 373–404. Simon Schaffer, 'Natural Philosophy and Public Spectacle in the Eighteenth Century', *History of Science*, 21 (1983): 1–43. Jan Golinski, 'A Noble Spectacle: Phosphorus and the Public Culture of Science in the Early Royal Society', *Isis*, 80 (1989): 11–39.

5 Iwan Rhys Morus, *Frankenstein's Children: Electricity, Exhibition and Experiment in Early-Nineteenth-Century London* (Princeton, NJ, 1998).

6 David N. Livingstone, *Putting Science in its Place: Geographies of Scientific Knowledge* (Chicago and London, 2003), pp. 25, 29.

7 Robert P. Crease, *The Play of Nature: Experimentation as Performance* (Bloomington and Indianapolis, 1993).

lay audience and the experts. They can discuss the importance of scientific matters and provide arguments as to whether Galileo or Cardinal Bellarmini was right, disregarding the difference existing between their scientific knowledge.[8]

This kind of scientific theatre facilitates dialogue between the general public and science, as it helps to surpass the 'deficit model' used in the field of public understanding of science. According to the deficit model, the lay audience lacks sufficient knowledge to form an autonomous opinion in matters related to scientific knowledge. Because of this asymmetry, experts and non-professionals cannot be equal partners in discussions of science-related problems. Therefore, according to this model, the public should learn more science. Hence, while showing the dilemmas of the main characters, Galileo or Heisenberg, the scientific side of their thinking – geocentric astronomy or exploitation of nuclear energy – is also explained in an easy, simplified manner.[9] So, these theatrical plays are of pedagogical significance, as they teach some scientific ideas together with political and moral issues connected with science. Having seen some dramas, the audience learns to think about and discuss these kinds of vitally important matters. Scientific theatre contributes to forming a contextual, participatory and deliberative communication between science and non-science.

These approaches consider science to be a homogeneous activity. Ludwik Fleck, however, has stressed that science comes in various forms in terms of its communication, and popular science is one of them.[10] He connects the types of science with the communicating groups, called 'thought collective' (*Denkkollektiv*), which have members who share a 'thought style' (*Denkstil*). The members of a thought collective can discuss ideas in mutual understanding because of their common thought style. The thought collective consists of an esoteric and an exoteric circle, the former includes the specialists, while the latter the non-specialist experts, the students who will become experts and the lay audience. Popular science speaks to the exoteric circle; hence it is different from all journal, handbook and textbook science, which speaks to the specialists, non-specialist

---

8 In an internet article, Judy Kupferman gives a list of science-related dramas and analyses their didactic significance. Judy Kupferman, 'Science in Theater', <http://www.jewish-theater.com/visitor/article_display.aspx?articleID=485&pagenumber=1> (accessed 12 February 2006).

9 Interesting analyses of this kind of scientific theatre were delivered at a conference on Frayn's drama. The texts of some papers are posted on the Internet homepage of the Niels Bohr Institute (<http://www.nbi.ku.dk>). The Niels Bohr Archive's History of Science Seminar: 'Copenhagen and Beyond: The Interconnections between Drama, Science, and History' held at the Niels Bohr Institute, 19 November 1999. See in particular Jan Golinski, 'Copenhagen as history of science narrative' and Robert Marc Friedman, 'Reflections of a historian of science', <http://www.nbi.dk/NBA/files/sem/copfried.html> and <http://www.nbi.dk/NBA/files/sem/symp/golinski.html> (accessed 23 March 2006).

10 Fleck's views on popularization are also analysed in Jonathan R. Topham's historiographical study. See Jonathan R. Topham, 'Rethinking the History of Science Popularization/Popular Science', Chapter 1 in this volume.

experts and students respectively. Neither are the circles closed, nor the types of science insulated from each other. Fleck supposed that knowledge circulates between these types.[11]

According to Fleck, popular science, unlike other types of science, is characterized by its certainty, simplicity and vividness. More recently, Richard Whitley has criticized 'the traditional view' of popularization for disregarding social and cultural factors related to the communication between science and a lay audience. While emphasizing the significance of a contextual view of the popularization of science, he added two relevant features of popular science to Fleck's characterization: popular science is authoritative and decontextualized.[12] Stephen Hilgartner considered the Fleckian approach to be a simplistic 'dominant view'. He explained that the boundary between popular science and genuine science is a matter of degree regarding both their content and practitioners.[13] In harmony with these criticisms, James Secord admonished the uncertainty of the boundaries between Fleck's esoteric and exoteric science, and popular and expert science.[14]

Fleck based his typology on printed text. He and many authors disregarded other forms of popularization, including the most widespread ones, namely popular scientific lectures and museums. Nevertheless, around the turn of the nineteenth century, scientific theatre contributed to the impression that genuine science radically differs from popular science. The audience considered the performance onstage to be an imitation of nature and an imitation of scientific practice, without assuming it to be reality. Scientific theatre could persuasively show science to be certain, simple, authoritative, decontextualized and, particularly, vivid.

Popularization as Business, Governmental and Civic Activity

At least tacitly, vividness must have also been in the minds of the initiators of the Urania Scientific Theatre, but not only that. A popular scientific venture was a more complex enterprise than mere education of the public. The Urania Scientific Theatre's first show, *An Excursion to the Moon*, was of a philanthropic and pilot nature because the organizers wanted to see whether this form of popularizing science had any future. The answer was positive. They decided to set up a public company to run the scientific theatre in the hope that the owners would make a

---

[11] Ludwik Fleck, *Genesis and Development of a Scientific Fact* (Chicago and London, 1979) [first published in 1935].

[12] See Richard Whitley, 'Knowledge Producers and Knowledge Acquirers: Popularization as a Relation Between Scientific Fields and their Publics', in Terry Shinn and Richard Whitley (eds), *Expository Science: Forms and Functions of Popularisation.* (Dordrecht, Boston, London, 1985), pp. 3–28.

[13] Stephen Hilgartner, 'The Dominant View of Popularization: Conceptual Problems, Political Uses', *Social Studies of Science,* 20 (1990): 519–39.

[14] James A. Secord, 'Knowledge in Transit', *Isis*, 95 (2004): 654–72.

reasonable profit. The shareholders established the company in 1899, under the name of the 'Urania Scientific Theatre Society Corporation', as was announced in several journals.[15]

A detailed description of the company's structure was published in the first issue of the *Journal of the Urania Hungarian Scientific Theatre*.[16] Four founding members started the company: Viktor Molnár, a councillor, later state secretary in the Ministry of Religion and Education, Gábor Várady, Member of Parliament, Nándor Somogyi, entrepreneur in the fields of arts and literature and Professor Radó Kövesligethy, astronomer, geophysicist and leading scientist of the time. Politics, business and science were represented in the group, revealing the wide variety of interests behind the establishment. The board of directors had 12 members including the four founding members. Viktor Molnár was selected to be president, and Nándor Somogyi executive director. They set up a supervisory board consisting of four members. The company had around 80 shareholders recruited from politicians, high-ranking priests, landowners, bankers, industrialists, artists, scientists and ordinary private individuals. The start-up capital amounted to 100,000 Hungarian crowns, a good start for a new cultural venture. Urania held annual meetings for its shareholders to account for activities and finances.

The company's goal was to run the theatre using its own staff. At first, the staff included an executive director, Nándor Somogyi, and a scientific director, Radó Kövesligethy, who was a highly esteemed authority in his field. The personnel consisted of about 20 people, including technical staff, a cashier, legal adviser and two actors.

It is telling that Viktor Molnár, a politician, became the real leader rather than a scientist, a teacher, an artist or a businessman. He worked for the Liberal Party that governed Hungary between 1875 and 1905. The Liberal Party supported the compromise between Austria and Hungary that led to the dual Austro-Hungarian monarchy in 1867. Because the Liberal Party ruled for such a long period, it was called the 'Governing Party'. The opposition, the nationalist Independence Party, opposed the compromise. It stood for an independent Hungary. In 1905, the Liberal Party lost the national election and dissolved into other political formations. Nevertheless, 'national liberalism', as it was called, survived for a while as the dominant ideology in Hungary. The influence of this ideology weakened radically with the outbreak of World War I. Molnár, as a politician, worked for the governing party in the field of education. Other politicians participating in the organization of Urania belonged to the same party. Characteristically, the wife of Molnár's boss, Count Albin Csáky, the Minister of Education, actively participated in Urania's activities. Molnár was a devoted supporter of disseminating knowledge to a

---

15 *Gyógyszerészi Hetilap* [*Pharmacist Weekly*], 38 (1899): 403, 693, 706. *Gyógyszerészi Értesítő* [*Pharmacist Bulletin*], 7 (1899): 525.

16 Unsigned article: 'A tudományos színházról' [On the Scientific Theatre], *Uránia Magyar Tudományos Színház Közlönye*, 1 (1899): 3–13, <http://www.filmintezet.hu/magyar/filmint/filmspir/23/urania.htm> (accessed 27 February 2006).

wide non-trained audience. Using his rising ministerial positions, he promoted the education of the non-schooled population, and considered Urania to be an important instrument in his actions. It was the extension of a government policy relying on private initiatives.

Because of Molnár's comprehensive education ambitions, Urania soon became a more complex entity. He initiated the establishment of another organization, the Urania Hungarian Scientific Society, started in 1899 with himself as president. The goal of this society was to support the scientific theatre intellectually, morally and financially, and to widen its activities to include other genres of popularization, extending it to all intellectual fields and reaching a less-educated, non-theatre-going audience. In realizing these goals, Urania set up 26 sections representing different intellectual fields that, for practical reasons, were soon reduced to eight. They comprised almost all fields – from astronomy, physics and chemistry to geography and medicine; from ethnography, history and philosophy to architecture, fine arts and technology. As a scientific society, Urania had members but no shareholders. Founding members had to pay 200 crowns and ordinary members 10 crowns as a membership fee for at least three years, while honorary members, well-known people, only lent their names to enhance the importance of the society. Even some enterprises and institutions such as schools and industrial associations supported Urania. All members received the journal for free and received 50 per cent reductions at the theatre box office. The government secured gradually increasing financial support after 1903.

The multifaceted intellectual landscape combined with many genres. Urania very effectively used the printed forms of popularization. In 1900 the society launched a new journal, entitled *Urania*, a colourful monthly magazine, and in 1901 a series of books, the Urania Library. The volumes, in reality booklets, summarized some fields of science and the arts. Another traditional genre of popularization that Urania favoured was lectures, through which the society intended to reach a wide audience. To increase its income, Urania opened a fine arts shop in 1906 that also functioned as a gallery. In addition, a new partner, called the Urania Visual Educational Aids Factory – photographer Mór Erdélyi and Partner – joined the organization in 1908 to meet the growing demand for pictures.

Besides politician Viktor Molnár, professor of physics Jenő Klupathy also played a leading role as secretary general; he was succeeded by Károly Szász, later President of Parliament in 1912, then Gábor Finály, a historian.

The society grew until around the time of World War I, and then rapidly declined. Its annual meetings were usually held together with the Theatre Society and the proceedings were published in *Urania*. From these reports, data of its growth can be determined. The number of Urania members was 400 in 1901, 1,313 in 1910 and 1,702 in 1917. In 1904, they delivered 259 lectures and performances, while in 1906 this number had risen to 653. The audience in 1906 amounted to 114,000, rising to 266,000 in 1910. The journal had 370 subscribers in 1908, excluding members who received it free. In 1913, 2,200 copies of the journal were printed, quite a big run compared with other contemporary scientific, literary or popular journals. The society's capital grew from the initial 5,000 crowns to 105,000 crowns by 1914. The corporation's income amounted to around 8,000–10,000 crowns, from which a dividend was paid.[17]

This growth lasted until the outbreak of World War I. After 1914, all numbers and enthusiasm declined. At the 1916 annual meeting, the corporation's members spoke out against Molnár because of the decreasing dividend.[18] After Hungary's defeat in the war, Urania made efforts to start a new life, but in the environment of social and economic collapse this ambition proved hopeless. In 1922, Urania was dissolved.

The Scope of Urania

Urania took the goal of disseminating a wide body of knowledge to a wide circle of the population seriously. In a similar way to the old idea of the encyclopedic worldview, the society intended to show human culture as a unity in its journal, in the lectures and in the theatre. The journal published long articles about scientific results from different disciplines and about some developments in law, medicine, fine arts, literature and music. Many articles were accompanied by illustrations. The long articles were supplemented by short ones on the latest discoveries, technical gadgets, vehicles, aeroplanes, useful medical advice, meteorological phenomena, announcements of events in intellectual circles and reports on the society's activities, including its annual meetings.[19] The argumentative and linguistic level of the articles was quite high. They targeted an educated audience, rather than the wider population. Molnár served as the chief editor, while the other editors were well-known experts in their fields.

Concerning character, a conspicuous difference was present between the articles on science and other fields. Novels, historical and, in general, humanistic writings were of a conservative, nationalist mentality, drawing idealized pictures of the Hungarian past and the superiority of Hungarian tradition, literature and spirit. They largely disregarded poets and novelists who belonged to the 'avant-garde', using new forms and speaking about contemporary life in a critical tone. In contrast, the articles on science were modern. They spoke of the latest inventions and research subjects, such as electricity, automobiles, telegraphy, radioactivity, aeroplanes, submarines, photography, astronomy, earthquakes and geographical phenomena such as salt lakes. The journal was also interested in spiritualism,

---

[17] The data was gathered from *Urania*. Ernő, Szabó, 'Az Uránia folyóirat, Színház Társaság' [Urania Journal, Theatre Company], I–II. *Népművelés*, 9 (1980): 26–7, 30–33.

[18] The accounts do not provide the names of the authors and have no special titles. Therefore, in the following I refer to the journal without reference to the author or title. *Urania*, 17 (1916): 324–30.

[19] These chronicles, as the journal called them, are the factual basis for my account, given the lack of archival sources that I could find.

hypnosis, telepathy and the like. Interesting articles were published on cell biology, evolution theory and the history of astronomy and chemistry. The contrast between the conservative nature of humanistic writings and the modernity of articles on scientific matters reflects the clash of an emerging modern, urban middle class with the traditional ruling, rural and land-owning elite.

Many short news items that appeared in other countries, mainly Germany, in popular scientific journals, daily newspapers or in *Naturwissenschaften*, were only published in translation. Other articles dealt with all kinds of practical problems such as cleaning, economic heating, perspiration – almost anything.

The layout was elegant and the beautiful pictures enhanced *Urania*'s attractive character. For the reader, science might have seemed an interesting, pleasant and sometimes useful genre that directed people's fantasies towards a new, exciting world.

Urania had no specific scientific authors. The articles were written by leading professors and high school teachers or non-Hungarian scientists whose articles appeared in translation. The style of the texts was mostly fresh, well written, smooth, easy to read and interesting, but they could only be understood by the educated non-specialist, non-scientist reader. Both the language and the background culture of the articles presupposed that the reader was schooled at a relatively high level characteristic of the urban, bourgeois middle class of the time.

Urania aimed to spread culture to a broad audience. Therefore, it started new actions to reach a population outside its initial audience. The 'rural action', as they called it, began in 1902, soon after the start of the society's work. Through this, Urania intended to extend itself towards the non-urban population, or at least that living in smaller towns and cities. Lectures were delivered outside Budapest based on texts devised in the capital city. The texts were published in the Urania booklets that could be bought at the site of the lecture. It was difficult to organize popularizing lectures in rural places because of the lack of infrastructure and lecturers. Urania often had to recruit lecturers from the local population. In villages, for instance, the local teachers read the printed texts aloud to the audience, and illustrated the lecture with photographic slides using a projector lent to them by the society. The 'rural action' proved to be an enormous success. At the annual meetings, reports accounted for the rapid growth of both the audience and the number of lectures. In 1903, 93 lectures were delivered in 38 counties. Because of its fast growing popularity, the society did not always have enough projectors to meet the increasing demand. In some places, people misused the name of Urania to advertise low-quality lectures in the hope of enhancing their income, proving that the brand had value.[20] Natural science was not among the subjects of the rural action lectures. They were mostly about history, literature, fine arts and geography – practically travel stories.

With the second action, Urania sought to extend itself towards workers. From 1907, Urania established a new genre called the workers' gymnasium (gymnasium

---

[20] *Urania*, 4 (1903): 460.

meant high school), an informal series of lectures given specifically to uneducated workers. This was a form of free education, a substitute for high school, lasting for three years and giving at least 88 classes per year. The subjects included history and literature, but also some 'useful' themes, such as economics, accounting and law. Later they added some physics, biology, technology and health to the curriculum. These informal schools gave a kind of diploma that had no practical value because it was not accepted as a certificate of having participated in formal education. In 1908, the society started to organize workers' gymnasiums in 56 towns, while in 1910, 126 towns announced their intention of participating in the movement. In fact, in the same year, 76 courses were held in 50 towns involving around 10,000 classes and 5,000 students. After that year the action slowly declined and after World War I it could not be revitalized.

Although the data is not reliable, as the whole organization had an interest in painting a flattering picture of its activities, Urania apparently became the most important association to attempt to draw a large circle of non-educated people into culture.[21]

The Work of the Scientific Theatre

Urania's widespread activity started with the idea of the scientific theatre, this less than obvious genre of science popularization. Theatrical shows disappeared the moment the curtain fell. Therefore, the way the Urania scientific theater worked can only be understood from descriptions given by *Urania* and some contemporary newspapers. Some surviving flyers and posters also provide short descriptions of the contents of the performances.[22]

The very first performance, *An Excursion to the Moon*, had three acts showing painted and photographic pictures of a total eclipse as seen from the Earth and from space. A lunar eclipse was then shown in the same style. Even the mountains on the surface of the Moon were visible. Actors read the written text in emotively pathetic tones. After each act, an orchestra played pieces by Bach, Schubert, Liszt and Mendelssohn. Professor Kövesligethy, the leading astronomer, directed the acting.

The corporation rented a remarkable building for the theatre. The house was built in 1893. The first level was a nightclub until Urania furnished it for its own purposes, and apartments occupied the upper floors. The beautiful main hall and the furniture were Moorish in style, with lots of gold, thick, dark vine-red velvet curtains and rich decoration to imitate the sultry sexual atmosphere of a

[21] See Ernő Szabó, 'Az Uránia folyóirat, Színház Társaság', p. 33.

[22] Domokos Dániel Kis, 'Az Uránia száz éve' [Urania's One Hundred Years], *Valóság* 46/1 (2003): 1–8.

Middle-Eastern 'caravan seraglio' inherited from the nightclub.[23] The hall had 18 boxes on the ground floor and first floor, and 306 seats on the ground and 94 on the first floor. Fifteen hundred electric lamps provided lighting for the hall, making it wonderfully modern. Four differently coloured (white, red, yellow and blue) electric lamps lit the stage from behind, by which specific lighting effects could be produced. The most attractive devices were the projectors showing large photographic images, even from microscopes.

Urania was a scientific theatre in a non-metaphoric sense. Its aim was to show the object of science onstage by entirely different means to Fleck's ideas of popular science. Nevertheless, the certainty, simplicity, authoritative and decontextualized character, and mainly vividness of popular science must have been present on the stage. The question is what kind of science could be put on stage and by what means.

According to Urania's own description, the performances normally started with a short lecture, a so-called 'scientific chat', lasting for 20 to 30 minutes. Then came the main attraction, the theatrical performance. Painted pictures and projected photos lit in different colours were shown while music played. Meanwhile, actors read the text that explained the pictures and the phenomena they represented. Pictures were also projected between scenes while the stage was reorganized.[24]

The premiere in the Urania building was *The Fight for the North Pole*, written by Jenő Cholnoky, later a professor of geography, and read by an actor, Balázs Dajka. The play consisted of three acts and 15 scenes. The pictures showed a storm around icebergs, the Northern Lights, huge ice floes, impenetrable fog and incoming ships, some of which sank. The physicist Jenő Klupathy presented a short talk on liquid air, proving that cold could make wonders. Besides chatting, he also performed experiments for the audience. Next, Miklós Konkoly Thege, a leading astronomer, produced a show using 40 pictures, mostly from the Prussian observatory in Potsdam. Then Radó Kövesligethy returned, this time with a performance on shooting stars in November. Another show on the ice cave in Dobsina proved highly successful. The walls of the cave were covered by a 30-40 metre thick layer of ice that could be seen in the photos taken by Loránd Eötvös, a professor of physics. The first performance of the play called *Spain* was given in 1900. Performed many times in many places, it was the biggest hit yet because of the witty text and the 200 projected pictures, some of which were moving, showing dancing and bullfighting, which were exotic for the time in Budapest.

---

23 The building was described in a detail by Ferenc Bor, Zoltán Fehérvári, 'Uránia Magyar Tudományos Színház: Uránia mozi' [Urania Hungarian Scientific Theatre: Urania Movie], <http://www.budapest-photo.hu/Urania%20Filmszinhaz_1.htm> (accessed 15 October 2007) and Erki Edit, 'Az Uránia a pesti mozik Nemzetije volt'. [Urania was the national movie theatre in Pest], *Népszabadság*, 3 June 1999, <http://www.filmintezet.hu/magyar/filmint/filmspir/21/lapoz.htm> (accessed 15 February 2006).

24 The description and the data were published in 'A tudományos színházról', see note 16.

The writer Gyula Pekár proved to be the best in the genre. He wrote several plays and movie scripts and became a specialist in scientific theatre. He was a prolific but not particularly significant writer and a right-wing Christian Party politician, state secretary and minister for a while.

It soon became clear that it was the projected photos that attracted the largest audiences. Therefore, the subject of the performances shifted towards fine arts, history and geography, which could be better illustrated by pictures of wars, interesting buildings and beautiful landscapes. There was major success with *Paris in 1900*. This three-act play showed such scenes as walking along the streets of Paris, the Champs Elysée, dancing at the Moulin Rouge, people on the Champs de Mars, returning from the Bois de Boulogne, in front of the Grand Palais, scenes in the Place de Concorde, the World Expo and the Hungarian pavilion at the Expo. This was Urania's greatest success for ten years.

The scientific theatre was based on dioramas, personal, visual and audio experiences supplemented by an explanatory text read in several tones and styles. The genre did not favour the deductive, theoretical kind of science, but rather the descriptive, experimental kind or natural history. Boltzmann's statistical theory of heat or Maxwell's electrodynamics equations could not raise the interest of the theatre audience in anything like the same way as icebergs or electric machines. Natural history led directly to art history or social history, travel, exotic scenes and the like. It was precisely that vivid and visual character that determined the fate of the scientific theatre.

With the growing significance of projected pictures, natural science had to leave the large main hall of the Urania building. Because the big audiences preferred stories about travel and history, scientists had to withdraw to the smaller rooms on the floor, but continued to present their subjects, to show experiments and use pictures as illustrations. In the basement of the building, a photo laboratory was set up to produce photos on 8.5×8.5 cm glass plates.[25] After 1908, the Urania Visual Educational Aids Factory was set up – a company owned and led by Mór Erdélyi, an eminent court photographer – to supply the theatre, rural action and workers' gymnasiums with pictures and other devices.[26] Indeed, Urania's photos were used in parallel with the Urania booklets to provide the illusion in rural areas that the audience could receive the same service as the Budapest audience. As electricity was not widespread, the projectors were often powered by spirit from wine lamps.[27]

---

25 Some series of photos were published on a CD-ROM attached to a study of Urania's pictures. Gyula Munkácsy and Melinda Bogdán, 'Az Uránia "decorative" képei' [Urania's decorative pictures], in *Fényszülte képek. Tudomány – technika – művészet. A fotográfiai innováció és művészet kapcsolata Magyarországon, 1839–2001* (Budapest, 2002).

26 Judit Baróti, 'Erdélyi Mór élete és munkássága', *Fotóművészet*, 40 (1997): 82–90.

27 By chance, around 25,000 photographic glass plates with old Urania pictures on them, all made by Erdélyi's company, were found recently in a basement: 'Többezer Uránia-dia menekült meg az enyészettől' [Several Thousands Urania Slides were

Meanwhile, the main room in the beautiful Urania house in Budapest became increasingly popular, where the lesser-educated but larger audience enjoyed pictures. This gradually shifted towards the new wonder of the period: the cinema, or movie theatre. In 1896, one year after the Lumière Brothers' first show in Paris, the first moving pictures were shown in Budapest in a coffee house, Royal, and in the same year the Sziklai Brothers opened the first movie theatre to show the Lumière Brothers' films. The very first Hungarian moving picture was made in 1901 by Béla Zitkovszky, Urania's photography technician, on the roof of Urania's theatre in order to make use of the natural light there. Gyula Pekár wrote the script. Zitkovszky constructed his camera by himself and first showed the film in the theatre, and later in many other places. He shot many other movies that were included in Urania's programmes.[28] Today, Urania is considered the national movie theatre, where mostly Hungarian films are shown.[29]

Movies were soon to become a huge success in Hungary. By 1908, 40 to 50 movie theatres were operating in Budapest and there were many more in other Hungarian towns. Urania was one of the first places in Budapest to show, produce and use movies. However, the attraction and success of these moving pictures contributed to the downfall of Urania as a place for science popularization. The competition became too strong for the corporation, income decreased and the shareholders became displeased.[30]

The Peripheral Character of the Scientific Theatre

The very idea of Urania was originally not Hungarian. The people who first suggested its establishment in Budapest were not politicians or business people, but scientists: Loránd Eötvös and Albert Berzeviczy, eminent people in scientific life. Eötvös was a physicist, famous for his extraordinarily precise measurement of the gravitational constant, a leader in the physics community and later president of the Hungarian Academy of Sciences. Berzeviczy was a serious historian, the author of many books, and Eötvös's successor as president of the Academy. He had a long and significant political career that started in the Liberal Party until 1905. After World War I, Berzeviczy served the nationalist, conservative Horthy regime. He always supported science and education, including the popularization of science. In 1897, emphasizing the major significance of spreading scientific knowledge,

rescued], *Népszabadság Online*, 26 March 2003 <http://; http://www.urania-nf.hu/sajtoszemle/030326_2.php> (accessed 10 February 2006).

[28] Among many works on the early history of Hungarian movies, including Magyar Bálint, *A magyar némafilm története* [History of Hungarian silent film] (Budapest, 1966).

[29] The movie theatre has a nice homepage with many images of the rooms, which have been restored. <http://www.urania-nf.hu> (accessed 10 February 2006).

[30] Viktor Molnar, 'Új közszellem' [New mentality], *Urania*, 15 (1916): 1–3; Report on the meeting of Urania shareholder company, *Urania*, 19 (1918): 22, 121–8.

he stressed the importance of theatre. In relation to the popularization of sciences, Berzeviczy mentioned an institution (in Berlin), 'Urania', that showed scientific phenomena and experiments on stage, supplemented by lectures explaining the show. In a newspaper article, he stressed his commitment to theatres used for teaching and mentioned some theatres in Paris that combined the show with lectures that commented on and interpreted the drama.[31] Eötvös, on the other hand, suggested they should set up a scientific theatre to perform scientific spectacles and experiments to a wide audience. This theatre would apply the example of Urania operating in Berlin, which he had seen and enjoyed, to Hungarian circumstances. He wanted to name the Hungarian version Universum to distinguish it from the original and planned to use it for the popularization of solely natural sciences.[32]

So, the source of the idea behind the Hungarian Urania lay in Berlin, Germany – the centre of the 'world-science' to which Hungarian science belonged. 'World-science' refers to a network of national or regional sciences sharing scientific ideas and practices devised at the centre of the network. This approach assumes that science is neither necessarily global nor uniform; rather it is divided into various networks operating simultaneously at a particular time. World-sciences have centres that determine the work of the network. Scientific culture in the countries that formed part of the German world-science learnt many things at the centre, including scientific ideas, methods and institutional arrangements.[33] Urania as an institution for popularizing science was born in Germany. The idea was carried from its birthplace at the centre to the periphery by interested people such as Eötvös.

The Berlin Urania Society (*Gesellschaft*) was established in 1888 by Wilhelm Foerster, an astronomer and the director of the Berliner Königlichen Sternwarte, an observatory.[34] He was a follower of Alexander von Humboldt, who always fostered the idea of disseminating scientific knowledge. Foerster met Max Wilhelm Meyer, an astronomer and a man devoted to popularization. Meyer published successful and popular scientific articles in the *Berliner Tagesblatt* newspaper. The two men decided to cooperate. From his childhood, Meyer had wanted to become both a writer and a scientist. He studied astronomy and worked in many places in Germany, Italy, Switzerland and Austria. To connect his interests in writing, astronomy, theatre, lecturing and popularizing, Meyer wrote a theatre play, *Pictures of the Star World* (*Bilder aus der Sternenwelt*), performed for the

[31] Berzeviczy's newspaper article is cited in 'A tudományos színházról', see note 16.

[32] 'Tudományos színház a fővárosban' [Scientific Theatre in the Capital City], *Gyógyszerészi Értesítő*, 5 (1897): 46.

[33] I detailed the world-science framework in relation to the centre–periphery problem in Gábor Palló, 'Internationalism in Soviet World-Science: the Hungarian Case' in Elisabeth Crawford et al. (eds), *Denationalizing Science*: *The Contexts of International Scientific Practice* (Dordrecht, Boston, London, 1993), pp. 209–32.

[34] See Gudrun Wolfschmidt, 'Popularisierung der Astronomie: Instrumente, Sternwarten, Planetarien' in Gudrun Wolfschmidt (ed.), *Popularisierung der Naturwissenschaften* (Berlin, Diepholz, 2002), pp. 107–14.

first time in Vienna, at Christmas 1884. He intended to use literature and fine arts to show natural phenomena on stage as if the audience were watching a show in a theatre.[35] Foerster and Meyer decided to establish an institution aimed at educating the general population (*Volksbildung*) in one house but in several genres.[36]

The idea received enthusiastic support from the industrialist, Werner von Siemens. He established the Urania Corporation (*Aktiengesellschaft*), starting out with 205,000 German marks. The Prussian government donated the land for the building. In July 1889, the house of Urania was opened, but soon proved too small as a result of its huge success (about a 100,000 visitors in the first year). Urania was moved to a new, larger building in 1896. Meyer became the first director of science and commerce.[37] Urania originally had five departments, each of which had its own director: 1. Astronomy, with an observatory for visitors, 2. Cabinet of physics, 3. Microscopy, 4. Precision mechanics, 5. Scientific theatre. The Cabinet of physics, Microscopy and Precision mechanics constituted a prototype for science museums, showing experiments, interesting phenomena (mostly electric phenomena), devices and models of machines that could be used by the visitors. All of these were supplemented by Urania's journal, *Himmel und Erde* (Sky and Earth), edited by Meyer himself. The Berlin Urania organized exhibitions, popular scientific lectures and theatre performances.[38]

Describing the theatre, one visitor remarked that it was like a Jules Verne novel in dramatized form. The most modern techniques were used onstage, including lighting effects, decorations, murals and projected pictures. The lectures on experiments presented such recent discoveries as, for example, Roentgen rays. Historian Arne Hessenbruch has pointed out that Roentgen rays were related with magic and spiritualist shows on the stage of the Berlin Urania, proving how apt scientific theatre was for communicating some special, non-mainstream kinds of scientific content to an audience.[39] The Berlin Urania used projected photos and

---

35 Eric Krug, 'Der Traum von der Urania: Dr. Wilhelm Meyers Leben und Werk', *Sterne und Weltraum,* 8 (1969): 8–11.

36 Andreas Daum placed Urania in the context of the contemporary German institutional system of popularization. Andreas W. Daum, *Wissenschaftspopularisierung im 19. Jahrhundert: Bürgerliche Kultur, naturwisseschaftliche Bildung und die deutsche Öffentlichkeit, 1848–1914* (Munich, 1998), pp. 178–84.

37 Jutta Aschenbrenner, 'Bildung und die Muse der Sternenkunde', *Berlinische Monatsschrift*, 7/1 (1998): 38–44. Probleme/Projekte/Prozesse, Das erste URANIA-Gebäude <http://www.luise-berlin.de/bms/bmstext/9807prof.htm> (accessed 15 February 2006).

38 Meyer, often called Urania-Meyer, detailed Urania's collection, activity, building and related activities in terms of numbers, names and pictures as they existed after its establishment. M. Wilhelm Meyer, 'Die Urania nach Ihrer Fertigstellung', *Himmel und Erde,* 2 (1890): 226–39, 268–86.

39 Hessenbruch, in his highly interesting study, provided a lively picture of the work of the scientific theatre in the context of contemporary Berlin culture, so thirsty was it for all kinds of modern thoughts, wonders and inventions. Arne Hessenbruch, 'Science as

later documentary movies in its performances. The plays impressed the audience in the same way as works by famous writers or movie stars.[40] The theatre had 700 seats in the second, larger building. In its first year, 85 performances were given in the theatre and its popularity grew at a rapid pace.[41]

Because of their success, Urania houses were opened not only in Budapest but also in several other cities. The idea spread all over the periphery of the German world-science. In Vienna, a Urania society aimed at popularizing science started work in 1897. It too used modern devices such as projected photos and later moving pictures in its lectures just like in Berlin, the model behind Urania. The Viennese Urania also included an observatory. Emperor Franz Joseph opened the Viennese Urania house in 1910. However, this house was destroyed, then rebuilt and the new building is still in operation. The Viennese Urania now has a movie theatre, a puppet theatre and an observatory.[42] Other Urania societies were also established in Zurich (1907), Jena (1909), Breslau and Magdeburg (1913), Prague (1917) and Graz (1919). Somewhat later, these were followed by similar institutions in Meran, Chemnitz, Moscow and St Petersburg.[43] From Budapest, Urania started spreading towards the Balkans. In a letter, the Bulgarian minister of culture asked Molnár about Urania's methods because he intended to set up a similar organization.[44] In Budapest, Urania ran a play by Adolf Strausz, entitled *Serbia* and there was a request for it to be shown in Belgrade. After the performance, Peter, the Serbian king, met Viktor Molnár to talk about the Urania experience.[45]

Concluding Remarks

The Budapest Urania, including the theatre, was but a part of a network operating throughout the German world-science in the late nineteenth and early twentieth centuries.[46] The idea of scientific theatre was devised in Germany, and the Budapest Urania followed soon after. In the theatre some dramas, texts and arrangements

Public Sphere: X-Rays Between Spiritualism and Physics', in Constantin Goschler (ed.), *Wissenschaft und Öffentlichkeit in Berlin 1870–1930* (Stuttgart, 2000), pp. 89–126.

40 Wolfschmidt, 'Popularisierung der Astronomie', p. 110.

41 The numbers are published in Aschenbrenner, 'Bildung und die Muse der Sternenkunde'.

42 <http://www.planet-vienna.com/spots/Urania/urania.htm> (accessed 15 February 2006).

43 Aschenbrenner, 'Bildung und die Muse der Sternenkunde'.

44 *Urania*, 6 (1905): 233.

45 Ibid.

46 Scientific theatres operated outside the Urania network much earlier. I thank Jonathan Topham for the information about the British scientific theatres playing at the West End Theatre in the 1820s, and the Royal Polytechnic Institution and Great Globe in the 1840s and 1850s.

imitated the German pattern. Even the text of the first performance and some others were borrowed from the Berlin Urania and translated from German to Hungarian by an enthusiastic countess, Lady Albin Csáky, wife of the Minister of Education and Religion. The form of the corporation, the additional journal and the mixing of several genres were all adopted.

On the other hand, on some points there were conspicuous differences between the Budapest and the Berlin Urania. The periphery did not, and could not, slavishly copy the original Urania operating at the center. Budapest learnt the idea in Berlin and adapted it to the local, peripheral cultural, political and economic circumstances. According to a contemporary Hungarian account, the Berlin Urania was incomparably richer than the Budapest version insofar as its building, instruments and exhibitions were concerned.[47] The idea of scientific theatre was the same in both places, but a fleeting glance at the buildings tells us that the realization was different. The dignified house in Berlin radiated the seriousness of its inhabitants, the instruments, electricity, astronomy and science, while the Budapest building looked like a place for amusement. It was part of a flamboyant, frivolous new city, where science was not such a grave, solemn matter. In addition, the building did not house the whole Urania. The gallery, the publishers of the journal and the Urania Visual Educational Aids Factory worked in other places in the town.

In the Berlin Urania, the organization was institutionally unified, while in Budapest the society and the theatre corporation differed. The Budapest Urania as a scientific society had more ambitious popularization goals, including its 'rural action' and the workers' gymnasium. Unlike in Berlin, the Budapest Urania had neither a museum nor an observatory, and astronomy did not play an outstanding role in its activities. Therefore, the name of the organization did not refer to the Greek muse of astronomy, but rather to the Berlin organization. The early usage of moving pictures in Budapest and the theatre's development into a movie house were also distinctive features. Although in the 1930s the Berlin Urania became the prime theatre for Universum Film Aktiengesellschaft (UFA) – German Art Cinema – documentaries, it did not change into a movie house. Another important difference was that the Budapest Urania had a more universal goal than its Berlin counterpart. It embraced a wide range of subjects, including fine arts, literature, practical matters, history, science and technology.

Urania represented an early version of modern communication that made parallel use of the oral, musical, printed and theatrical forms with photographs and painted images, and with an observatory and a museum. This mixture of many genres proved highly successful in popularizing science in the prewar period. The Budapest Urania functioned as an independent element in a network that was present in the area dominated by the German centre.

The Budapest Urania Scientific Theatre placed the content of science on stage, mainly natural history and history, instead of the moral, personal dilemmas

---

[47] Kálmán Kapitány, 'A berlini Uránia', *Urania*, 5 (1904): 41–3.

of scientists.[48] Scientists or the history of science played no role in its shows. Characteristically, it separated science and scientists. Despite its vividness both in Fleck's and the narrow sense, the scientific theatre embodied the traditional idea of popularization. In its arrangement, nature's drama in a non-metaphoric sense took place on the stage that was viewed by the audience.

Urania as an institution in a network of science popularization used several channels to communicate science in a modern way using modern devices. In doing so, Urania became popular, and proved that science could be more than useful. It could be amusing and entertaining.

48 I found one exception in the programme. In 1903, Emil Haberfeld's play *The Doctor in the Old Days and Today* spoke about old doctors and history and showed pictures taken by X-ray. *Urania*, 4 (1903): 308.

Chapter 9
The Popularization of Astronomy in Early Twentieth-Century Sweden: Aims and Motives

Johan Kärnfelt[1]

In the winter of 1922 the Swedish astronomer and director of the Uppsala observatory, Östen Bergstrand (1873–1948), wrote a despairing letter to his former student and now junior colleague, Knut Lundmark (1889–1958). Bergstrand was struggling with a lengthy manuscript for a major popular astronomy handbook, and in the letter he complains about his toils. The work is difficult and time-consuming, and the publisher's deadline has long since passed. Cautiously he asks for Lundmark's assistance on one of the more specialized chapters. Lundmark had recently defended his thesis on the extragalactic distance to the Andromeda nebulae.[2] His work won international recognition and he was to become one of the major advocates of the view that the spiral nebulae were situated far outside the Milky Way and were distant star systems in their own right.[3] At the time of Bergstrand's letter he was working on a scholarship at the famous Greenwich Observatory in London, England.

If Bergstrand was reluctant to put an extra burden on the young astronomer's shoulders, Lundmark for his part did not hesitate to accede to the request. Parallel to his academic studies, he had embarked on a career in popular astronomy and was also to become one of Sweden's most celebrated popular science authors and lecturers. Yet, at this early stage, he apparently felt the need to defend his choice:

1 I would like to express my gratitude to Aant Elzinga for smoothing out the language in this text.

2 Knut Lundmark, *The Relations of the Globular Clusters and Spiral Nebulae to the Stellar System: An Attempt to Estimate Their Parallaxes* (Stockholm, 1920).

3 On Lundmark's role in this debate, see Robert W. Smith, *The Expanding Universe: Astronomy's 'Great Debate' 1900–1931* (Cambridge, 1982), pp. 105–9, and Gustav Holmberg, *Reaching for the Stars: Studies in the History of Swedish Stellar and Nebular Astronomy, 1860–1940* (Lund, 1999), chapter 4.

> Naturally it will take some time to write the article, but I can't work with science alone. Such is my nature. When I have had 10–12 hours workday here, I will gladly do something else during a couple of hours.[4]

Lundmark's statement raises some interesting questions. The most striking in my view is that the letter reveals a scientist who was working long and burdensome hours six, sometimes seven, days a week at the observatory doing reductions and calculations, but who still managed to find both the time and energy to continue into the evening writing popular astronomy. Evidently the same could be said about his professor. By the time Bergstrand's book, *Astronomi*, was eventually published in 1925, he had invested hundreds and hundreds of evening, weekend and holiday hours to fill the over 700 pages.[5] The obvious question is why. Why did so many Swedish scientists, in this case astronomers such as Lundmark and Bergstrand, engage in popularization? It did nothing to promote their academic career. It was time-consuming and tended to interfere with the all-important task of research. So what did they hope to achieve? Lundmark's reference to his 'nature' could be just part of the answer; however, in a historiographical inquiry we have to broaden the approach.[6]

The Swedish Genre of Popular Science

Before entering into the realm of popular astronomy, it is necessary to provide a short outline of the history of the Swedish genre of popular science. It has some unique features that are important for the argument to be put forth below, and it adds to contextualizing and situating the astronomers with regard to popularization.

Let me thus begin with a linguistic remark. When below I address the genre of popular science in general, it must be noted that the English concept of popular science is a rather awkward, even misleading translation of the Swedish word *populärvetenskap*. Whereas the term 'science' denotes the natural sciences, the Swedish word *vetenskap* – just like its German counterpart *Wissenschaft* –

4 'Naturligtvis tar det en del tid [att] skifva uppsatsen men jag kan ej enbart arbeta med vetenskap. Min natur är sådan. När jag haft 10–12 timmars arbetsdag här så sysslar jag gärna med något annat under ett par timmar' My translation. Knut Lundmark to Östen Bergstrand, 14 July 1924, in The Östen Bergstrand Archive, Uppsala University Library, Uppsala.

5 Östen Bergstrand, *Astronomi* (Stockholm, 1925).

6 The motives behind the Swedish astronomer's engagement in popularization were my main interest in a research project completed a couple of years ago. The end result is the monograph Johan Kärnfelt, *Till Stjärnorna: Studier i Populärastronomins Vetenskapshistoria Under Tidigt Svenskt 1900-tal* [*To the Stars: Studies in the History of Popular Astronomy in Early Twentieth Century Sweden*] (Gothenburg, 2004). The aim of this article is to give a summary of my work for an English-speaking audience.

covers the natural sciences, the social sciences and the humanities. Accordingly, *populärvetenskap* is a much broader concept than popular science. There is no easy way out of this dilemma, but for the sake of brevity I will henceforward use the concept of popular science in this broader sense.

Turning then to the development of a Swedish genre of popular science, it can first be noted that its history stretches back to the 1880s.[7] There were predecessors, especially in the form of the 'useful findings' (*nyttiga rön*) extensively circulated primarily by the Royal Swedish Academy of Sciences from the mid-eighteenth to the early nineteenth century; however, a recognizable modern form of popular science came at a later date. A short detour is necessary to clarify this point.

The main difference between the 'useful findings' and the later form of popular science can be found in their respective epistemological underpinnings. When the Royal Swedish Academy of Sciences started to circulate useful findings in their Proceedings and Almanacs, the relationship between scientists and the public was not yet fully developed.[8] Even if the Academy had some famous and highly productive Fellows such as the botanist Carl von Linnaeus (1707–78) and the astronomer Anders Celsius (1701–44), it was not self-evident that they and their colleagues had the exclusive right to produce reliable knowledge. In their contemporary culture of knowledge, a useful find could have its source not just amongst the Fellows of the Academy but equally likely amongst the general public. What gave it credibility was not that it issued from a professional scientist working in accordance with scientific method, rather that it came from someone that the Academy held to be trustworthy. A farmer, for example, who had found a new and useful way to nurture crops, or a new and useful plough design, could thus get the 'discovery' or invention published in the Academy's Proceedings.[9]

The modern form of popular science is structured around a very different culture. It does not involve this two-way communication pattern of useful findings; rather it presupposes a culture where the public is deemed passive and

[7] I have studied the development of the Swedish genre of popular science in my thesis: Johan Kärnfelt, *Mellan Nytta och Nöje: Ett Bidrag till Populärvetenskapens Historia i Sverige* [*Between Utility and Delight: A Contribution to the History of Popular Science in Sweden*] (Stockholm/Stehag, 2000). The following is a summary of the main arguments of the thesis.

[8] I discuss this at length in Johan Kärnfelt, *Mellan Nytta och Nöje*, chapter 3. For a more general discussion of this issue, see Steven Shapin, 'Science and the Public', in R.C. Olby et al. (eds), *Companion to the History of Modern Science* (London, 1990), pp. 990–1007, and Bernadette Bensaude-Vincent, 'A Genealogy of the Increasing Gap Between Science and the Public', *Public Understanding of Science*, 10 (2001): 99–113.

[9] It should be noted that trustworthiness in this setting did not necessarily presuppose the status of gentleman as discussed in Steven Shapin, *A Social History of Truth: Civility and Science in Seventeenth-Century England* (Chicago, 1994). When the Academy evaluated a useful find, the social status of the person behind it sometimes did come into play, but it seems that they could just as easily have their say on the reasonability and general interest of the find as such.

ignorant and the production of knowledge is exclusively limited to a small group of professional scientists. Popularization therefore becomes the dissemination of specialized scientific knowledge in a form that is accessible and enjoyable for a general audience.

In countries like Germany, France and Great Britain, this brand of popular science was fully developed by the beginning of the nineteenth century. Sweden, however, a small and peripheral country, lagged behind. Even if Swedes in general were literate by the turn of the century, a culture wherein scientific issues would be of any interest to the general public did not yet exist. Although publishers did make some earlier attempts at translating and publishing foreign authors, it was not until the 1880s that popular science found its audience in Sweden. The writings of authors such as Camille Flammarion in particular – I will return to him later – were now circulated in extensive editions. At the same time, we also find a clearly discernible and locally produced genre of popular science.

The defining feature of the establishment of a Swedish genre of popular science is that it cannot be reduced to the output of a purely scientific enterprise. Scientists did, willingly and sometimes enthusiastically, fill the form with content, but they did not create the form as such. Instead, the genre of popular science – seen as an infrastructure of spaces for popularization (such as book and lecture series, specific journals, newspaper features and so on), channels of distribution, rhetorical and pedagogical approaches, ideological goals as well as a culture wherein science became important to the public mind – originally emerged in close association with the Adult Education Movement of the late nineteenth and early twentieth centuries.[10]

This movement was immensely important at the time. Olof Palme, former Swedish Prime Minister, once said that Sweden is a 'study circle democracy'.[11] With that statement he acknowledged the profound impact on Swedish society that the movement once had. The study circle was one of a number of techniques developed within the Adult Education Movement around the turn of the century to popularize a scientific and scholarly worldview, to educate an entire people and

[10] With adult education – *folkbildning* in Swedish – we once again run into a translation problem. First, the English concept refers to the education of adults. This is reasonable, especially since *folkbildning* targeted adults, but it must be noted that the Swedish word *folk* actually translates to people, or folk. *Folkbildning* is education of the people. Second, we have the notoriously difficult concept of *bildning*. The translation to education works quite nicely, yet with it we lose some central connotations in the process. *Bildning* has its roots in the German concept of *Bildung*. When introduced into Swedish in the early nineteenth century, it preserved the double meaning of the German language, in which it points on the one hand towards a lifelong process of cultivation or education, and on the other towards the end result – to be (or not to be) cultivated or educated. This divide opened up for a number of uses, especially in a political setting, that is difficult to capture in the English concept of adult education.

[11] Palme coined the expression in a speech given at the Social Democratic Party convention in 1969.

in the process to transform a traditionally hierarchical and agriculturally oriented society into a modern industrial democracy.[12]

In European history, the Adult Education Movement is peculiar to the Scandinavian countries; in certain respects, however, it may be compared to the British University Extension Movement.[13] The obvious similarity is that both movements tried to disseminate academic culture and a scientific worldview to people who lacked academic training and tradition. At the same time, the two movements differed from each other. To begin with, while the Extension Movement issued from inside the universities, the Adult Education Movement had its roots in a multitude of autonomous organizations founded from the 1880s and onwards outside the academe. The Swedish universities did eventually, and rather reluctantly, jump on the bandwagon, but by that time the scene had already been shaped by organizations such as the Stockholm's Workers Institute (*Stockholms arbetareinstitut*), the Swedish branch of the International Order of Good Templars, the Workers Educational Association (*Arbetarnas bildningsförbund*) and student organizations such as Verdandi and Heimdal.

Another difference was that the Extension Movement literally wanted to extend the outreach of the universities to new groups in society, whereas the Adult Education Movement claimed no such ambition. Situated entirely outside the academic system, it was, rather, content to utilize scientific and scholarly knowledge as a major tool in what ultimately must be regarded as a political campaign. Speaking first and foremost for a rapidly growing working class, its chief aim was to overthrow a political system that effectively denied workers political influence.[14]

---

[12] Invented in 1900 by Oscar Olsson, teacher at and later head of the Swedish branch of the International Order of Good Templars, the study circle was a clever way to organize adult education in a manner that reduced the costs for literature and at the same time did away with the need for teachers. Five members were required to form a study circle. The circle decided on a subject – for example, astronomy, Swedish history or contemporary literature – and then bought a number of *different* books on the area in question. The books were distributed in the group and the members read the one at hand. Every second week or so, the group met and discussed the readings. When the meeting was over they switched books and every individual now had a new book to read for the next meeting. In this way the participant's insight in the subject matter slowly developed and matured, not in relation to a teacher's agenda but in relation to the dynamics of the group. In Olsson's view, the study circle was not actually education *of* the people, but education *through* the people. (This is one of the instances in which the double meaning of *bildning*, mentioned in the previous footnote, comes into play.)

[13] On the University Extension Movement, see J. Rée, 'Idealism and Education', *History of Education*, 9/3 (1980): 259–63.

[14] The Swedish parliamentary system at the time gave influence to citizens in proportion to the amount of property they owned. This gave a political voice to the wealthy, the nobility, clergy and bourgeoisie, yet also to the peasants, who in most cases owned their farms.

The massive campaign launched by the Adult Education Movement was vital to gaining universal suffrage in Sweden. Still, the efforts did not end after the relevant bills were passed (1909 for men and 1921 for women). In the wake of its political success, the Movement's aims were reformulated. Leading ideologists within the movement knew that it was one thing to have political rights, but a very different thing to actually be able to put those rights to use. The educational level amongst workers at the time was limited to a couple of years of primary school, which was not enough. Given a new political reality, adult education was then construed as a vital part of the efforts needed to empower and encourage workers to actually engage in political life.

The close association with the Adult Education Movement thus gave the genre of popular science an explicitly political twist. The key groups in the movement were the Liberals and Social Democrats. Despite their ideological differences, they were able to unite in the struggle for an extended democracy, in the belief that suitable adult education could only be achieved by popularizing an ideologically unbiased scientific worldview. Even more importantly, they shared a common fear induced by left-wing agitators. Powered by tidings from abroad, it was believed that such agitators might rouse the anger of the working class and in so doing throw society into turmoil. Under these circumstances, adult education, and with it the popularization of science, was seen as a necessity: To begin with, it was supposed to present the workers with an objective and politically neutral worldview that drew on all brands of science and learning. Firmly grounded in such a worldview, the workers would have a stable platform from which to observe the political landscape. With this knowledge, it was argued, workers could ponder the issues at stake and form their own opinions in a more rational manner.

Adult education, however, was not limited to the dissemination of scientific knowledge as such. A further premise was that popular science could and should promote insights into scientific method. With elementary knowledge of the workings of science – that is some familiarity with its empirical foundations, rationality, objectivity and disinterestedness – the workers, it was believed, would be equipped with a reliable tool that enabled them to judge and deal with the arguments presented by the political alternatives, especially the left-wing agitators. Popular science was actually supposed to function as an antidote to all kinds of revolutionary tendencies. Armed with the ethos of science, workers were supposed to recognize an unsound argument when they met one; and, further down the road, they were destined to make an enlightened choice when the time came to visit the ballot box.

Swedish Astronomy

With the foregoing context in mind, we now return to the question raised at the outset of this article. Why did Swedish scientists engage in popularization? It is safe to say that strong cultural and political imperatives were at work. Indeed, approximately 50 per cent of all professional scientists engaged in popularization of one sort or another. Evidently some were more active than others, but it is still an impressive figure. Why was there such a high degree of participation? Did they share the ideological and political goals that nurtured the Adult Education Movement and that in a fundamental way shaped the genre of popular science? Did scientists on average share the belief that the pathway to a stable, modern democratic society was via an enlightened working class? Before we try to answer questions like these, it is necessary to introduce my informants, the Swedish astronomers active during the heyday of the Adult Education Movement.

Swedish academic astronomy in the early twentieth century was a rather small enterprise. The country only had three observatories: two of these were run by the universities in Lund and Uppsala respectively, and one was under the Royal Swedish Academy of Sciences in Stockholm. By comparison, it is interesting to note that there were 14 professional observatories in France, 19 in Britain and 34 in Germany.[15] Yet in spite of the limitations, Swedish astronomers had managed to keep pace with their colleagues abroad, at least throughout the nineteenth century. During the first half of the century they were especially active within geodetic surveying, participating in several expeditions to measure the curvature of the earth. Later on, some of them achieved international renown for their work in celestial mechanics.

By the turn of the twentieth century, however, the earlier success showed signs of decline. A new generation was about to take over, and its heritage in terms of material resources was literally withering. Time had taken its toll, and the Swedish observatories were in quite bad shape. Built in a pre-industrial society, they were all situated in the hearts of the cities, where heavy traffic had now begun to disturb the sensitive instruments. The electrification of cities also brought increasing light pollution, which was gradually making high-quality observations impossible. Still more troublesome was the fact that the observatories had been constructed during an earlier era of astronomy. Thus, they were all designed for the positional astronomy that had served celestial mechanics during the nineteenth century, and as such they were incapable of rising to the demands of the new astrophysics. Created around the groundbreaking technology of spectroscopy, astrophysics involved entirely new questions – questions about the interiors of stars, about the processes that make them shine. It was a branch of astronomy that would ultimately change the foundations of the mother discipline.[16] Nor did astrophysics merely offer a new and exciting view on celestial bodies; it also called for a very different kind of instrumentation, and even a different kind of observatory. When astrophysics was introduced in Sweden at the end of the nineteenth century, it constituted the major challenge facing a new generation.

---

15 Data from Knut Lundmark, 'Astronomer och Observatorier' [Astronomers and Observatories], *Populär Astronomisk Tidskrift*, 13 (1932): 54–76.

16 For the history of Swedish astronomy, see Holmberg, *Reaching for the Stars: Studies in the History of Swedish Stellar and Nebular Astronomy, 1860–1940*. On the impact and development of astrophysics, see Owen Gingerich (ed.), *The General History of Astronomy: Astrophysics and Twentieth-Century Astronomy to 1950: Part A* (Cambridge, 1984), vol. 4.

One consequence resulting from the institutional limitations was a low number of professional astronomers. Counting a handful of graduate students and astronomers working for the government with geodesics and as surveyors, there was a total of about 20 professionals at any given time during the first half of the century.

The modest size of the Swedish astronomical community has permitted me to use a specific historiographic approach. If we want to understand the aims and motives leading scientists to engage in popularization, there are at least two routes to take: we can either focus on printed material such as popular articles, books and the like, or alternatively we can try to get behind the public scene, drawing on other kinds of materials. In the Swedish setting, the first option has two major flaws. First, given the very strict style of popular science that flourished at the time (I will return to this below) there is just not all that much information to be found in popular astronomy as such. Second, when scientists sometimes did say something about their intentions, they were circumscribed by an arena defined, and to a large extent controlled, by the Adult Education Movement. Accordingly, they tended to say what they were supposed to say.

Given this background, the second route is preferable. In their private communication, astronomers were not bound by the restraints that held together the genre of popular science. And since there were not that many of them, it has been possible to systematically work through archival material such as correspondence, minutes and the like. (The well-established Swedish academic culture of donating private archives to the university libraries has been most helpful in this case.) As it turns out, and as the reader might already have guessed, the views exchanged in such sources contrast sharply with the rhetoric of the Adult Education Movement. However, to be able to appreciate the multitude of uses that astronomers found for popular science, we must start elsewhere, namely with the new style of popular address mentioned above.

The Celebrated Style of Popular Astronomy

The close connection between the genre of popular science and political life also had an impact on views of how science should present itself. The gravity of the political struggle had as its counterpart a very strict style of popular science created around the turn of the century. It was a question of a co-production of popular science and political order. The dominant style of popularization emerged as an integral part of the Adult Education programme, but was adapted and developed in detail by a variety of authors in different contexts. Here I will confine the discussion to the version that held sway in popular astronomy.

The new style of Swedish popular astronomy was very much the result of extensive work done by Östen Bergstrand (Figure 9.1). Bergstrand was one of the leading Swedish astronomers of his generation. Trained in Uppsala under the supervision of his predecessor and pioneer in spectroscopy, Nils Christoffer Dunér (1839–1914), he had an interest in astrophysics right from the outset.

Figure 9.1 Östen Bergstrand (1873–1948). Pen drawing by Carl Benedick.

He did his most important work on methods of reducing data from photographic plates, and when he became director of the observatory after Dunér's retirement, he gradually turned the institution into a Swedish hub for the new astrophysics. His influence on Swedish astronomy can also be seen in the fact that it was his students – Knut Lundmark and Bertil Lindblad (1895–1965) – who eventually came to occupy the astronomical chairs at the observatory in Lund and at the Royal Academy of Sciences in Stockholm.

Bergstrand formulated his rather elaborate programme for the popularization of astronomy in a couple of articles published in some of the leading adult education journals just after the turn of the century.[17] Later on, he supplied his own

[17] Östen Bergstrand, 'Vetenskap och Folkbildning: Några Reflexioner om "Populär Astronomi"'['Science and Adult Education: Some Reflexions on "Popular Astronomy"'], *Social Tidskrift: Organ för Folkbildningsarbetet och de Sociala Frågorna*, 3 (1905): 89–93, and 'Svensk Populär Litteratur i Astronomi: En Översikt' ['Swedish Popular Literature in Astronomy: An Overview'], *Folkbiblioteksbladet*, 5 (1907): 10–14. The first article contains the main arguments and his more programmatic views on the popularization of astronomy,

examples in a number of popular astronomy books, for example in his Swedish edition of Simon Newcomb's well-known *Astronomy for Everybody* and in the aforementioned *Astronomi* (1925).[18] In his articles, Bergstrand explicitly cites the colourful and poetical prose of the well-known French astronomer and author Camille Flammarion (1842–1925) as the bad example.[19]

Flammarion's books had been translated and published for a Swedish audience since the 1860s, and several generations of Swedes had learned about astronomy from his work. But by the turn of the century, his prose had gone out of style. In other European countries he was still seen as the archetype of a successful popularizer and an author that had largely created the genre of popular science. An example of this attitude was the royal reception Flammarion himself received during a visit to Spain in connection with the solar eclipse in May 1900.[20] On the Swedish scene, however, dominated as it was by the Adult Education Movement, the situation was very different. Within the movement, Flammarion could still be honoured as a *pioneer* of popular science and adult education, but that was all. Despite his acknowledged ability to reach out, to create interest and to communicate, leading ideologists claimed that his approach did not meet the standards of an adult education optimized for a democratic society built upon the rationality of science. His beautiful paintings of the heavens, his elaborations of the cosmic mysteries and his political views on more earthly matters were declared misleading and even dangerous.[21]

In his articles, Bergstrand gives an astronomer's voice to these views. The fact that Flammarion's books still were being reissued was disturbing according to Bergstrand, especially since some of them were reprinted in cheap pulp editions intended for a mass audience; still more worrisome was that his typical style continued to be exploited by Swedish authors of lesser dignity and for commercial purposes. This did not tally with the elevated seriousness that, in Bergstrand's view, was fundamental to the mission of science in general and astronomy in particular.

Bergstrand illustrated his main argument with the quote below from Flammarion himself.[22] In one of the chapters in his famous *L'Astronomie Populaire* (1880, first

---

and in the second article Bergstrand delivers a very critical review of the few items of popular astronomy in circulation at the time.

18 Newcomb's book was printed as Simon Newcomb, *Astronomi för Alla* (Stockholm, 1909).

19 On Flammarion, see Bernadette Bensaude-Vincent, 'Camille Flammarion: Prestige de la Science Populaire', *Romantisme*, 65 (1989): 93–104, and Danielle Chaperon, *Camille Flammarion: Entre Astronomie et Littérature* (Paris, 1997).

20 See Pedro Ruiz Castell, *Astronomy and Astrophysics in Spain, 1850–1914*, DPhil dissertation (University of Oxford, 2006).

21 See, for example, Oscar Olsson, *Folkbildning och Självuppfostran* [*Adult Education and Self-Cultivation*] (Stockholm, 1921), pp. 94–5.

22 Östen Bergstrand, 'Vetenskap och Folkbildning: Några Reflexioner om "Populär Astronomi"', p. 91.

Swedish edition, 1897) Flammarion enters into a lengthy discussion of binary star systems. Amongst other things, he tries to capture their specific beauty. He claims that they shine with

> the most admirable contrast of colours, where the eye is surprised to see the tint of the emerald united with the fire of the ruby, the topaz with the sapphire, the diamond with the turquoise, the opal with the amethyst, with one word they twinkle in the colours of the rainbow.[23]

Bergstrand was not impressed by this typical example of Flammarion's poetic style, and his main argument was quite simply that binary stars do not look like this. They do not shine like gemstones; they bear no resemblance to rainbows whatsoever. Viewed through a refractor telescope, they are merely specks of light amongst other specks of light.[24] This means that Flammarion deliberately misleads his readers. From his own experience, Bergstrand found this to be a dangerous game. Now and then, members of the public got their hands on a telescope, or even visited observatories; and they sometimes got extremely disappointed when the colourful skies of the famous Frenchman were nowhere to be seen. And even worse, Bergstrand claimed, there is an impending risk that their disappointment does not target the source, but rather spills over onto the discipline of astronomy, making astronomers look like liars and cheats. The moral, and the central doctrine in Bergstrand's programme, was that popular astronomy should not make *any* claims that science could not back up.

With his point of departure in a critique of Flammarion, Bergstrand tried to formulate a new approach. Since astronomy is a very serious enterprise, popular astronomy must reflect this. Accordingly, it should be cleansed of all mysteries, poetry and drama. Flammarion's colours must be replaced with reliability, objectivity and matter-of-factness. And with a hint to all the non-professionals who gladly lectured and wrote on astronomy in a manner that, in Bergstrand's view, replaced genuine scientific insights with a Flammarian style, he also demanded that the author have first-hand knowledge regarding the area in question. The result became an impersonal, highly technical, quite sober and actually rather boring type of prose, a prose that as far as possible mimics the prose of academic astronomy. (The present-day reader might well wonder who was actually supposed to read some of the more technical pieces that were published following Bergstrand's ideal.)

---

[23] 'de mest beundransvärda färgkontraster, där ögat förvånas att se smaragdens färgton förenas med rubinens eld, topasens med safirens, diamantens med turkosens, opalens med ametistens, med ett ord de tindra i regnbågens alla färger'. My translation. Camille Flammarion, *Populär Astronomi* (Stockholm, 1897), p. 575.

[24] Here we might have to remind the reader that Bergstrand and Flammarion wrote at a time when photography had just been invented and people had not yet seen the fabulous colour photography of modern astronomy.

Bergstrand's views on the popularization of astronomy had a profound impact on the astronomical community, especially since Swedish astronomers arrived rather late on the popular science scene compared to scientists in other fields. When astronomers started to publish for a larger audience in the early twentieth century, the genre of popular science had already been well established and was rapidly growing. For astronomers, by contrast, all the questions relating to popularization were new ones: Whom to address? In what language? And last but not least, for what purpose? In this setting, Bergstrand's work – both his programme for a new kind of popular astronomy and his own books – paved the way.

In addition, Bergstrand's recommendation that only professional scientists should engage in popularization did not appeal only to astronomers; it also resonated well with similar views heralded by leading voices within the Adult Education Movement. One consequence was that non-professionals wanting to contribute to the genre were more or less denied access. There were, of course, some authors who managed to get their manuscripts published, but in general it was astronomers, and thereby the views of Bergstrand, that came to dominate the subgenre of popular astronomy, at least up until the 1950s.

Reasons for Popularization

Popular science is a very flexible genre. It can easily be adapted to different audiences and different media; it can be used to disseminate widely varying kinds of scientific knowledge. However, combined with the stylistic ideal successfully promoted by Bergstrand, it also becomes, and as we now shall see, a very powerful tool in the hands of the professional scientist.

It is quite clear from my study that it was this instrumental utility, rather than the agenda hailed by the Adult Education Movement, that Swedish astronomers recognized and appreciated in popular science. Astronomers regularly exploited the popular science infrastructure created by the movement – they gave public lectures, wrote newspaper articles, crafted study circle materials and so on – but judging from their correspondence, they were unaware of or even disinterested in the aims underpinning the genre. 'Enlightenment', 'democracy', 'for the public good' – buzzwords connected to the genre and to the Adult Education Movement – were never mentioned in their letters. Instead, the correspondence reveals that astronomers actually had a very different agenda driven not by democratic ideals and political aims, but by much more down-to-earth materialistic motives. In this section, I will discuss some of the more important of these motives.

Private Economy

Surprisingly the most common rationale for popularization was the meagre level of the average astronomer's private income. Since the Adult Education Movement engaged hundreds of thousands of Swedes every year, there was, at least during the early decades of the twentieth century, a constantly growing demand for different kinds of popularizations. And since the demand was larger than the supply, such work consequently paid quite well. Axel Corlin (1897–1965), a young graduate student at the Uppsala observatory, wrote popular astronomy articles for the press for a couple of months following the appearance of *nova Aquilae* in the summer of 1918. With the extra income so earned he was able to support himself for the next semester at the university.[25] Knut Lundmark, as mentioned before, wrote popular science throughout his whole career, but it should be noted that his correspondence is full of complaints about having to write to be able to pay, for example, the next instalment on his student loans. Another telling example is the Greenwich incident that opened this essay. The letters reveal that it was not Lundmark's 'nature' that made him accept Bergstrand's plea, but rather the prospect of a small economic remuneration that would help him through some acute financial troubles.[26]

The most striking example, however, concerns Östen Bergstrand himself, and his popular monograph *Astronomi*. The project was initiated in 1920 during a time of severe economic difficulties for both Sweden as a state and the Swedish universities. Government funding was reduced and inflation crippled academic salaries. To maintain the standard of living academics had become accustomed to, they were pressed to find supplementary sources of income. In many cases, the gap was filled by commissions from a rapidly expanding Adult Education Movement. So when Bergstrand's publisher approached him to suggest that he write a popular handbook that could replace Flammarion's outdated *Populär astronomi* (1897), he immediately accepted the offer. During negotiations, his first bid for the assignment called for 45,000 Swedish crowns, a sum that roughly amounted to three times the annual salary of a university professor at the time.[27] His publisher was unable to satisfy this demand, but when the contract was signed Bergstrand actually obtained half of what he asked for – still quite a handsome sum of money.[28]

Summarizing, and judging from the correspondence, it must be said that the astronomers' engagement in popular astronomy was first and foremost a way of supporting themselves during hard times.

---

25 Axel Corlin to Knut Lundmark, 16 June 1918 and 2 October 1920, in The Knut Lundmark Archive, Lund University Library, Lund.

26 Knut Lundmark to Östen Bergstrand, 20 July 1924, in The Östen Bergstrand Archive, Uppsala University Library, Uppsala.

27 Albert Bonnier to Östen Bergstrand, 26 January 1920, and Östen Bergstrand to Albert Bonnier, 31 January and 15 February 1920, in The Albert Bonnier Archive, Albert Bonniers förlag, Stockholm.

28 *Kontrakt: Bergstrand, Östen*, 3 July 1920, in The Albert Bonnier Archive, Albert Bonniers förlag, Stockholm.

Gathering and Educating Amateurs

Considering now the institutional level, we find a number of other reasons for popularization. Amongst them, amateurs' cultivation of astronomy became a major one. Abroad, on the continent, in Great Britain and the United States, amateur astronomy had a long and influential history; but as late as 1920 Sweden still lacked such a culture, at least on a more organized scale.[29] Nevertheless, Swedish astronomers were aware of the need. As stated above, Swedish astronomy had a very limited number of active astronomers, and consequently as a community they were unable to cover the entire range of celestial phenomenon properly, including observations of variable stars, meteors and the aurora borealis. The nature of these phenomena demanded continuous series of observations made either over long periods of time or at different geographical sites. Astronomers knew that they could not cover this terrain by themselves; but they also knew from colleagues abroad that the deficiency might be compensated for by involving a widespread group of suitably equipped and properly trained amateurs. Accordingly, they set to work.

An important event was the formation of the Swedish Astronomical Society in 1919. Founded on the initiative of the astronomer and later insurance broker Nils Nordenmark (1867–1962), its main purpose was to serve as a meeting ground for professional astronomers, amateurs and other members of the public interested in astronomy. Even if the more programmatic formulations in its charter were meant for a wide audience, in practice the society focused on a much more exclusive task. The main ambition, as revealed by the correspondence between the founders, was to put actual or potential amateurs in contact with professional astronomers and, most importantly, to do so in order to benefit astronomy.

Given this context, the firm belief amongst astronomers was that amateurs should be kept on tight reins. Armchair amateurs – or 'astronomical enthusiasts' as they, rather condescendingly, were also called – were of no import. The objective of a proper amateur was to deliver scientifically useful observations, and therefore they could not be allowed to follow their own devices. They had to be tailored to fit the demands of the professionals. With its ambition to gather, educate and discipline amateurs, popular astronomy became a very important instrument, especially during the 1920s and 1930s. The subgenre – to a large extent supervised by the Swedish Astronomical Society – thus introduced areas of research well suited to amateur engagement. Driven by the style promoted by Bergstrand, it taught the basics of instrumentation and observation techniques and, last but not least, it provided recommendations on how best to draft a report so that it would be of use to professional astronomers. In general, popular astronomy therefore

---

[29] See, for example, Allan Chapman, *The Victorian Amateur Astronomer: Independent Astronomical Research in Britain, 1820–1920* (Chichester, 1998), and Marc Rothenberg, 'Organization and Control: Professionals and Amateurs in American Astronomy, 1899–1918', *Social Studies of Science*, 11 (1981): 305–25.

sought to fuel amateurs' interest, keep them up to date on recent developments in astronomy and, of course, keep them on the right track – that is, capable of delivering useful observations.

Popular Science as a Scientific Outlet

Associated with the preferred style of popular astronomy, we also find a number of reasons for popularization that were more closely tied to actual research. Since the reigning ideal sanctioned highly technical prose that in many respects was supposed to mimic proper scientific texts, it also allowed scientists to use the genre of popular science as a kind of scientific outlet. It could sometimes be used to publish preliminary research results, to debate technical questions, to inform colleagues and even to claim priority.[30] Knut Lundmark was one of the astronomers who seem to have systematically blurred the boundary between popular astronomy and astronomy proper (Figure 9.2). For example, he never failed to mention – primarily to remind his Swedish colleagues – that he was one of the first to measure extragalactic distances to the spiral nebulae, and many of his popular articles look more like reviews for technical journals than something intended for a broader audience.

Figure 9.2 Knut Lundmark (1889–1958). Unknown photographer.

[30] This aspect of popularization is also important to the argument in Terry Shinn and Richard Whitley (eds), *Expository Science: Forms and Functions of Popularisation* (Dordrecht, Boston and London, 1985).

Another case is Axel Corlin and his sighting of the 1918 nova. After his first popular articles on the nova, he was forcefully attacked in the press by a member of the public claiming priority. It was only after an extended battle, conducted in popular articles and involving him and a number of senior astronomers, that he managed to establish himself as the first Swedish discoverer of *nova Aquilae* 1918.

Furthermore, as indicated above, popular science also served as an arena for debate amongst scientists themselves. At the time, it was standard procedure within popular biology to attack one's scientific opponents. This was due to a well-established newspaper format, 'the biological causeries' (*biologiska kåserier*), used by Swedish biologists to develop rather elaborate forms of criticism.[31] Astronomers did not use this specific format, and when controversial issues were discussed in popular astronomy it was normally in a more reserved manner. Yet it is quite clear that popular astronomy was an important arena when it came to positioning oneself in relation to other colleagues in, for example, the important debate on the island universe theory.

Another use of popular science emerged when astronomers felt the need to extend their theoretical frameworks into areas of science where they could not claim scientific authority. Speculating, for example, on the possibility of life on other planets – an issue that in the aftermath of the debate whether the supposed channels on Mars was located in a no man's land between astronomy and biology, or even relegated to the genre of science fiction – was one thing; but it was an entirely different matter to get an article on the subject accepted by a serious scientific journal. Here, popular science served as an alternative forum. It allowed the author to move freely through scientific space at the same time that the accepted format permitted extensive technical reasoning. Knut Lundmark, again one of the more vivid examples, wrote a major popular book on life in the universe in which, while popularizing the subject, he also put forward his own scientific claims.[32]

Donations

The final motive for popularization that I would like to mention here also has to do with economic matters. As already noted, the material resources of Swedish astronomy were quite miserable, and the need for fresh funding was inevitable. Swedish astronomers were, however, well aware of developments in the United States, where private donations had facilitated construction of the world's largest

---

31 The biological causeries are discussed in Kaj Johansson, *Den Torgförda Biologin: Studier i Populärvetenskapens Problem och Tematik*, [*Biology in the Agora: Studies on Problems and Themes in Popular Science*], dissertation (Gothenburg, 2003).

32 Knut Lundmark, *Livets Välde: Till Frågan om Världarnas Beboelighet* [*The Dominion of Life: To the Question of the Habitability of the Worlds*] (Stockholm, 1935). Another example can be found in Svante Arrhenius, *Världarnas Utveckling* [*The Evolution of the Worlds*] (Stockholm, 1909) where he discusses his panspermia hypotheses.

telescopes. Swedish astronomers were convinced that it was the impact of popularization that ultimately triggered such sizeable donations.

Pinning their hopes on donors might sound a bit naïve, but Sweden had a well-developed culture of public patronage and a number of very active, generous and wealthy families. It was well known that a large number of public buildings and institutions that had been inaugurated during the first decades of the twentieth century were financed by funds raised from private sources. Furthermore, a number of polar expeditions had also been privately endowed.

Consequently, Swedish astronomers adopted the American strategy. They advertised their needs in popular articles published in magazines and daily newspapers. They could, for example, elaborate on the successful past of Swedish astronomy in contrast to its decline caused by the contemporary state of instrumentation. They were also quite frank about it, literally begging the wealthy to lend them a helping hand. One example is a sweeping campaign launched by Axel Corlin in Gothenburg in 1918 which, in the fading light of the dramatic nova, sought private funding for a professional observatory in the city. Even if Corlin was young and inexperienced, he was well connected and he managed to rally a number of senior astronomers to the cause. Interesting, too, was his realization that the strict and sober style of popular astronomy heralded by Bergstrand was counter-productive for his purpose. Thus, he intentionally gave the campaign an explicit Flammarian touch. He was well aware that he was running the risk of annoying Bergstrand, his professor, but he believed that if one wanted the wealthy to open up their pockets one had to speak to their feelings and imaginations rather than to their intellects. Unfortunately, the rather elaborate and well-orchestrated campaign – running during the autumn and winter of 1918 with public lectures, biweekly articles in the press, meetings and so on – largely turned out to be a failure. It resulted in only a minor donation, just enough for a small public observatory built in 1926 on a hill in one of the city's parks.

Astronomers had better luck in Stockholm later that same year. In this case, the result depended not on a huge campaign but on a single popular article written by Nils Nordenmark, secretary of the Swedish Astronomical Society. The article focused on the history of Swedish astronomy, but it also discussed the financial problems and outmoded instruments that threatened the discipline. In line with the direct approach mentioned above, Nordenmark ended his argument with a plea:

> One thing is for sure; something has to be done. And one thing I am convinced of: if the men lived, that at that time got Stockholm an observatory [1753], they would for certain once again make sure that we got a new and up-to-date one worthy of the city. I am almost tempted to wake them up from their graves.[33]

---

[33] 'Ett är visst; något måste göras. Och en sak är jag övertygad om: levde nu de män, som den gången skaffade Stockholm ett observatorium, så skulle de helt visst åter se till, att vi finge ett nytt, tidsenligt och staden värdigt observatorium. Det är nästa så, att jag skulle vilja väcka dem upp ur graven.' My translation. Nils Nordenmark, 'Huru Stockholm Fick

Nordenmark never needed to go all out. A couple of days later the famous banker Knut Wallenberg (1853–1938) read the article. Maybe inspired by the patrons of science of bygone times, he immediately decided to donate 1.3 million crowns for a new observatory. The new facility, designed specifically for astrophysical purposes from the outset, was later built in Saltsjöbaden outside Stockholm and inaugurated in 1931.

Wallenberg's donation was a major breakthrough not only for Swedish astronomy, but also for the practice of popular science advertising its needs. In the early twentieth century, potential donors were definitely very important targets when it came to the popularization of astronomy. Wallenberg's donation was one of a kind; but one should not forget all the lesser ones, donations that could, for example, finance a solar eclipse expedition, a new instrument or even something as basic as photographic plates. In this respect, the popularization of science literally paid off.

Conclusion

For a number of years, the scholarly study of popular science has been fired by an extensive and successful critique of what is sometimes referred to as 'the standard model of popularization'.[34] According to this model, the process of popularization starts with scientists doing science; in their laboratories, observatories and offices for that matter, they produce new knowledge. After this knowledge has first been thoroughly scrutinized, evaluated and approved by the scientific community, is it disseminated to the public. Expressed in another way, popularization starts when research has come to a close. Then techno-jargon, formulas and complexity are reduced; metaphors, simplifications and narrative are introduced. Thus made accessible and turned into an 'end product', it is consumed by the public. This means that the standard model regards research and popularization, scientists and the public, as distinctly separate categories.

Critics of the standard model have focused on a number of issues: the model is far too simple and dichotomous to account for empirical reality; it does not allow for any kind of feedback; and the public is reduced to a passively waiting

---

ett Observatorium: Ett Kapitel ur den Svenska Astronomiens Historia' [How Stockholm got an Observatory: A Chapter from the Swedish Astronomical History], *Populär Astronomisk Tidskrift*, 7 (1926): 112.

[34] See, for example, Shinn and Whitley (eds), *Expository Science*; Maurice Goldsmith, *The Science Critic: A Critical Analysis of the Popular Presentation of Science* (London, 1986); Stephen Hilgartner, 'The Dominant View of Popularization: Conceptual Problems, Political Uses', *Social Studies of Science*, 20 (1990): 519–39 and Massimiano Bucchi, *Science and the Media: Alternative Routes in Scientific Communication* (London and New York, 2004). Also see Shapin, 'Science and the Public' and Bensaude-Vincent, 'A Genealogy of the Increasing Gap between Science and the Public'.

monolith. In the process, the critique has demonstrated how the popularization of science is actually a more interesting phenomenon than commonly acknowledged. A better understanding of the processes guiding popularization also gives a better understanding of the workings of science, as both an academic enterprise and a cultural artefact.

The aim of my study has been to contribute to the newer picture. The focus on the astronomer's whereabouts, especially contrasted with the aims and motives behind the Adult Education Movement, has revealed the process of popularization in all its complexity. For example, it is apparent that astronomers intentionally targeted a number of different and quite exclusive groups within the general public – such as colleagues, potential donors, amateurs and so on – and that, at least during the first half of the twentieth century, they had very little interest in the greater public. It is also clear that an item of popular astronomy intended for a specific audience – let us say as part of a campaign to enlist potential amateurs – could at the same time be used within the Adult Education Movement addressing a very different audience and for very different reasons.

The dynamics of popular science is also evident in the multiple uses that the astronomers recognized. Here I have briefly discussed some of the more important ones that are visible in my material. However, given another approach from the one I have used, the list can easily be extended. Evidently astronomers sometimes engaged in popularization out of vanity and an ambition to be publicly recognized; but we should also remember the simple desire to share one's thoughts or take part in an ongoing cultural debate. The outcome of my work is also structured by the choice of focus group. A study of early twentieth-century biologists, for example, would have yielded a somewhat different picture. Biologists engaged in popular science for similar reasons as astronomers; but a striking difference is that, whereas Swedish astronomers never used popular science to enter into the political debate, biologists frequently did – and with a great deal of authority. Parallel to the development of the Adult Education Movement, Sweden also saw the birth of social engineering and the first steps towards the welfare state; and here biologists but not astronomers were important players.[35]

Yet even if a change of focus would yield partly different answers to the questions here asked, it would not change the general conclusion: In Sweden, popular science definitely cannot be reduced to a genre that brings enlightenment to the needy; Swedish scientist did not work within the field of popular science as a kind of charity destined to serve an uneducated general public – an ideal sometimes promoted by the Adult Education Movement. My study shows that

[35] Popular science as a forum where biologist could engage in political debate is discussed in Kaj Johansson, *Den Torgförda Biologin: Studier i Populärvetenskapens Problem och Tematik*. A comparative study on the role of eugenics in the development of the welfare state in the Scandinavian countries can be found in Gunnar Broberg and Nils Roll-Hansen (eds), *Eugenics and the Welfare State: Sterilization Policy in Denmark, Sweden, Norway and Finland* (East Lansing, MI, 2005).

the scientists' engagement in popularization was proportional to, and in most cases reducible to, needs *within* science itself. Furthermore, since the engagement actually did pay off, it is quite clear that the dichotomy between science and the public, research and popularization, has to be questioned. In the light of my study, I would instead suggest that the relationship is extremely complicated and multilayered, and that in important respects popularization cannot be distinctively decoupled from the actual formation of scientific knowledge. What we are faced with is a matter of making sense of a co-production of science and popular science, as well as social order.[36]

[36] Cf. Sheila Jasanoff (ed.), *States of Knowledge: The Co-Production of Science and Social Order* (London and New York, 2004).

Chapter 10

Physicians as a Public for the Popularization of Medicine in Interwar Catalonia: The *Monografies Mèdiques* Series

Enrique Perdiguero, José Pardo-Tomás and Àlvar Martínez-Vidal[1]

The *Medical Monographs* collection *Monografies Mèdiques* (*Monografies* hereafter), a series published in Barcelona between 1926 and 1937, provides the optimum conditions for studying the relationship between science, in this case medicine, and its public within the context of Catalonia, which, in the interwar period, might be considered on the periphery of European science.

In the first place, in its pages the creator and editor of the collection, the physician and politician Jaume Aiguader i Miró (1882–1943), explicitly articulated countless thoughts on the process of popularizing medicine in general, as well as his interest in winning over physicians as the primary target for the *Monografies* popularization efforts. While we also find expressions of his interest in offering *Monografies* to the 'educated public' (that is, non-physicians), our hypothesis is that Aiguader's contributions championing access to wider audiences – upheld, as we shall see, by some contemporary *Monografies* reviewers in the press of that time – were merely rhetorical. Not only was the definition of that other audience poorly defined, but its access to *Monografies* was also problematic given the technical medical matters covered in every issue, the way they were addressed and the advertising inserted in its pages.

In the second place, some general traits of the *Monografies* publishing project serve as an indication of a popularizing programme, some of whose aims involved political agitation and shaping a Catalan scientific culture in a nationalistic sense. Readers' mobilization in this regard began with the unambiguous backing of the consolidation and diffusion of a modern scientific Catalan language, and ended with the no less obvious endorsement of a series of expectations that medical Catalanism had espoused in order to strengthen and sanction the social status that physicians considered their due.

[1] This research was carried out within the project 'Los públicos de la ciencia, la técnica y la medicina en la España contemporánea. 1: La divulgación médica en la Cataluña noucentista', financed by the Science and Education Ministry (BHA2002-04611-C03-01). We wish to thank Manuel Martí Martínez (Universitat de València) for his help regarding some aspects of this study.

Finally, and closely related to the preceding paragraphs, to the extent that it was a project of high popularization among physicians, *Monografies* was a product and, at the same time, a support for the consolidation of medical specialties in Catalonia. In fact, within the local publishing world, *Monografies* was a unique exponent of the phenomenon of the birth of medical specialities, thriving in the interwar period and radically transforming medical knowledge and practice.[2]

Our aim in this text will therefore be to approach one case in the general phenomenon of science popularization through the analysis of *Monografies*, whose meaning is revealed when set in relation to these three particular contexts in which it was developed: science popularization, European nationalist movements and the birth of medical specialties. To this end, we have considered it essential to start from the specific historiographic fields concerning those concrete contexts.

Three Specific Contexts to Situate a Publishing and Scientific Venture

We are interested in underlining the different audiences for science popularization, for, as mentioned above, *Monografies* targeted a particular audience which it intended to mobilize for specific aims.

The so-called 'dominant view' of science popularization has, of course, been critically reviewed over the past two decades. Such a standpoint focused on producers, creators of 'genuine' scientific knowledge and leading actors of its 'diffusion' to the audience, which was regarded as an undifferentiated mass, a passive recipient of a more or less simplified version of 'true' knowledge. Any other agents and the process of communication itself were, according to this 'dominant view', merely a source of 'distortion' of the science to be popularized.[3] Nevertheless, criticism of this 'dominant view' has failed to lay it to rest.[4] In fact, the traditional way of understanding the relationship between science and its public is still present in texts seeking to solve what is considered a problematic bond, and

2 For a new periodization and a comparative perspective of the process, see George Weisz, *Divide and Conquer. A Comparative History of Medical Specialization* (New York, 2006), mainly pp. 189–256.

3 Some well-known expressions of these criticisms may be found in Richard Whitley, 'Knowledge Producers and Knowledge Acquirers. Popularisation as a Relation Between Scientific Fields and Their Publics', in Terry Shinn and Richard Whitley (eds), *Expository Science: Forms and Functions of Popularisation* (Dordrecht, 1985), pp. 3–28, especially pp. 3–10. Stephen Hilgartner, 'The Dominant View of Popularization: Conceptual Problems, Political Uses', *Social Studies of Science*, 20 (1990): 513–39.

4 An especially clear example can be found in the statement 'There is a general problem about the popularization of science. Science is developing very fast, yet public understanding of science is only growing slowly [...] there is now an increasing divergence between the advancement of science and the general scientific understanding of the vast majority of human beings': A.G. Howson and Jean-Pierre Kahane, 'A Study Overview', in Howson and Kahane, *The Popularization of Mathematics* (Cambridge, 2003), p. 1.

which lay at the root of a far-ranging political agenda aimed at improving 'the public understanding of science'.[5]

In our opinion, the kind of relationship *Monografies* tried to achieve between medicine and its public may be adequately understood if, taking into account the criticism of the 'dominant view', we accept that any exposition[6] or communication of science is popularization to a greater or lesser extent, provided it modifies knowledge when communicated in a different context from that in which it was produced and initially validated. Whitley has highlighted that the increasing scientific specialization starting in the second half of the nineteenth century diluted the existing clear distinction between experts and laypeople,[7] in that 'expertise' became restricted to an increasingly reduced circle. All the remaining immense area of knowledge did and does require science 'exposition' processes among scientists which, one way or another, are 'popularization'. They are intended to make certain knowledge available to others for a variety of purposes: to update, to seek professional cohesion, prestige and financial support, as a strategy for handling problem situations or as a way to redraw the boundaries between different specialties. The significance of this kind of intra-scientific communication lies in there being different ways of conveying problems, and therefore different ways of constructing them. Thus, in many different fields, especially science communication studies,[8] new conceptualizations are being put forward in order to focus even further on the relationships established between a complex plurality of audiences and knowledge claiming a scientific status for itself.[9]

---

[5] An interesting assessment is in Ulrike Felt, 'Sciences, Science Studies and their Publics: Speculating on Future Relations', in B. Joerges and H. Nowotny (eds), *Social Studies of Science and Technology: Looking Back Ahead* (Dordrecht, 2003), pp. 11–31. On other reviews of the 'Public Understanding of Science' agenda see: Brian Wynne, 'Public Understanding of Science', in Sheila Jasanoff, Gerald E. Markle, James C. Petersen and Trevor Pinch (eds), *Handbook of Science and Technology Studies* (Thousand Oaks, CA, 1995), pp. 361–88; and Sergio Sismondo, 'Expertise and the Public Understanding of Science', in Sismondo, *An Introduction to Science and Technology Studies* (Oxford, 2004), pp. 162–72.

[6] We take the concept from Terry Shinn and Richard Whitley, 'Editorial Preface', in Shinn and Whitley (eds), *Expository Science*, p. viii.

[7] For a detailed discussion, see Richard Whitley, 'Knowledge Producers', pp. 11–16, and Joske Bunders and Richard Whitley, 'Popularisation within the Sciences: The Purposes and Consequences of Inter-Specialist Communication', in Shinn and Whitley (eds), *Expository Science*, pp. 61–77.

[8] An overview can be found in Bruce V. Lewenstein, 'Science and the Media', in Jasanoff et al., *Handbook of Science and Technology Studies*, pp. 343–60.

[9] Massimiano Bucchi, 'Can genetics help us rethink communication? Public communication of science as a "double helix"', *New Genetics and Society*, 23 (2004): 269–83. A more general articulation of his ideas on scientific communication can be found in Massimiano Bucchi, *Science and the Media. Alternative Routes in Scientific Communication* (London and New York, 1998); and more recently Massimiano Bucchi, 'Communicating

Thus, the consideration of science popularization as a *continuum* comprising both intra- and inter-specialist communication[10] allows us to properly situate the *Monografies* publishing venture, as well as to understand the varied appeals to its readers constantly appearing in its pages. In this sense, viewing audiences as active elements that are able to modify the context of the production of scientific knowledge becomes especially interesting in order to properly understand the *Monografies* publishing venture as a whole. Such an endeavour sought to win readers from a very specific professional field, and it recruited its writers from this same sector. Meanwhile, it constantly proclaimed its lay influence and the fact that it welcomed a wide range of audiences in order to boost both the prestige and the support of physicians and medicine in Catalonia.

There is no doubt that historical studies regarding nationalist movements, and especially the mechanisms of constructing national identities, have undergone considerable change in the last 20 years.[11] They have chiefly focused on historians' role in the construction of different national imageries.[12] On the other hand, it is harder to find the same kind of considerations as regards scientists and the different scientific disciplines which were beginning or developing precisely in the same period as the consolidation of modern European nations. In this sense, the historiography of the so-called 'liberal professions' is extremely useful when suggesting specific case studies, such as ours.[13] We think a publishing endeavour such as *Monografies* must be linked to the essential question regarding the role of professional groups in the construction of the Catalan national identity in the last decades of the nineteenth century and the first few decades of the twentieth

Science', in *Science in Society: An Introduction to Social Studies of Science* (London and New York, 2004), pp. 107–24.

[10] This classification may be found in Michel Cloître and Terry Shinn, 'Expository Practice: Social, Cognitive and Epistemological Linkage', in Shinn and Whitley (eds), *Expository Science*, pp. 31–60. A slightly different nomenclature can be found in Rudolf Stichweh, 'The Multiple Publics of Science: Inclusion and Popularization', Institut für Weltgesellschaft. Universität Bielefeld, Working Paper, October 2003 <http://www.uni-bielefeld.de/soz/iw/pdf/stw_science_popular.pdf>.

[11] To give a starting point on which there seems to be certain agreement, see Ernest Gellner, *Nations and Nationalism* (Oxford, 1983). For an overview of the discussion Gellner's arguments has prompted, see John A. Hall (ed.), *The State of the Nation: Ernest Gellner and the Theory of Nationalism* (Cambridge, 1998).

[12] A recent example is Christian Amalvi, *Les lieux de l'histoire* (Paris, 2005). For Spain, see Carolyn P. Boyd, *Historia Patria. History, Politics and National Identity in Spain, 1875–1975* (Princeton, 1997).

[13] A historiographic outlook can be found in Maria Malatesta, 'Uno sguardo agli studi sulle professioni', in Angelo Varni, *Storia delle professioni in Italia tra Ottocento e Novecento* (Bologna, 2002), pp. 21–49; Maria Malatesta (ed.), *Society and the Professions in Italy 1860–1914* (Cambridge, 1995).

century. This, indeed, is what we sought to describe when we coined the expression 'medical Catalanism'.[14]

Recent historiography presents Catalan nationalism as a cross-class movement within which professional sectors and wage-earners in the tertiary sector or in factories had a clear influence.[15] These social groups posed an alternative to the politics of Spanish oligarchic liberalism through the period known as the Restoration (1875–1923).[16] The Catalan League initially managed to provide this option with a conservative orientation, yet it was established as a Catalanist party through a democratic political approach and practices which broke with the oligarchic tradition of Spanish liberal nationalism. The rebellion of the middle classes was channelled through economic and professional corporations, which were in turn strongly influenced by the updating of the premises of political Catalanism. In this way, the first viable political expression of a non-Spanish nationalist project was realized from the perspective of a conservative social force which defended only moderately nationalist claims. At a later stage, especially through the years 1920–36, the initial conservative hegemony fell apart, yielding left-leaning Catalanist political proposals,[17] of which Jaume Aiguader was one of the leading exponents.

We are currently interested in showing how the Catalanist questioning of Spanish oligarchic liberalism was spearheaded by a social sector whose critical entry in the public sphere took place through educators, writers, journalists, civil servants and professionals in the fields of science, technology and medicine (physicians, pharmacists, veterinary surgeons, engineers, architects and so forth).

---

[14] José Pardo-Tomás, Àlvar Martínez-Vidal and Enrique Perdiguero, '"Per la Ciència i per la Pàtria"': medical catalanism (1898–1936)', in Social Science History European Conference, Amsterdam, 22–25 March 2006: <https://digital.csic.es/handle/10261/7884>.

[15] Three basic references for this historical reinterpretation of Catalan nationalism are Joan Lluís Marfany, *La cultura del catalanisme. El nacionalisme català en els seus inicis* (Barcelona, 1995); Jordi Casassas, *Els intel·lectuals i el poder a Catalunya (1808–1975)* (Barcelona, 1999); and Enric Ucelay-Da Cal, *El imperialismo catalán. Prat de la Riba, Cambó, D'Ors y la conquista moral de España* (Barcelona, 2003).

[16] On the construction of the Spanish national identity, see José Álvarez Junco, *Mater Dolorosa. La idea de España en el siglo XIX* (Madrid, 2001). By this author see also 'The Formation of Spanish Identity and Its Adaptation to the Age of Nations', *History and Theory*, 14 (2002): 13–36. A convincing discussion of some of his premises can be found in Ferran Archilés and Manuel Martí, 'Un país tan extraño como cualquier otro: la construcción de la identidad nacional española contemporánea', in María Cruz Romeo and Ismael Saz (eds), *El siglo XX. Historiografía e historia* (Valencia, 2002), pp. 245–78. By these same authors: 'Ethnicity, Region and Nation: Valencian Identity and the Spanish Nation-State', *Ethnic and Racial Studies*, 24 (2002): 245–78.

[17] Josep R. Llobera, *Foundations of National Identity: From Catalonia to Europe* (New York and Oxford, 2004); Xosé M. Núñez Seixas, *Historiographical Approaches to Nationalism in Spain* (Saarbrücken, 1993); Clare Mar-Molinero and Angel Smith (eds), *Nationalism and the Nation in the Iberian Peninsula: Competing and Conflicting Identities* (Oxford and Washington DC, 1996).

These groups' penchant for nationalism was the expression of undefined anti-system political aspirations – common behaviour in Europe – which had a strong sense of 'status incongruity', according to the expression insightfully coined by Alberto M. Banti.[18] The exclusion from the construction of a Spanish liberal state, coupled with the sense of participating in trends of European modernity while lacking the political, institutional, economic and cultural instruments needed to cope with them, paved the way for the attempt at nationalization through politics and cultural agitation. The key pieces in this process were the consolidation of Barcelona as an urban centre able to polarize Catalan social and cultural life, the coding of a symbolic heritage for a modern national culture,[19] the political claims of the Catalan language and the demand for a space for its own cultural and scientific progress.[20]

Barcelona, a large city with almost one million inhabitants, attracted considerable labour contingents both from the rest of Catalonia and increasingly from the neighbouring Spanish regions (Aragon, Valencia and Murcia) to the industrial centres in its metropolitan area. As regards the field of healthcare, the accelerated increase in new knowledge and the development of diagnostic and therapeutic techniques, requiring longer and more expensive training, took place in parallel with the consolidation of a demand from the middle classes for more sophisticated and efficient medical care. All these circumstances prompted the development in Barcelona of a dynamic healthcare sector, where clinics and health centres, mercantile and labour mutual benefit societies, and pharmaceutical, chemical and biological analysis laboratories proliferated.

In this context, some characteristic features of the process constituting the first medical specialities emerged during the first third of the twentieth century. This process reproduced the social division of labour within the medical profession and consisted of the gradual differentiation and identification of each field of medical knowledge and practice. Around this time, the first expert societies appeared within the medical circles of Barcelona, with their own rules, meetings

---

[18] Alberto M. Banti (ed.), *Storia della borghesia italiana. L'età liberale* (Rome, 1996) and Alberto M. Banti (ed.), *Immagini della nazione nell'Italia del Risorgimento* (Rome, 2002). See also Alberto M. Banti, 'Italian professionals: markets, incomes, states and identities', in Maria Malatesta (ed.), *Society and the Professions*, pp. 223–54.

[19] Benedict Anderson, *Imagined Communities: Reflections on the Origin and Spread of Nationalism*, Revised Edition (London and New York, 1991).

[20] Vicente Cacho Viu, *El nacionalismo catalán como factor de modernización* (Barcelona, 1998). Miroslav Hroch's studies are still essential to understand the historical conditions of the construction of national identities without starting from pre-existing states. See Miroslav Hroch, *Social Preconditions of National Revival in Europe: A Comparative Analysis of the Social Composition of Patriotic Groups among the Smaller European Nations* (New York, 2000).

and journals.[21] Such societies fought to develop and define their respective areas of interest, as well as introduce adequate training mechanisms. In the long term, they also sought to impose monopolistic systems on their specific fields. This process was not devoid of tensions between opposing forces, some aimed at fragmenting the professional community, while others reinforcing cohesion.[22]

For instance, in 1926, almost concurrently with the publication of the first issue of *Monografies*, the Catalan Society of Paediatrics was instituted in Barcelona with the support of most Catalan corporations. Besides the adoption of Catalan as the primary language for scientific communication, its aims included joining the Academy and Laboratory of Medical Sciences of Catalonia, which came to fruition in 1931 after arduous negotiations. This served as a model for all the other Catalan specialists associations, so that the aforementioned Academy became an authentic society of societies, a backbone for medical specialities.[23] The *Monografies*, which offered up-to-date texts on specific topics written by local experts, represented a coherent response to the tensions arising from the emergence of these specialities.

The *Medical Monographs* as a Publishing Venture

The *Monografies*, as a project aimed at connecting medicine with its publics, was linked to other initiatives from the same core of people and institutions spearheading medical Catalanism. The pages of the *Monografies* are both a reflection and a symptom of this constellation of projects, as is shown by the constant inclusion of news about other Catalan publications, and courses on new medical developments being held in clinics and hospitals in Barcelona, the Physicians Union of Catalonia and the Academy and Laboratory of Medical Sciences. In a variety of instances, the role of the *Monografies* was explicitly asserted as the first within an array of publishing initiatives with greater ambitions.[24]

Moreover, its publishing endeavour was complemented by other, diverse and more sweeping popularization projects which were encouraged both by Aiguader, who was closely involved in all of them, and his main core of contributors. These

---

21 The Psychiatry Association of Barcelona and the Otorhinolaryngology Society of Barcelona were founded in 1911, the Catalan Society of Dermatology and Syphilography in 1915, the Ophthalmologic Society of Barcelona in 1925, the Catalan Society of Radiology and Electrology and the Surgery Society of Barcelona in 1926; the Obstetrics and Gynaecology societies were founded in 1927 and the Urology Society in 1929.

22 Jan Goldstein, 'Psychiatry', in William Bynum and Roy Porter (eds), *Companion Encyclopedia of the History of Medicine* (London and New York, 1997), pp. 1350–52.

23 Inmaculada Pastor-Cubo and Àlvar Martínez-Vidal, 'L'Acadèmia entre la dictadura i la II República (1923–1939): cap a una societat de societats', *L'Informatiu. Acadèmia de Ciències Mèdiques de Catalunya i de Balears*, 15 (2002): 25–8.

24 Jaume Aiguader, 'Després de dos anys', *Monografies*, 14 (1928).

were products that, with differing degrees of success, were published by the same company, Arnau de Vilanova (often printed in the typographic workshop 'Cosmos') and promptly advertised by Aiguader in the pages of the *Monografies*. Thus, for instance, a Catalan edition of *Introduction to the Study of Experimental Medicine* by Claude Bernard, translated and annotated by Jaume Pi-Sunyer i Bayo in 1935, was published in the 'Collection Miguel Servet'.[25] Aiguader's work on physicians and medicine, *Praise of Physicians and Medicine*, was also published in this collection.

Yet another example of these projects prompted by Aiguader was a more ambitious 'scientific encyclopedia in Catalan'. This work was designed to address workers and arts and crafts students, but its introductory text also included a rhetorical allusion to the 'educated man' as a potential reader.[26] The most direct political agitation, an essential piece of Aiguader's popularizing effort, was also present in the collection 'La Sageta',[27] which he also edited, as he did the *Monografies*, and in which articles signed by Aiguader himself appeared under such meaningful titles as *Catalonia and the Revolution*.[28]

The *Monografies* came out at a critical moment, during the dictatorship of General Primo de Rivera which, with the consent of the monarchy, lasted from 1923 to 1930. Democratic institutions were suspended, there was severe press censorship and publications in Catalan were put 'on probation'. This political situation came about, it is worth remembering, after a certain degree of political autonomy during the few years of the Mancomunitat de Catalunya (1917–23).[29] This experience had given those physicians and scientists more or less involved with political Catalanism an accurate idea of the extent of their leadership chances and agenda of social transformation, which was only possible with a recovery and expansion of political autonomy.

Aiguader's overt opposition to the dictatorship – which brought him, among other things, a six-month prison term between 1926 and 1927 – cannot be considered apart from the initial public reception of the *Monografies*. In this

[25] Claude Bernard, *Introducció a l'estudi de la Medicina Experimental. Pròleg de J.F. Fulton i prefaci de Paul Bert. Traduïda i anotada per Jaume Pi-Sunyer i Bayo*, 2 vols (Barcelona, [1935]).

[26] The plan was to produce a collection of 50 to 60 handy and reasonably priced volumes on physics, chemistry and natural sciences: *Monografies*, 31 (1929).

[27] In the collection, published by 'Arnau de Vilanova', appeared the following monographs: Joan Puig i Ferreter, *Vida interior d'un escriptor* (Barcelona, 1928); Carles Rahola. *Els jueus a Catalunya* (Barcelona, 1928); Jaume Aiguader i Miró, *La lleialtat a l'època* (Barcelona, 1929); Jaume Aiguader i Miró, *Amb Catalunya i per Catalunya* (Barcelona, 1930); Jaume Aiguader i Miró, *Catalunya i la Revolució* (Barcelona, 1931).

[28] Aiguader, *Catalunya i la Revolució*.

[29] Antoni Roca Rosell, 'Ciencia y sociedad en la época de la Mancomunitat de Catalunya (1914–1923)', in José M. Sánchez Ron (ed.), *Ciencia y sociedad en España* (Madrid, 1988), pp. 223–52.

sense, the publication became a vehicle for opposing the regime, just like almost all the cultural Catalanist publications at the time.[30]

As a vehicle for popularization, therefore, the *Monografies* raises an interesting question concerning its relationship with its readership. Although circumscribed initially by their professional activities, the readers received a product that was clearly defined from a political standpoint. Initially, this might have sparked rejection within some sectors. However, given the adverse political context for Catalanism, it did provide the *Monografies* with greater support derived from the phenomenon of 'nationalist cultural resistance', which is always able to create far-ranging complicities.

Figure 10.1 Jaume Aiguader i Miró, newly elected mayor of Barcelona (centre).

In the polls of April 1931, which led to the proclamation of the Republic, Aiguader was elected mayor of Barcelona (Figure 10.1), a role which probably kept him away from the day-to-day management of the *Monografies*. In addition to the successive and increasingly wider political responsibilities taken on by Aiguader (a congressional representative in the three Republican terms, and twice minister in 1936), this allows us to distinguish two clearly differentiated periods for the *Monografies*:

[30] In reference to *Monografies*, once the dictatorship had fallen, Aiguader wrote that perhaps its main aim was to protest against the humiliations suffered by the Catalan language and Catalonia: Jaume Aiguader, 'Monografies Mèdiques traduïdes al francès', *Monografies*, 37 (1930).

1. A first period encompassing slightly more than half of the issues (40 of 52 instalments) over five years (May 1926–April 1931), marked by the strategy of achieving viability for the publishing endeavour within a hostile political environment.
2. A second period with an independent publishing company which fully fell within a different political framework but was interrupted by the outbreak of the Civil War, when 77 issues of 104 instalments had already been published.

The *Monografies*, in its monographic format, broached a wide range of topics. Forty-six issues, that is, almost half the total, focused on internal medicine, especially emphasizing three fields: digestive disorders, endocrine gland ailments and nutrition. Space was also given to disorders of the respiratory and circulatory systems as well as blood-related ones. In the context of healthcare's transition towards an epidemiological pattern dominated by degenerative diseases, infectious pathologies did not constitute the primary focus of the *Monografies*, except for tuberculosis and syphilis. However, increasingly important diseases such as diabetes, hypertension, cardiac insufficiency, hepatic conditions, rheumatic diseases and cancer were dealt with on a regular basis. Tuberculosis received special attention from the surgical point of view. In all, there were 19 issues devoted to different aspects of surgery. Women and children were the focus of 14 issues, and 23 more were devoted to other specialities, most notably psychiatry and radiology. It is also worth emphasizing that the focus was on diagnosis much more than on therapeutics, except in the case of tuberculosis and certain digestive tract disorders.

Regarding the structure of the monographs, each issue had two parts, neatly differentiated in both their presentation (layout, typography, quality of paper) and their content. The monograph itself, usually of between 50 and 100 pages, was in the middle of the issue, with its own specific cover, no advertising and printed on higher-quality paper (Figure 10.2). The front and back pages (printed on slightly thicker, grey-coloured paper featuring a reproduction of the internal cover of the monograph and generally including the index of both back and future issues in the series) were sewn to the central booklets. There were between 12 and 32 pages containing all the publicity that, together with subscriptions, constituted the main source of funding for the collection.[31]

However, the most interesting matter is that, from the beginning, Aiguader always inserted text in the spaces between advertisements. The typographical layout and the structure of these pages clearly show how these texts were subordinated to the size, layout and form of the publicity. These texts included press reviews (with titles such as '*Our Press*' and '*Catalan Medical and Scientific*

[31] As far as the footnotes in this chapter are concerned, it is very important to bear in mind that until 1931 the non-monograph pages were not numbered. From August 1931 (issue 54), these pages were numbered by issue; after 1935 the pages were numbered according to an annual sequence.

Figure 10.2 The cover of one issue of the *Monografies Mèdiques*.

Press'), announcements for courses for students and graduates, conferences, appointments, obituaries and other matters of interest mainly for professionals in the field of medicine. Nevertheless, the main section, the 'News Bulletin' from the first issue, was always at the beginning. In fact, more than a news report, it was an editorial, almost always written by Aiguader himself and accompanied by texts by friends and contributors, often excerpted from their publications in the Catalanist press (*La Veu, La Publicitat, La Nau*) and sometimes from the regional press as well (*Reus, Lleida*).

These texts are the best illustration of Aiguader's ideal for the *Monografies*, not only because of the content of his writings, but also because of the selection of news and excerpts that were included. They are, also, the best source we have so far[32] of reliable evidence on the reception of *Monografies*, an aspect that we believe deserves attention in view of the aims of this chapter.

A Preliminary Look at the Reception of the *Monografies*

As described above, the *Monografies* was an explicit and self-reflective effort to connect medicine and its audiences. Its main objectives as a publishing project were to make available to non-specialist physicians, in a summary and original form, the new developments generated by the gradual fragmentation of medical knowledge;[33] to promote reading and writing on medical issues in Catalan; and to generate an opportune environment for the development of science in Catalonia, showing its peculiarities and potential compared to a weakened Spain. However, was it successful? All evidence seems to indicate it was. Although we have very little relevant evidence available, the fact that the series lasted a decade, the existence of reissues of certain issues and some translations into Spanish[34] and even translation projects into French[35] are nevertheless far from negligible.

---

32 Several attempts to locate administrative documentation on the *Monografies* (mail with advertisers and subscribers, the archive of Arnau de Vilanova's publishing company, the Cosmos typographic workshop, etc.) have not yet yielded positive results. The same is true of Aiguader's personal archive, recently donated by his descendants to the Catalan National Archive.

33 In the two-year review of the *Monografies* published in the first issue of 1928, this idea is clearly emphasized, pointing out the need to write original books in Catalan that synthesized new medical knowledge.

34 As far as we know, in addition to the translation and publication of the first issue in Spanish – Leandro Cervera, *Fisiología, fisiopatología y clínica de las glándulas de secreción interna* (Barcelona, [1933?]) – there is at least one more translated into Spanish: August Pi i Sunyer, *Las distrofias por retardo* (Barcelona, [1934?]). The translation was by Jaime and Carlos Pi-Sunyer, the author's son and nephew. In both cases the publisher was 'Arnau de Vilanova', the same as in the Catalan version.

35 A plan apparently devised by Aiguader himself together with the Occitan physician Ismael Girard, a frequent participant in the meetings of the Catalan-speaking Physicians

The *Monografies* had several sources of financing for its furtherance: patronage, publicity, subscriptions and direct sales to the public. Regarding patronage, it seems[36] that shortly after the publication was launched, financial support was secured from lawyer and landowner Paulí Geli i Maymí, elected representative for the *League* in the *Mancomunitat* for the electoral district of Figueres in all three elections (1917, 1919 and 1923) that were held before the dictatorship. However, its success in attracting advertisers, subscribers and direct buyers soon allowed the journal to do without patronage yet still break even.[37] According to the assessment made public by the editor-in-chief three years after it was launched, the *Monografies* was self-financing, without either profit or loss, although Aiguader proclaimed himself in favour of making it profitable.[38]

Concerning advertising, a review of the pages devoted to publicity in every issue of *Monografies* shows the importance and abundance of this means of financing (Figure 10.3). The advertisements were mainly for pharmaceutical products, although ads for finance companies, magazines, dietetic products, medical equipment and private clinics where outstanding figures in medical Catalanism were working were also included.[39] The advertising was, thus, clearly targeting physicians as consumers, therefore confirming them as the main audience of the *Monografies*, the rhetoric appeal for a non-medical 'educated public' notwithstanding.

---

and Biologists Association (1913–36). In an article devoted to the plan, it was asserted that the *Monografies* would spread throughout the world from his publication in the heart of Europe: Jaume Aiguader, 'Monografies Mèdiques traduïdes al francès'. He discusses this matter again in *Monografies*, 40 (1930) and in *Monografies*, 43 (1930), where the announcement appears of the publication *Monographies Médicales*. It appears that no issue of the *Monografies* was actually published in French.

36 Josep M. Poblet, *Jaume Aiguader: una vida amb Catalunya i per Catalunya* (Barcelona, 1977), p. 46, without any reference to the source of the information.

37 According to an exceptional witness, given his close involvement with Aiguader and his editorial projects, over time, *Monografies* was an important source of income for Aiguader and his family, after the first years of hard effort and no profits: Jaume Pi-Sunyer, 'Prefaci', in Jaume Aiguader, *Miquel Servet* (México, 1946).

38 Aiguader considered that in order to sustain a culture, the culture itself must maintain the authors: Jaume Aiguader, 'Balanç espiritual i econòmic de Monografies Mèdiques', *Monografies*, 29–30 (1929).

39 Aiguader's account of the winning of advertisers was very explicit. He commented that the majority of advertisers greeted the publication of medical works in Catalan with enthusiasm. However, others showed no interest in supporting the new publication and increasing their costs. Finally, a few refused to collaborate because of the doubtful circulation of a publication in Catalan. Aiguader admitted the birth-pains of the first issue: Jaume Aiguader, 'Balanç espiritual i econòmic'. The difficult beginnings regarding publicity are also revealed by the request included in the first issue asking the readers to support the sponsors.

Figure 10.3 Some of the advertisements published in the *Monografies Mèdiques*.

In relation to subscribers, there were already 800 in 1929, three years after the collection was begun. This is a far from negligible figure when we consider that there were almost 3,000 physicians in Catalonia. If we are to believe Aiguader's account, some 200 new buyers acquired the complete collection of the *Monografies* each year and remained subscribers. The bulk of them were from Catalonia, although there were also subscribers from the Balearic Islands and Valencia, the result of an advertising campaign which, though weak, was consistent with the pan-Catalanist rhetoric typical of Catalan nationalism.

In Aiguader's opinion, most subscribers were 'practitioners' whose interest 'in getting up to date' with many different topical issues related to medical practice was primarily fulfilled by the *Monographs*. Moreover, the *Monografies* claimed the added advantage of using the vernacular language of those readers 'lacking the time' to acquire that knowledge through books or specialized journal articles, which were mostly written in languages other than Catalan and were often 'fragmentary, contradictory and prone to quick enthusiasm or to disappointment'.[40]

[40] Aiguader was fully aware of the contemporary phenomena of growth and obsolescence of scientific literature, as well as of the problems and prospects of scientific communication at the time. He found journal articles too detailed and books too prone to obsolescence and too lengthy. Thus, he sought to publish an intermediate product aimed at medical practitioners who needed to apply new knowledge immediately. Aiguader, ibid.

Medical students, who, according to different accounts, 'constantly asked for back issues of the *Monografies*', were apparently the most significant group of non-subscription readers.[41]

Certainly the usual appeal to the layperson was never missing in the annual reviews on the progress of the *Monografies* that Aiguader or his contributors used to publish in both the daily press and the opening pages of the publication. Thus, by mid-1929, to show that the *Monografies* was reaching an audience wider than just physicians, Aiguader maintained that direct sales – in the Llibreria Catalònia (Catalònia Bookstore) situated in Plaça de Catalunya, the city's nerve centre – were about 200 copies per issue. Moreover, he claimed that issue 8 (*Current questions in the treatment of syphilis*, by Antoni Peyrí), which had come out at the beginning of 1927, had a print run of 2,000 copies (500 more than usual) but had nevertheless sold out.[42]

The constant allusion to the success of the endeavour from the pages of the *Monografies* itself was combined with reviews published in both the daily press in Catalan and the medical press. This was most likely a strategy to mobilize related writers that was aimed at dispelling any doubts about a project born amidst much uncertainty about its viability. Thus, in its second issue, Aiguader already asserted that 'our first monograph has been a success' and discussed the reviews of the first issue that had appeared in the two daily newspapers which served as the mouthpieces for political Catalanism: the more conservative *La Veu de Catalunya* (30 June 1926, signed by physician Manuel Moragas)[43] and the more left-leaning *La Publicitat* (20 June 1926, signed by 'C.', an initial behind which hid 'one of the most illustrious names of our scientific world' according to Aiguader).

Months later, the *Monografies* reproduced Jesús M. Bellido's review in *La Publicitat*, where he insisted that it was 'a scientific success' and 'a bookstore bestseller'. However Bellido, one of the leaders of the Institute of Physiology, insisted that the *Monografies* was aiming at two kinds of audiences: 'all physicians' and 'the curious educated reader, albeit layman in medicine'. He wrote of the qualities of its popularizing drive: 'they are easy to read and are full of practical information and healthy doctrine [with a] clear and careful style'.[44] The reactions

---

41 Ibid. Students as an essential audience to explain the success of the *Monografies* turned up again when the project of the *Encyclopedia Arnau de Vilanova* was presented. Jaume Aiguader, 'Altres raons de l'Enciclopèdia Arnau de Vilanova', *Monografies*, 43 (1930).

42 In fact, a 'second corrected edition' would be made in 1936, when the author was already head of the Service for Anti-venereal Fight of the Catalonia Government.

43 Moragas's entire review was reproduced in *Monografies*, 3 (1926). Likewise, the editorial from *La Publicitat* entitled 'Els nostres escrits científics' [Our scientific writings] was reproduced in its entirety in *Monografies*, 5 (1926), which was mostly devoted to praising the editorial commitment of the *Monografies*.

44 Bellido considered the *Monografies* an effort beyond 'vulgarization', emphasizing its practical value for medical doctors. For the educated general public, the *Monografies* was a means of updating their knowledge of the progress of medicine: Jesús M. Bellido, 'Monografies Mèdiques. L'Electrocardiografia per Cristià Cortés', *Monografies*, 7 (1927).

to the publication of the *Monografies* in the Catalanist press were not limited to welcoming the publication of the first issues. In 1935, for example, when the tenth anniversary of the *Monografies* was approaching, several articles on the series were published in the general press[45] and professional journals.[46] However, it is important to point out that after ten years, appeals to the general, educated reader were very unusual; medical doctors and medical students had become accepted as the target audience for the *Monografies*.

In order to complete these general comments on the reception of the *Monografies*, we believe it is essential to consider the contributors who joined this popularization project, since the almost 70 authors of the *Monografies* undoubtedly came from the body of physicians who were also readers and subscribers. In his articles in the *Monografies*, Aiguader repeatedly encouraged readers to write and participate in the project; he also sometimes complained about the difficulties involved in mobilizing contributions on topics he deemed necessary to examine, and of the problems caused by some authors who seemed unable to attain the style the *Monografies* sought in presenting a given topic to its readers.

An important fact in order to understand the background of the *Monografies* authors is that almost half of them (31 out of 66 altogether) became part of the teaching staff at the Faculty of Medicine in Barcelona during the period of the Generalitat (1932–39) – Catalonia's Republican government. Given that from the beginning of the century one of the aspirations of medical Catalanism was to attain a specific and autonomous teaching and research environment – an 'authentically Catalan' university which did not depend on the oppressive centralism of the Spanish university system – the realization of political autonomy within the Republic led the University of Barcelona to become what was called the Autonomous University. While only three *Monografies* authors had been university professors in the period preceding the Autonomous University (another four had been 'assistant professors' and three more 'course supervisors' in the previous university model), 11 *Monografies* authors were professors on the faculty of the new university, and 20 more were hired as teachers for the so-called 'monographic advanced courses' which were intended both for specialization and the PhD level. Moreover, of the eight freelance professors (*professors lliures*) of medicine hired directly by the Autonomous University, three had already been *Monografies* authors. The political concordance between the editorial project of the *Monografies* and the Republican regime helps to explain the huge impact of the exile (the fate of a great majority of the authors published in the *Monografies*) and repression that followed Franco's victory in the Civil War.[47]

---

[45] For example, Martí de Riquer, 'La tasca de les editorials catalanes: les Monografies Mèdiques', *La Publicitat*, 7 May 1935; and Nicolau Battestini, 'Monogràfies Mèdiques', *El Diluvio*, 9 August 1935.

[46] Salvador Vives, 'Monografies Mèdiques', *Butlletí del Sindicat de Metges de Catalunya*, 183 (1935): 457.

[47] Jaume Claret, *El atroz desmoche. La destrucción de la universidad española por el franquismo* (Barcelona, 2006), pp. 251–86; quotation on p. 285.

A Model for an 'Ideal' Public, a Precise Level of 'Popularization'

As we have seen, Jaume Aiguader stated clearly and repeatedly what the *Monografies* wished to represent. In this way, he outlined an 'ideal' public that was active both scientifically and politically and far above the presumed passiveness of the general public, as regarded by the dominant view of popularization mentioned above.

In September 1928, Aiguader published 'The Cultural Moment', one of the clearest texts on the role he wished for the readers of the *Monografies*, with cultural and political agitation as essential features. He wrote, among other things: 'We physicians cannot afford to be inactive spectators of the current cultural movement: we have much to teach the people.' The whole text is emblematic of his way of understanding the relationship between medicine and its audiences. Physicians must consider the popularization of medicine addressed to the general public, and it is worth noting that Aiguader himself used the term popularization: 'la popularització dels principis elementals de la nostra Ciència' [the popularization of the basic principles of our science]. The editor of the *Monografies* advised his authors to limit the scope of this popularization of medicine to notions such as the basics of anatomy, physiology and histology: the most suitable subjects for people wishing to learn. Together with the appeal to the general public, Aiguader emphasized the need for doctors to get involved in the project as authors and readers of a new scientific vision of the world.[48] He intended, through the *Monografies*, to create a specific scientific culture that was both new and in Catalan:

> Our collection is not entertainment for idle people working in isolation just for themselves; our collection, besides its scientific task, its teaching mission, its concern for researchers and our publicists, also wishes to be the beginning of a process to particularize a new scientific culture. In all modesty and humility, and yet with great pride, it intends to be the catalyst of a great deed: to give our vernacular language a scientific personality.[49]

In order to reach that ideal public, the precise level of 'expository science' needed for the *Monografies*'s popularizing endeavour was constantly emphasized. The editorial published in *La Publicitat* in November 1926, reproduced in issue 5 of the *Monografies*, is very explicit: both 'simplistic popularizations' and overly specialized vehicles were liable to be rejected, and an intermediate space had to be found, a 'third way' that would make the *Monografies* 'understandable for the educated public'.[50] A 'synthetic style' had to be adopted as a result of the authors' creative work and the requirements of the nascent medical specialities.

---

48 Jaume Aiguader, 'El moment cultural', *Monografies*, 20 (1928).

49 Jaume Aiguader, 'Després de tres anys', *Monografies*, 24 (1929).

50 'Els Escrits Científics', *La Publicitat*, 12 November 1926, article reproduced in *Monografies*, 5 (1926).

Every medical monograph had to be a brief treatise of the main facts involved in a medical speciality. The authors, therefore, were required to put together all the relevant research and their own knowledge to achieve a text with theoretical and practical unity. Publishing their own research, a series of medical records or descriptions of rare clinical cases were explicitly discouraged. Aiguader, aware of the intended 'exposition' style, pointed out that writing this kind of article would yield a distinctive scientific personality and would allow participation in the world of scholars 'as masters, not as apprentices'.[51]

How was the ideal level of popularization achieved? Uneasily. In fact, an analysis of the exposition styles shows how the aims of the editorial project launched by Aiguader in 1926 were achieved via constant negotiation between the editor, the authors, subscribers and readers (mainly medical doctors and medical students). Regarding the 'styles of popularization', although some issues of the *Monografies* were closer to the undesirable 'too specialized' end of the scale, in view of the disparity of the popularization levels of the texts, it seems Aiguader gave authors a great deal of leeway. While some authors expressed their concerns with the popularization enterprise, others had no such qualms. Aiguader did, however, have problems finding authors and convincing them to meet deadlines and adapt to the educational aim of the series.

In our opinion, the authors most closely identified with medical Catalanism expressed their concerns about the popularization styles of their texts more clearly. Frequently, they explicitly declared their desire, despite the difficulties, to become accessible to the readership, usually medical doctors, and in Catalan. For example, Leandre Cervera, writing on the inner secretion glands, clearly stated that he was addressing readers with some basic knowledge of gland physiology.[52] Likewise, Salvador Goday, writing on the feeding of infants,[53] expressed in the preface his decision to survey the foundations of modern infant dietetics in order to avoid a simple practical treatise on infant feeding that would let down his readers, Catalan medical doctors. In his double issue on psychoanalysis, Emili Mira claimed that divulgation meant giving answers to basic questions about new developments.[54] However, to fulfil this aim he took two decisions. First of all, he preferred to avoid oversimplifying psychoanalytical theories. Second, when speaking about controversial aspects of psychoanalysis, he used quotations from Freud in order to allow no room for interpretation. This decision did not prevent Mira from giving his opinions on psychoanalysis while admitting they were not part of the popularizing effort. Indeed, the need to clarify the way the text has been constructed was also present in other monographs. Although it is difficult to generalize, the risks of

51 Jaume Aiguader, 'Després de dos anys', *Monografies*, 14 (1928).

52 Leandre Cervera, 'Fisiologia, fisiopatologia i clínica de les glàndules de secreció interna', *Monografies*, 1 (1926).

53 Salvador Goday, 'Alimentació del nen durant la primera infancia', *Monografies*, 19 (1928): 3.

54 Emili Mira, 'El Psico-anàlisi', *Monografies*, 3 (1926).

oversimplification feared by Mira were also present in the work of other authors, together with the desire to offer useful advice to practitioners.[55]

Other authors such as Caralps, in his issues on the surgery of lung tuberculosis, also attempted to define his potential reader.[56] He sought to explain the advances in lung surgery to the general practitioner, and with this in mind he attempted to write an inter-specialist text.[57]

At the other end of the wide spectrum of popularization styles, the *Monografies* presented several monographs which avoided discussions on how best to address the reader and used a highly technical style which would only be accessible to specialists, thus quite implicitly adopting an intra-specialist approach to explaining medicine.[58] Other authors, such as Domingo writing on immunology and vaccines,[59] produced highly technical texts; yet he felt the need to justify the dryness of his monograph and appeal to the 'culture' of Catalan medical doctors.

Furthermore, we need to bear in mind that all the *Monografies* authors were also subscribers or readers of the series, so they probably used their insights from reading the monographs when they coped with the challenge of writing a text.

Naturally, not all the *Monografies* authors used their participation to project or defend their way of understanding popularization and how achieve it. The examples of Mira, Domingo and Goday are perhaps the most enlightening, although their efforts in this vein were not restricted to their involvement in the *Monografies*. Likewise, and even more obvious, is the work of the aforementioned Leandre Cervera (author of the first issue of the *Monografies*, in 1926) who, starting in 1933, established his own popularization journal: *La Medicina Catalana. Portantveu de l'Occitania Mèdica* [*Catalan Medicine. The Voice of Medical Occitania*]. Over the five years of its existence, *La Medicina Catalana* was mainly devoted to the summarized translations into Catalan of texts from the leading medical journals of Europe and America.

---

55 See, for example, Salvador Goday and M. Carbonell, 'Transtorns nutritius del lactant', *Monografies*, 62–3 (1932).

56 A. Caralps, 'Tractaments quirúrgics de la tuberculosi pulmonar II', *Monografies*, 93–4 (1935).

57 This kind of surgery text addressed to non-surgeon readers is also represented by Manuel Corachan, 'La Cirugia en el processos abdominals aguts', *Monografies*, 5 (1926). However, inter-specialist texts are also the norm for texts addressing diagnostic procedures; see, for example, Ramon Torres and P. Bosch-Solà, 'Semiologia Röntgenològica de l'aparell respiratori intratràcic', *Monografies*, 51–2 (1931) or Frederic Duran-Jordà, 'Anàlisi i tècnica coprològica', *Monografies*, 53 (1931).

58 Examples include Josep Ducuing, 'Flebitis, trombosis i embòlies postoperatories', *Monografies*, 44 (1930) and Jordi Guasch, 'El Tractament de les anèmies', *Monografies*, 49 (1931).

59 Domingo San Juan, 'Inmunitat i vacunoteràpia preventiva', *Monografies*, 4 (1926).

A last point to highlight regarding popularization styles is that, given the publication time span of the *Monografies* and the political changes Catalonia underwent during this period, some of the medical students reading the first issues achieved the status of author by the end of the *Monografies* endeavour. Once more, this double role had a clear impact on the way some authors attempted to popularize the medical matters published in the series.

Concluding Remarks

This initial examination of the *Monografies* has left aside other matters needing further exploration. One of them of great interest is the role played by the *Monografies* in the world of medical publications in Catalan. From the chronological standpoint, it is clear that when the *Monografies* began being published in 1926, it was one of the very few medical texts published in Catalan. We also know that the *Monografies* was the most successful product of Aiguader's editorial initiatives. However, in order to properly determine the importance of the *Monografies*, we need to assess its relevance in relation to the dozen and a half, at least, medical journals published in Catalan immediately before the outbreak of the Civil War. We have to bear in mind that during the first part of the 1930s there was a veritable explosion of medical journalism in Catalan. As mentioned above, the *Monografies* was a well-established publication; but in order to get a true picture of the scale of the popularization discourses, we need to consider whether it was a competitor or an ally of the other medical publications in Catalan. Another aspect we need to explore in the future is the role of the radio, a means to disseminate medical knowledge also used by Aiguader, in the popularization of medicine.

Taking into account that the end of the *Monografies* project was brought about by the Civil War, another possibility is to explore the role of the *Monografies* in the libraries at the front organized during the conflict. We know there were issues of the *Monografies* among the reading list sent to the front from Barcelona, but this aspect requires further exploration.[60]

However, the characteristics of the *Monografies* studied show, in our opinion, the plurality of aims of this popularizing vehicle and the interest of studying it. The *Monografies* tried to become, successfully it seems, an instrument of 'expository science' among both the nascent medical specialities and specialist and general practitioners for whom other vehicles of scientific communication were inconvenient. To achieve this, a popularization product in Catalan was created that was utterly self-reflective and developed as a weapon of political agitation and

---

[60] The paper by Teresa Huguet 'Reading in the war: Catalan "Library Services of the Front" as a means for the popularization of science and technology, 1937–1939', given at the STEP meeting held in Mahon (Minorca, Spain) in 2006, contains data on the books that were included in libraries at the front, allowing us to know the four issues of the *Monografies* that were included.

national construction. The *Monografies* sought to contribute to the building of 'a Catalan medical culture',[61] to give prestige to Catalan science, to provide scientists with resources and to spread summary and original works in order to transcend the periphery and penetrate the academic centres.

Obviously, the *Monografies* was deeply marked by the scientific, social and political milieu of Catalonia in the 1920s and 1930s. The dynamism of the nascent medical specialities in the Barcelona medical setting, together with the political goals shared by a majority of the Catalan middle classes, shaped the *Monografies* as a product addressed to inter-specialist communication in Catalan. The *Monografies* was an enterprise of political resistance, yet also an attempt to offer a new means to popularize medicine and retain a medical identity in order to cope with the fragmentation brought about by the growth of medical specialities.

[61] Salvador Vives, 'Monografies Mèdiques', *Butlletí del Sindicat de Metges de Catalunya*, 183 (1935): 457; article reproduced in *Monografies*, 93–4 (1935): 261–4; quotation on p. 264.

Chapter 11
With or Without Scientists: Reporting on Human Genetics in the Spanish Newspaper *El País* (1976–2006)

Matiana González-Silva[1]

The media was the source of most of the new scientific information acquired by European citizens in the late twentieth century. During those years, television, magazines and newspapers were key elements in the public awareness of scientific discoveries, and in the process by which new knowledge became 'common sense' for certain social groups.[2] In a parallel process, science in the media became an increasingly more important subject of scholarly work. From Media Studies to Social Psychology, Science and Technology Studies, Sociology and History of Science, numerous scholars have studied a wide variety of aspects of the relationship between science and its publics.[3]

---

1 This chapter is based on research funded by the Mexican Consejo Nacional de Ciencia y Tecnología (Conacyt). I am grateful to the Department of History of Science of the Institució Milà i Fontanals-CSIC in Barcelona (and in particular to Jon Arrizabalaga, supervisor of my PhD dissertation) for offering me the ideal environment for my work. I also very much appreciate discussions with members of the research group La ciència i els seus públics at the Centre d'Estudis d'Història de les Ciències (CEHIC), Universitat Autònoma de Barcelona.

2 John Durant, Martin Bauer and George Gaskell (eds), *Biotechnology in the Public Sphere* (London, 1998), p. 240.

3 Research has focused on very different aspects of scientific popularization. The political and professional agendas that it has served, the relationship between the actors that intervene in the communication process, the image of science fostered by the media and the way in which different audiences appropriate scientific knowledge made publicly available through certain channels are among the most interesting.

For general overviews of science in the media, see, among others: Bernardette Bensaude-Vincent, 'In the Name of Science', in John Krige and Dominique Pestre (eds), *Science in the Twentieth Century* (Amsterdam, 1997), pp. 319–38; Roger Cooter and Stephen Pumfrey, 'Separate Spheres and Public Places: Reflections on the History of Science Popularization and Science in Popular Culture', *History of Science*, 32 (1994): 237–67; Bruce Lewenstein, 'Communiquer la science au public: L'émergence d'un genre américain. 1820–1939', in Bernardette Bensaude-Vincent and Anne Rasmussen (eds), *La Science Populaire dans la Presse et l'Édition. XIXe et XXe siècles* (Paris, 1997), pp. 143–53; Christopher Dornan, 'Some Problems in Contextualizing the Issue of Science in the

The aim of this chapter is to contribute to this research, focusing on the forces that shaped a particular way of reporting on scientific news: that of the leading Spanish newspaper of its time, *El País*. It analyses reports about human genetics published in *El País* from the year it was launched, 1976, to 2006, and centres its enquiries on the image of science the newspaper conveyed to its readers.

Following social psychologist Serge Moscovici, a pioneer in the study of the public image of a certain discipline in the general press, this chapter takes newspaper articles as the 'objectified traces' of the communication processes within a particular social system.[4] Newspaper reports are also conceived to be a fundamental 'source of representation' of science in the public sphere: the basis for the social construction of the meaning of science in a certain social context.[5]

The chapter focuses on *El País*'s coverage of scientific issues, based on the idea that journalists not only speak about the surrounding world, but also help shape it. Journalists play an active role in the construction of science's public image. They do so not only by means of explicit statements, but through many other aspects too, which together constitute a particular journalistic style: the framing of the news, the preferred journalistic genres or their position regarding controversies. Journalists decide which participants to give a voice to, what questions they will ask and how they will put any research into context. In so doing, they are influenced in numerous ways, and in turn influence the general evolution of science and technology in the local context in which they operate.[6]

It has generally been stated that the more peripheral the context, the more uncritical popularization becomes. This would be true not only from the geographical point of view (remoteness from the scientific *centre*), but also

Media', in Eileen Scanlon, Elizabeth Whitelegg and Simeon Yates (eds), *Communicating Science* (London, 1999), pp. 179–205. The bibliography of particular case studies is too broad for review in this chapter.

4 Serge Moscovici, *La psychanalyse, son image et son public* (Paris, 1961).

5 Durant et al., *Biotechnology in the Public Sphere*. Jurdant has gone further, stating that more than transmitting scientific knowledge to a general audience, the fundamental epistemic function of the popularization of science is to anchor science to the real world. This would operate through a 'reality effect' mechanism in which journalists communicate certain facts to the public. These facts are subsequently commented in social arenas, not as dealing with science, but with the reality itself. Baudouin Jurdant, 'Popularization of Science as the Autobiography of Science', *Public Understanding of Science*, 2 (1993): 365–73.

6 This relates to more general considerations of the role of the media in contemporary societies. According to Thompson, 'struggles for recognition have increasingly become constituted as struggles for visibility within the non-localized space of mediated publicness. The struggle to make oneself heard or seen (and to prevent others from doing so) is not a peripheral aspect of the social and political upheavals of the modern world; on the contrary, it is central to them.' John B. Thompson, *The Media and Modernity. A Social Theory of the Media* (Cambridge, 1995), p. 247.

regarding the audiences at which popularization is addressed.[7] Historically, 'science popularisation in the periphery played an important role as a strategy for the legitimisation of the main values and ideas of the scientific culture of the centre.'[8] This would have led to the obscuring of critical statements for fear that they could undermine the prestige of science and its practitioners.

Journalistic coverage of human genetics in *El País* very much fulfils remoteness in the two aforementioned senses. On the one hand, we are dealing here with articles published in a daily newspaper addressed at the general public, socially situated at a great distance from the place where science is generally produced.[9] On the other hand, during the last decades of the twentieth century, Spain was not at the international forefront of human genetics, the major achievements of which were mainly located in the United States, Britain and France.

In order to visualize the way the peripheral condition of Spanish science influenced the various ways in which *El País*'s journalists approached scientific subjects, the chapter focuses on the differences between the coverage of certain areas of human genetics in which Spanish participation was virtually zero, and others where a local community of geneticists was consolidated over the study period. This will highlight the crucial role that the presence or absence of a local community of geneticists played in the definition of the model of scientific journalism adopted by *El País*, and will question the very fact that scientific popularization on the scientific periphery leads to an uncritical and hagiographical account of research.

Restored Spanish Democracy, Science and *El País*

The period and place covered by this chapter offer a splendid opportunity to show how science interacts with the broader context in which it evolves, as it studies the journalistic style of the leading Spanish newspaper in parallel with the fundamental changes in late twentieth-century Spain on the one hand and the evolution of the local scientific scene on the other.

---

7 Harry Collins has stated that 'science only looks certain when one moves away from the "core-set", either in "sociometric space" or in space. … Since the general public is by definition maximally distant from the research front, their position ought to give rise to a model of science as a process that produces certainty.' Harry Collins, 'Certainty and the Public Understanding of Science: Science on Television', *Social Studies of Science*, 17/4 (1987): 689–713, p. 692.

8 Agustí Nieto-Galan and Faidra Papaneloupoulou, 'Science, Technology and the Public in the European Periphery. A report of the 5th STEP Meeting (1–3 June 2006, Mahon, Minorca)', *Journal of Science Communication*, 5/4 (2006): 2–6.

9 Mass communication is defined as such due more to its accessibility to a 'plurality of receptors' than to the quantity of people that conform its audience. Thompson, *The Media and Modernity*, p. 24.

After almost four decades of dictatorship, during the last 25 years of the century Spain experienced a process of profound social and political transformation. The so-called 'transition to democracy' began with the death of the dictator Francisco Franco, in November 1975. In the early years of this process, political parties were legalized, as were divorce, contraception and public demonstrations. A new Constitution was written and Spaniards voted for the first time in more than 40 years. In a very short time, Spain passed from intense political activism to a pleasant environment of consolidated institutions; it joined the European Community (Union), and experienced impressive economic growth.

During these decades, the emergence of a new 'public opinion' was one of the most important phenomena in Spain. Franco's death opened the door to a flood of fresh air in the media, which was now completely free from censorship. In this context, the emergence of the newspaper *El País* was quite an event in the Spanish public sphere. First published in May 1976, less than a year after Franco's death, it soon became the most read, and the most influential, newspaper in the country.[10]

In the general Spanish media, *El País* rapidly established itself on the left wing of the political spectrum. Although it was generally open to the expression of the most varied opinions by its collaborators, the newspaper had a very fluent relationship with the Socialist Party that gained power in 1982. The alliance was even more noticeable after the triumph of the right-wing Popular Party that won the general elections in 1996.[11]

In the years following the end of the Spanish dictatorship, *El País* played a crucial role 'in the formation of a democratic public opinion and in the introduction to social discourse of a whole set of debates about subjects that, until that time, had not had a place in public discourse'.[12]

Science was certainly among those subjects. During the 1980s, politicians began to conceive science as being an impulse for the general development of the country,[13] and the amount of scientific news in the pages of *El País* increased several times over. In 1985, *El País* launched a weekly supplement, *Futuro*, devoted entirely to the subject. The assignment of a predetermined space for scientific news also led to the progressive specialization of certain journalists and the establishment of collaboration agreements with different US newspapers and with *Nature* magazine's news service.

Scientific coverage within *El País* was in the hands of a relatively small group of journalists. A constant presence since the mid-1980s was Malen Ruiz de Elvira,

---

10 *El País* became the head of an extremely powerful entrepreneurial group with interests in broadcasting and other editorial sectors. María Cruz Seoane and Susana Sueiro, *Una Historia de* El País *y del Grupo Prisa. De una aventura incierta a una gran industria cultural* (Barcelona, 2004).

11 Ibid.

12 Gérard Imbert, *Le discours du journal* El País (Paris, 1988), p. 5.

13 Xavier Roqué, 'Premsa i cultura de la Ciència a Catalunya', in: *Actes de les III Trobades d'Història de la Ciència i de la Tècnica* (Barcelona, 1995), pp. 47–60, p. 53.

trained both in engineering and journalism, and general scientific editor of the newspaper from 1986. In 1990, Ruiz de Elvira published a paper in which she stated how she thought journalists should cover scientific news better. Reporters, she stated, should not perform scientific 'vulgarization', but rather provide 'journalistic information about scientific subjects'. This idea entailed covering not only scientific discoveries, but also frauds, disputes or the use of scientific arguments for commercial or political purposes. Considering that scientists often saw the media as a means of 'personal promotion', Ruiz de Elvira insisted on the need to maintain journalists' autonomy in the definition of the informative hierarchies. She was also in favour of allowing a plurality of actors to express their opinions about science in the pages of the newspaper.[14]

Ruiz de Elvira's editorial points of view contrasted, however, with those of other scientific journalists who joined the editorial team in the 1990s. The most relevant case is that of Javier Sampedro, who covered a large percentage of the genetic stories after he was hired in 1995. Sampedro was not only a major admirer of the Anglo-Saxon way of popularizing science, traditionally hagiographical and committed to the promotion of science, but was also very close to the scientific community, since before becoming a journalist he was himself a molecular biologist specializing in *Drosophila* genetics.

Science not only interested journalists, but the readership as well.[15] With the passing of time, scientific subjects in general, and human genetics in particular, gained visibility and fostered an increasing number of opinion columns, editorials and letters to the editor. Public awareness of the importance of science ran parallel with the new political impulse it received through the launch of the first National Plans of Research and Development during the eighties,[16] and the creation of new research infrastructures. Many other Spanish newspapers also launched regular specialized scientific pages, while in the following decade research institutions included the distribution of their results among their political aims.[17]

---

14 Malen Ruiz de Elvira, 'Las fuentes de la noticia en ciencia', *Arbor*, 534 (1990): 93–102.

15 In 1995, the ombudsman of *El País*, Juan Arias, stated that over the last year he had received more complaints related to social and scientific issues than to political or cultural information. Juan Arias, 'Un año después', *El País*, 19 November 1995. Page numbers of the print edition in which articles originally appeared are not specified because texts cited in this chapter were taken from *El País*'s Internet archive at <http://www.elpais.com/archivo/buscador.html>. This tool allows keyword searching on everything published in the newspaper since 1976, but it retrieves articles in a simple text format without the original page design.

16 Luis Sanz-Menéndez, 'Policy Choices, Institutional Constraints and Policy Learning: The Spanish Science and Technology Policy in the Eighties', *International Journal of Technology Management*, 10 (1995): 622–41.

17 Carlos Elías, 'Estudio cuantitativo de las fuentes en el periodismo español especializado en ciencia', *Revista Latina de Comunicación Social*, 38 (2001) <(http://www.ull.es/publicaciones/latina/2001/latina38feb/122elias3.htm>.

On the general Spanish scientific scene, biology was particularly strong, thanks to the relatively successful development of biochemistry and molecular biology during Franco's regime, 'compared to many other scientific disciplines which remained underdeveloped in Spain'.[18] The consolidation of both disciplines was favoured by the fact that the Spanish biochemist Severo Ochoa won the Nobel Prize for Medicine in 1959. Many young scientists studied in Ochoa's laboratory in the United States and returned later to Spain, while Ochoa's prestige prompted the Spanish scientific authorities' interest in these branches of research.[19]

In this context, Spanish human genetics developed irregularly over the last three decades of the twentieth century. While the study of the genetic bases of different diseases and the development of genetic diagnoses expanded fairly successfully,[20] there was almost no local research that dealt with the mapping and sequencing of the human genome.[21] As we will see later, the diverging degree of development of the different areas of human genetics led *El País* journalists to approach them in completely different ways.

Genes, Diagnoses and Diseases[22]

Being in charge of some of the most successful research programmes in Spain related to human genetics, scientists working on the genetic aetiology of different diseases were responsible for locally spreading the very idea of a disease defined in terms of the DNA of the people carrying it.[23] However, this was not a popular perspective during the first years of *El País*'s existence. In the late seventies and early eighties, most of the reports that mentioned the relationship between genes and diseases depicted it as ambiguous, remote and having no traces of determinism;

18 María Jesús Santesmases and Emilio Muñoz, 'Scientific Organizations in Spain (1950–1970): Social Isolation and International Legitimation of Biochemists and Molecular Biologists in the Periphery', *Social Studies of Science*, 27 (1997): 187–219, p. 211.

19 María Jesús Santesmases, 'Severo Ochoa and the biomedical sciences in Spain under Franco, 1959–1975', *Isis*, 91 (2000): 706–34.

20 Emilio Muñoz, *Imágenes Actuales de la Ciencia y la Tecnología Españolas* (Madrid, 2002).

21 With the exception of two experts on bioinformatics who participated in the visualization of the human genome sequence of Celera Genomics, no Spanish name could be found in the papers on the sequence of the human genome published in *Nature* and *Science* in February 2001.

22 A detailed revision of *El País*'s coverage of the genetic aetiology of diseases can be found in Matiana González-Silva, 'Del Factor Sociológico al Factor Genético. Genes y Enfermedad en las Páginas de *El País* (1976–2002)', *Dynamis*, 25 (2005): 487–512.

23 Mariano Barbacid, one of the international leaders in the study of oncogenes and the development of oncochips, returned to Spain in 1998 after having worked for 24 years in the United States. Other research groups developed tests for Alzheimer's disease and the genetically determined excess of cholesterol in blood.

local scientists used to stress instead the social and economic influences on human diseases, conceiving genetics as merely a 'milieu' or general constitution that conferred 'certain predispositions' to certain pathologies.

From a journalistic point of view, the young newspaper did not report on particular discoveries or scientific breakthroughs. Instead, it published long stories showing the different approaches to the same problem that coexisted on the scientific scene, indicating the cultural determinants of science and their potential consequences for the broader social sphere. According to this general trend, specific reports on human genetics were very rare; this science was only seldom mentioned in its own right, and mainly appeared in reports as one example among others of the kind of science experts were performing at the time. In the context of the social and political turmoil that characterized the Spanish 'transition to democracy', reports used to discuss the implications that adopting a determined scientific explanation would have on different issues considered urgent at the time.[24] Statements about the ideologies underlying certain scientific positions were therefore common, usually linking 'biologicist' explanations of behaviour and diseases with extreme right political stances, and rejecting them as scientifically poor and socially harmful.

This situation lasted until the early eighties, when the discovery of oncogenes by two Spanish geneticists settled in the United States prompted a completely new approach to human genetics by *El País* journalists: Reports on this breakthrough were the first to make a priority of technical details, with statements such as 'oncogenes, in their non activated form of proto-oncogenes, can be found in every eukaryotic live organism'. At the same time, promises for future therapeutic benefits began to appear. It being the first time that two national scientists had taken centre stage in international scientific news, the discovery also generated the first public controversies among local scientists about the real value of a genetic discovery.[25]

During the eighties, the depiction of the role of genes in human disease gradually shifted from that of a general dysfunction of the cell to a much more direct link. Most of these reports discussed foreign research and were written outside the walls of *El País*. But local reports maintained a particular style: The few Spanish geneticists interviewed in the newspaper insistently spoke about the complexities of the organic process that occurred between the genetic condition and the illness. When reporting on local research, *El País* journalists devoted a

---

24 In 1978, the elaboration of a national plan for the prevention of 'subnormality' prompted many discussions about the origin of such a condition. During these controversies, genetics was sometimes mentioned; but the real concerns were the perinatal cases of mental retardation and the social conditions in which affected individuals lived. Other areas that led to mentions of genetics were education and health policies.

25 'El País. Descubrimiento de españoles relacionado con el mecanismo original del cáncer', *El País*, 17 July 1982. 'Tribuna: Ysidro Valladares. Los oncogenes y la investigación actual sobre el cáncer', *El País*, 25 July 1982.

lot of space to recounting the path that had led to a certain discovery (hypothesis, experiments, interpretation of the data and so on), among other aspects that together fostered the image of science as a complex activity in which subjectivity was also important. The eighties can be considered a kind of 'transition period'. Despite the particularities found in news written by journalists within *El País*, genetic research related to diseases began to be framed in an increasingly more technical way. In accordance with the general context of scientific and technological promotion in which science gained importance as a newsworthy event, the public and political implications of science which had been common in the early years of the newspaper were set aside.

This situation was consolidated during the nineties, which could be considered the 'glory years' of discourses that inevitably linked genes with particular diseases. Reports on the latest discovery, identification or localization of the widest variety of genes followed on from each other at a frenetic pace, accompanied by promises of the new diagnoses and treatments that would necessarily follow. At first, *El País* reported on this branch of research as if Spanish scientists were not actively participating in it. It was not until 1994 that the newspaper first reported on a local research study related to the genetic aetiology of diseases. The very title – 'A team of scientists in Barcelona identifies a gene related to a hereditary disease'[26] – showed that the fact that Spaniards had finally entered the international trend of identifying genes was perceived as more relevant than the actual results of the research.

The arrival of Spanish scientists on the international human genetics scene led to the transformation of their discourses, which became virtually indistinguishable from those of their international colleagues. Within the *El País* editorial department, this coincided with a drastic change in the way they reported on human genetics. Long stories, which in the previous years had allowed journalists to explain complex situations and varied approaches to the same problem, were replaced by ever shorter texts that left no room for anything other than the bare data provided by the scientists, whose claims and promises were accepted without filters or questioning.

In those years, *El País* even reported on the most insignificant of results related to the genetic aetiology of diseases. As it did so, it completely ignored the process of creating scientific knowledge that, without hypotheses or experiments, only seemed to be determined by the truth. The reiterate use of short texts also contributed to the fading out of controversies, portraying the image of science as a whole as being unique and unquestionable, a neutral endeavour whose achievements the newspaper could only reflect, as clearly noticeable in a 1994 text published in the 'news in brief' format, announcing the discovery of a gene related to breast cancer:

---

[26] Sebastián Serrano, 'Un equipo científico de Barcelona identifica un gen relacionado con un mal hereditario', *El País*, 1 April 1994.

> A group of US scientists has isolated the gene responsible for a hereditary form of breast cancer, which means an enormous advance in the study of the prevention and treatment of this deadly disease. For four years, scientists all over the world have known about the existence of the gene, called BRCA1, and have been trying to isolate it. After this discovery, a blood analysis for detecting its presence may be developed.[27]

The implementation of genetic diagnosis programmes in adults offers the best example of how the appearance and consolidation of a local community of geneticists contributed to modifying *El País*'s coverage of science. Early reports on genetic diagnoses available in foreign countries depicted them as being a highly conflictive tool that posed severe ethical problems, causing as they did strong psychological stress to patients diagnosed with diseases for which there were no treatments. But as soon as these genetic tests became a local reality, in the second half of the nineties, *El País*'s attitude was completely uncritical when it reported on the statements made by local scientists. Reports stressed the optimistic perspectives the tests had allegedly opened up, particularly the promises that people diagnosed as carriers of a mutated gene would be the first to receive treatments for their illnesses (once these treatments had become available, of course).[28]

Geneticists involved in the launch of genetic test programmes began a campaign to gain legitimacy for the new technologies, on the grounds of the causal link between genes and diseases. Compared to other Spanish geneticists, those related to genetic diagnoses offered the most geneticized discourses of those found on the pages of *El País*, stating for instance that some genetic mutations conferred 'between 70 and 100 per cent of probabilities' of developing cancer,[29] or even that 30 per cent of the cases of Alzheimer's disease thought to be sporadic were due to the combination of the father's and the mother's genes.[30]

The extreme genetization found in many reports on new diagnosis tests generated some of the greatest controversies regarding human genetics that ever took place within the pages of *El País*. In each case, the newspaper supported geneticists against their detractors, not only in its tone and the space devoted to

27 'Reuters. Científicos aíslan el gen causante del cáncer de mama', *El País*, 14 September 1994.

28 Other promises included periodical checkups for carriers of determined oncogenes. Mayka Sánchez, 'Un hospital español estudia la detección precoz de un tipo de cáncer de colon', *El País*, 22 May 1995. Mayka Sánchez, 'Diagnóstico precoz de la mala absorción hereditaria del hierro', *El País*, 5 May 1997.

29 Oriol Güell, 'Barbacid inicia en Madrid un plan para detectar el cáncer genético', *El País*, 3 December 2001.

30 The main promoter of the genetic idea of Alzheimer's disease was Ramon Cacabelos, director of a private institute devoted to the research of this illness, and later promoter of an international programme for its 'genetic prevention'. Javier Sampedro, '4.000 familias españolas con casos de Alzheimer tienen ya su "ficha" genética', *El País*, 2 July 1997.

them, but also overtly in some editorials. When reporting on the new tools or programmes, journalists obscured every disagreement regarding their reliability and pertinence. They also avoided any mention of the potential problems that the private funding of research centres could raise – quite the opposite of what had occurred in the preceding years when mentioning the origin of the funding of some studies somehow introduced an element of suspicion to the reliability of the scientific results.[31]

Compared to the early period of the newspaper, limits were also put on the people allowed to express their opinions on controversial issues. *El País* only gave a voice to experts within the field, directing every argument towards scientific issues (the origin of Alzheimer's disease, the predictive capacity of oncochips and so on), and leaving aside both the ideological determinants of science and its impact on other areas of social life. When a controversy arose, critics of the genetic approach to diseases were relegated to the marginal platform of letters to the editor. *El País* even published the telephone numbers of the centres where potential patients could get tests performed, and supported the direct causal link between possessing certain genes and the risk of becoming ill, as stated in the 2001 title of a report dealing with the launch of a new diagnosis tool for familiar hypercholesterolemia: '80,000 Spaniards are at a high risk of precocious heart attack due to a defective gene'.[32]

In short, when reporting on research related to the genetic aetiology of diseases and their prediction through genetic diagnoses during the nineties, the scientific journalism of *El País* was very close to what Hilgartner has called the 'dominant model of scientific popularization',[33] a model traditionally linked to scientific promotion, which emphasizes technical details and obscures the network of relationships in which science is embedded.

Covering the Human Genome Project from the Periphery

During the years covered by this analysis, *El País* not only reported on the discovery of certain genes linked to diseases, or on the way to detect them. Those were precisely the years in which the mapping and sequencing of the human genome boomed, and even though hardly any Spanish geneticists were involved in this branch of genetics, it received extensive attention from *El País* journalists.

[31] This was the case, for instance, with a study funded by the leading Spanish tobacco company that suggested genes were almost entirely responsibility for lung cancer. Alina Quevedo. 'Tabacalera financia la investigación de mutaciones genéticas inducidas por el humo del tabaco', *El País*, 3 October 1990.

[32] Javier Sampedro, '80.000 españoles tienen un alto riesgo de infarto precoz debido a un gen defectuoso', *El País*, 2 July 2001.

[33] Stephen Hilgartner, 'The Dominant View of Popularization: Conceptual Problems, Political Uses', *Social Studies of Science*, 20 (1990): 519–39, p. 519.

The first reports on the Human Genome Project (HGP) appeared in the late eighties. They discussed a new plan structured in the United States, which seemed neither necessary nor revolutionary. In the early period of the initiative, *El País* gave a voice to scientists who were highly critical of it through concern that funding would be diverted from other branches of biology that they considered much more interesting.[34] This led to the introduction of scientific priorities in reports on the HGP from the very beginning of its journalistic visibility. From then on, *El País* continued to focus more on issues related to the economics, geography and politics of the project, rather than on its technical aspects, its promises and results.

The 'sociological approach' to the HGP was very well outlined as early as 1988, on the occasion of an international conference on the human genome held in Valencia, Spain. The newspaper's science editor, Malen Ruiz de Elvira, was sent to the conference as a special correspondent, where she stressed the lack of agreement among scientists regarding a possible moratorium for experiments implying the genetic manipulation of human gametes and embryos, and the 'geopolitics' of human genetics. She also published a short text entitled 'Image operation', where she stressed how hard it was to understand why such a conference was being held in Spain, where research on the human genome was completely irrelevant. In every case, the 'scientific' details of the HGP, as well as the promises of the medical benefits it would bring, were pushed into the background.[35] These were precisely the pillars of the promotion of the project in the countries that were actively participating in it.[36] *El País* promoted instead discussions of the ethical, legal and political issues surrounding the HGP; it did so in the conviction that, as was once stated in an editorial, 'science is no more neutral than any other human activity'.[37]

Throughout the nineties, *El País* acted as an important public forum in which philosophers, scientists, medical doctors and regular columnists expressed their points of view regarding the study of the human genome. Most of these columns were critical of the HGP, particularly regarding the identification of the human genome with 'human nature', one of the subjects that Spaniards dealt with most frequently. Conversely to discussions in countries that actively participated in the HGP, where the main concerns were related to the way in which the project would affect the everyday life of scientists and common citizens,[38] public discussions in

---

34 Gonzalo Casino and François Jacob, 'Para curar el SIDA y el cáncer hay que recurrir a la clonación', *El País*, 12 November 1987.

35 M.R.E. 'EE UU cooperará con otros países en el proyecto del genoma', *El País*, 25 October 1988. Malen Ruiz de Elvira, 'Los científicos reunidos en Valencia rehúsan poner límites a la manipulación genética', *El País*, 27 October 1988; 'Una operación de imagen', ibid.

36 José Van Dijck, *Imagenation: Popular Images of Genetics* (New York, 1998).

37 'Editorial. Ciencia y ética', *El País*, 1 November 1988.

38 In the United States reports focused on pragmatic issues, such as the confidentiality of genetic information, genetic discrimination and scientific secrecy. Van Dijck, *Imagenation*.

Spain took a very similar path to philosophical anthropology, as if the practical consequences of the HGP had nothing to do with the local reality.[39] According to that tendency, *El País* housed reflections such as that of philosopher Fernando Savater, who explored the links between genomics and potential manipulation and discrimination. According to Savater, separating science from its application was a 'reasonable way of saying nothing, as irrefutable as stating that ballistic calculus has never killed anyone'.[40]

Particularities of *El País*'s approach to the HGP can be appreciated in reports dealing with the patenting of human genes, a controversy that first arose in the United States in the early nineties. At that time, *El País* published texts taken directly from *The New York Times*, *Los Angeles Times* and *The Washington Post*, which situated the main consequences of the measure in the inner organization of science, particularly in how it could affect cooperation among scientists.[41] But the newspaper also interviewed local scientists, whose concerns proved to be completely different. As opposed to their US counterparts, Spanish geneticists focused on more abstract aspects, such as the legitimacy of patenting genes, given that genes were not created by man, but existed naturally, and because it was against the 'philosophy of scientific progress'.[42] In any case, the issue was presented in *El País* as a matter that only concerned scientists in the United States, and one which was not that important as it did not pose any 'new moral problems',[43] apparently the main concern regarding the HGP in Spain.

Stating that patents were a commercial problem but not a scientific one could be considered a way of supporting the HGP; the same can be said of the reiterate separation of basic knowledge from its practical applications, which are both arguments that were often used in editorials commenting on the advancement of the project and its potential dangers (privatization of knowledge, genetic manipulation, eugenics and so on). At every crucial moment, editorials defended the pertinence and importance of the HGP. What should be pointed out is that, as opposed to what happened in other branches of human genetics, sympathy for the HGP did not prevent *El País* from showing the complexities of genomics, or even from publishing some fierce criticisms directed at its very scientific core.[44] Awareness

---

39 See, among others: 'Tribuna: Jesús Mosterín. La lectura del genoma', *El País*, 8 January 1997. 'Tribuna: Vicente Verdú. Felicidad genital', *El País*, 20 February 1997; 'Tribuna: Manuel Vicent. Sueños', *El País*, 28 November 1999.

40 'Tribuna: Fernando Savater. Problemas clónicos', *El País*, 12 December 1993.

41 See, among others: NYT. 'EE UU quiere patentar genes humanos antes de saber para qué sirven', *El País*, 23 October 1991.

42 Alicia Rivera, 'La patente de genes humanos indigna a los científicos', *El País*, 24 October 1991.

43 'Editorial. Alarmismo ético', *El País*, 5 November 1991.

44 Many of the critical voices came from France, a significant scientific reference for *El País* until the mid-nineties. Sebastián Serrano, '"El conocimiento completo del genoma humano no es urgente", dice el Nobel François Jacob', *El País*, 14 July 1993.

both of the plurality of opinions regarding the relevance of the HGP, and of the uncertainties scientists were facing, in turn offered journalists the chance to show the social determinants of scientific research. It was therefore very common to find references to methodology, presuppositions or the differences between the French and the US 'scientific cultures' regarding the management of scientific results, the funding of research and the role of the public authorities in the promotion and regulation of science. Reports always tried to contextualize the state of the question; for instance, the 1998 text that stated that

> the human genome is far from being heritage of Humanity as stated by the Universal Declaration adopted last Thursday by United Nations. Genes engraved with our ways of living and getting ill have owners, public and private companies embarked on a meteoric race to find and patent the potentially curative pieces of DNA.[45]

The first specific results achieved by the HGP in the late nineties fostered a boom in reports on the initiative. These ran in parallel with an increasing percentage of texts taken directly from US newspapers, which in turn caused a growing technification and medicalization of genetic discourses published in *El País*. Over those years, the newspaper echoed the US geneticists' optimistic promises, and aligned itself with the journalistic agenda of that country. Pragmatic issues such as the availability of scientific results gained prominence in public debates, to the detriment of ethical and philosophical considerations.[46] Controversies among scientists were also obscured: for instance, when the European Union authorized the patenting of genes with a medical application, it seemed there was a broad consensus on the issue, with only the exception of a few organizations depicted as radical, Greenpeace among them.[47]

These were ambiguous years, in which highly technical and laudatory texts coexisted alongside other more moderate ones, mainly written from within the walls of the newspaper. Local journalists continued to explain the economics and politics of human genetics, hoping to help the readers understand what was really going on in an extremely new and confusing realm. They talked about the technological innovations that were being used in 'new biology', outlined the theoretical presuppositions of the genetic approach to biomedicine and reintroduced some sociological aspects of science, such as the variety of actors and interests that were changing the general organization of scientific research. On a local level, the benefits of the HGP were situated not only in the realm of medicine, but also

---

45 Lucía Argos, 'Las multinacionales se adueñan de los genes', *El País*, 13 December 1998.

46 Rachel Smyly, 'El Proyecto Genoma logra un hito al descifrar el primer cromosoma humano', *El País*, 2 December 1999.

47 Amaya Iríbar, 'Europa permite patentar material genético de origen humano', *El País*, 2 September 1999.

in the understanding of human evolution or human migrations, offering an image of human genetics as the source of basic knowledge, rather than merely a search for medical applications.

After some years in which foreign news services led *El País*'s reports on the HGP, during the final phase of the project Spanish correspondents in the USA got involved in the regular coverage of this scientific initiative. The newspaper once again took over the establishment of its own journalistic priorities and consolidated its own style, with the publishing of long, independent stories that always explained something other than the mere statements made by scientists: where the commercial interest in the human genome was leading, what kind of political negotiations were being held before the publication of its sequence, how knowledge was being managed or the geneticists' concerns about the damage that the human genome race could cause to the public image of their discipline. Despite supporting scientists and praising their achievements in sometimes extremely laudatory terms, *El País* seems to have been convinced that the technical aspects of the HGP were insufficient for proper coverage of such a major new endeavour. As they once stated in an editorial, progress in genetic knowledge demanded 'a parallel progress in knowledge and social abilities, which would allow us to assimilate it in a natural manner'.[48]

After the publication of the human genome sequences, in 2001, some interviews went so far as to question the scientific value of the human genome sequence due to the differences between private and public results; while others threw light on the geneticists' promotional strategies, introducing suspicion about the reliability of their statements due to the increasingly financial interests of researchers. The sociological framing of genomic news was also reinforced; it was particularly noticeable in reports on scientific conferences or celebrations, occasions that offered *El País* journalists the opportunity to make direct contact with the main international figures in the field of genomics. Instead of focusing reports on technical achievements and on the medical promises that were endlessly repeated by the leaders of the HGP, journalists stressed lobbying, alliances and even gossip among scientists, in texts that more resembled a parliamentary chronicle than a traditional scientific story focused on scientific truths.

The chronicle on the presentation of the sequences of the human genome is a good example of this particular journalistic approach:

> The truce agreed last June between public and private researchers allowed for the common presentation of the human genome map to be made in Washington yesterday. But the truce was so fragile that it exploded during the event and in the presentations held by the international organization in Paris, Berlin, London and Tokyo.

---

[48] 'Editorial. Cromosoma 22', *El País*, 6 December 1999.

So stated a report that also detailed how, when scientists from both teams exchanged business cards, 'it was clear that they did not know each other'.[49]

The end of the sequence of the human genome also fostered a boom in columns commenting on more general issues, such as genetic determinism, the comparison between the human essence and the human genome or the way in which a geneticized vision of the human being could affect moral responsibility or the concept of individuality. Once again, the social, political and economic approach to the HGP adopted by *El País* journalists does not mean that it did not echo the rhetorical statements of the genome leaders. After the fulfilment of the project, the newspaper not only gave a voice to the main actors involved in the sequencing, but also gave a privileged position in terms of space and visibility to local columnists that praised the scientific achievement. Some local reporters published some extremely geneticized texts, situating the future of all biology in the knowledge of genes. But despite this resemblance to the main international trends, *El País* also included other aspects of scientific research which as a whole offered a picture of genomics as being a much more complex endeavour than the mere publication of a sequence of nucleotides composing human DNA.

In the months that followed the fulfilment of the HGP, local scientists tried to take advantage of the prevailing genomic enthusiasm for obtaining new funds for this field of research. This initiative had a significant influence on the editorial policies of *El País*. On the one hand, the campaign obliged the newspaper to incorporate scientific management and local scientific policies in its journalistic agenda. On the other, it prompted a much more promotional style when reporting on local genomic research.[50]

Conclusions

This very general depiction of the way in which *El País* reported on human genetics shows that the newspaper's model of scientific journalism varied greatly depending on which branch of this science was being dealt with. In reports on the genetic aetiology of diseases and genetic diagnosis tools, journalists adopted a technical and uncritical approach, merging into the prevailing international journalistic trends of the last quarter of the twentieth century. However, reports regarding the sequencing and mapping of the human genome reported not only on scientific discoveries, but also on many other aspects of scientific life. The differences are particularly striking given that one of these two modes of

---

[49] J. del Pino and M. Ruiz de Elvira, 'Los científicos piden acceso libre al genoma', *El País*, 13 February 2001.

[50] Matiana González-Silva, 'Funding Through the Press: Genomics as a New Political Issue in the Pages of *El País*', in Néstor Herran, Josep Simón, Ximo Guillem-Llobat, Tayra Lanuza Navarro, Pedro Ruíz Castell and Jaume Navarro (eds), *Synergia: Jóvenes Investigadores en Historia de la Ciencia* (Madrid, 2008).

reporting on science (the one used to inform about the HGP) matches very well with what was considered 'good scientific journalism' by *El País*'s scientific editor, Ruiz de Elvira – something that cannot be said of reports on the link between genes and diseases.

We are therefore dealing with the same newspaper, the same journalists, the same 'ideal' model of scientific journalism and the same time span. Can we also say, however, that it was exactly the same context? This leads us to the main hypothesis of this chapter: that the presence or absence of a local community of researchers was crucial in the definition of the journalistic approach to science by *El País*, and one of the key explanations for the differences found in the journalistic approach to these two branches of human genetics during the nineties.

The coverage of both scientific areas diverged more clearly during the last decade of the twentieth century; previous shifts in the newspaper's model of scientific journalism were common to every branch of human genetics, and thus accept common explanations.[51] Further hypothesis must be ventured, however, to explain why *El País*'s style of scientific journalism diverged so strongly during the nineties depending on the subject. More specifically, the question is what led *El País* journalists to drift away in certain cases from the programme stated by the scientific editor of the newspaper in 1990, and what led them to maintain it in certain other cases.

With regard to reports on the genetic aetiology of diseases, the Spanish context very much fulfils the main features associated with the so-called 'dominant view of popularization', namely, its role in the strengthening of the authority of both scientists and scientific journalists. The technification and medicalization of these stories in *El País* coincided on the one hand with the consolidation of a local community of geneticists devoted to this area of research, and on the other with the specialization of journalists in scientific issues. Geneticists working on the identification of certain human genes were among the few 'national heroes' of Spanish science in the nineties, and were treated as such by local journalists. Within *El País* it is significant that two of the journalists who dealt most with this area had emerged from the core of the genetics community or had been very close to it. They therefore shared the values and worldview of the scientists they interviewed, and were likely to promote it. The proximity of journalists to the local community of geneticists led to the merging of the news values of both communities, a circumstance that can be considered one of the causes of the generally positive and uncritical tone geneticists were treated with in the newspaper.

---

[51] For instance: the abandonment of the political arena in which the consequences of scientific knowledge were situated in the seventies can be attributed to the general climate of stability that Spain gained through the restoration of democracy. Meanwhile, the general shift towards more conservative positions both in the general Spanish population and within *El País* would be the origin of the oblivion of ideological criticisms of human genetics. González-Silva, 'Del Factor Sociológico al Factor Genético'.

Another factor that certainly contributed to the shaping of this model of scientific journalism was the growing influence of the United States. The leading geneticists who returned to Spain during the nineties brought back not only a scientific perspective, but also a culture of science promotion based on the promises of medical benefits and the stressing of the neutrality and independence of scientific research. Local research centres also became aware of the importance of public relations.[52] Altogether, it can be said that the Spanish scientific community exerted major pressure on *El País* to report on the branches they were working in, and in the most convenient way for them.

Things were different in other branches of human genetics in which this local community was non-existent. Although *El País* sometimes published some extremely technified texts regarding the study of the human genome, and even though it echoed the promises and rhetoric of the very powerful international community of geneticists, it never abandoned its intention of offering a more 'journalistic' than promotional approach to the initiative, trying to do more than merely transcribe the scientists' claims and results. In so doing, it followed the main lines stated by Malen Ruiz de Elvira in her 1990 paper.

It could be ventured that the emphasis on the sociological aspects of the HGP was due to the lack of specific results to report on for most of the time the project lasted. But this shortage of scientific data was not at all a local circumstance, and certainly did not lead to the same style of science journalism in the countries leading the mapping and sequencing of the human genome, where newspapers highlighted the promises of future treatments and diagnoses that would be derived from the knowledge of the human genome, and offered highly detailed descriptions of the technical aspects of the HGP.[53] As opposed to what happened in reports on the discovery of genes related to diseases, an area in which *El País*'s journalistic style during the nineties was almost indistinguishable from that adopted by anglophone newspapers, in everything regarding the HGP the Spanish newspaper maintained an approach defined in accordance with its own news values.

It seems that the absence of local scientists working on the mapping and sequencing of the human genome gave the newspaper the freedom to put into practice the kind of scientific journalism it thought was best. Deprived of the 'moral obligation' of promoting local researchers, and taking advantage of the

---

[52] This was the case, for instance, with the main Spanish scientific institution, the Consejo Superior de Investigaciones Científicas (CSIC), which in 1996 launched a communication campaign with weekly press conferences on research carried out in its centres. Carolina Moreno Castro, *La biotecnología en la prensa diaria (1988–1998): Análisis y tendencias*, PhD dissertation (Madrid, 2001). An analysis performed on six Spanish newspapers during the first semester of 1998 showed that more than the half of the scientific information proceeded from scientists working at the CSIC, although they represented only 4.25 per cent of scientists in the country. Elías, 'Estudio cuantitativo de las fuentes'.

[53] Van Dijck, *Imagenation.*

perspective given by distance, *El País* journalists maintained a more detached and critical attitude to the HGP than the one they showed towards other branches of human genetics. The shift towards a more positive tone of reports on genomics was noticed after 2001, coinciding with the local campaign to gain more support for this science, reinforcing the hypothesis that the presence of a local community of geneticists was a key influence that favoured the technical and hagiographical framing of human genetics in *El País*.

This hypothesis contradicts, however, the general assumption that popularizing science on the periphery normally leads to an oversimplified image of it, detaching it from its external links. In the case of reports on human genetics published in *El País*, the more peripheral the condition, the less uncritical scientific journalism became. This can be appreciated not only by comparing them to reports published in the scientific 'centres', but also to the same newspaper's texts dealing with different areas of human genetics with different degrees of local participation. Being on the periphery was therefore not the most determinant factor in the adoption of the 'dominant view of popularization' by *El País* journalists. It is best to seek the origin of such a model precisely in the existence of a local community of geneticists with enough power to exert their influence on the newspaper.

Spanish scientists certainly saw the media as a means of gaining legitimacy, and were ready to use the public arena in their own interest. But this does not mean *El País* shared the exact same objectives, or that other factors did not also contribute to shaping the model of scientific journalism adopted by this newspaper. In this sense, the intention of providing 'good journalism' must not be underestimated. Another influence was the image the editorial team had, in the sense that reporting on science contributed to the legitimating of media as a vehicle for the transmission of knowledge.[54] For a newspaper that aspired to be the main international reference for Spanish journalism,[55] the HGP was a very good opportunity to demonstrate its capacity to widely cover news generated far from its head office, and to do so in an original and independent way.

Other aspects could also have contributed to the sociological framing of the study of the human genome. The recurrent analysis of the ethical and philosophical consequences of the HGP was actually an approach that *El País* shared with other Spanish actors.[56] A plausible explanation is that this approach offered Spaniards an

---

[54] Milagros Pérez Oliva, 'Valor Añadido de la Comunicación Científica', *Quark. Ciencia, Medicina, Comunicación y Cultura*, 10 (1998): 58–69. Pérez Oliva was at that moment responsible for health news in *El País*.

[55] Seoane and Sueiro, *Una Historia de* El País.

[56] It is relevant to remember that the second international conference on the human genome, held in Valencia in 1990, was especially aimed at discussing ethical issues. Other similar conferences were held in Santander (1990), Madrid (1992) and Bilbao (1993). The University Chair on Law and the Human Genome at the University of Deusto also responds to similar interests.

alternative path into the international scientific public arena, given that they could not make major scientific announcements from Spain.

In brief, being on the scientific periphery during the last decades of the twentieth century did influence *El País*'s model of popularization, but not in the sense that might have been expected. In the area of research in which Spain was most peripheral – that is, the cartography and sequencing of the human genome – the lack of a local group of geneticists led scientific journalists to adopt a detached attitude to science and its practitioners; while the pressure of an increasingly powerful community of Spanish geneticists working on the identification of genes related to diseases, and the launch of genetic diagnosis, prompted a model of scientific journalism that fostered a more authoritative and non-controversial image of science.

Concluding Remarks

Faidra Papanelopoulou, Agustí Nieto-Galan and Enrique Perdiguero

In this book, we have dealt with science and technology communication processes, which for several reasons have occupied a double peripheral position in the historiography of our discipline. The introductory chapters of this volume showed that the 'popularization of science and technology' has been considered a marginal field for a long time; a field focusing on processes of the diffusion of scientific knowledge and a mere complement to studies of the production of 'science proper'. In addition, our focus on countries that have not played a leading role in the emergence of modern science and technology seems to extend this marginality geographically.

From the standpoint of this double 'periphery', how do the case studies that appear in this book add to the historiography of science popularization? Is there anything new or original to add to the general framework provided by Jonathan Topham and Paola Govoni in the opening chapters of this collective volume? In other words, is it worth doing new research and publishing new case studies of the history of the popularization of science, or in a broader sense, of the history of science communication in countries that have often played a secondary role on what concerns science and technology? Will we perhaps end up with an antiquarian collection of erudite stories in which obscure peripheral figures go to great efforts to diffuse 'good' scientific ideas from the centre to 'uneducated' publics in the periphery? Far from that narrow and limited approach, our intention from the outset was to extract new historiographical lessons from our case studies.

Perhaps one of the first obvious conclusions is that the history of the popularization of science and technology in the European periphery is a more promising academic enterprise than the history of the great luminaries. A broader approach to scientific culture in local contexts is perhaps more rewarding than a history based on the search for internationally acclaimed figures. Therefore, it implies a necessary recovery of an enormous heritage of as yet unknown primary sources which are stored in libraries and archives across the Continent. The primary sources used in this book are a good starting point in this direction: Italian books on popular science, Portuguese botanical poems, physics textbooks circulating around the Continent and the British Isles, scientific articles in the Spanish daily press, chronicles of the Urania shows in Budapest, reports and articles on exhibitions in Denmark, Belgian encyclopedias, medical journals written in Catalan, articles and books about thermodynamics in Spain, Swedish books on popular astronomy.

Although we have to admit that some of the main aspects of these communicative practices can also be identified in countries and in cultural contexts that we could consider to have been scientific centres during several periods of history, there are some aspects that are particularly associated with the peripheral factor. One of our first conclusions is that the case studies show the prominent role of minor scientific figures when we examine different levels and practices of popularization. The popularizers studied in this book would never appear in a *Dictionary of Scientific Biography*, and in some cases not even in national biographies and encyclopedias. Some of them were prestigious professors in their own countries, who for several reasons devoted a good part of their efforts to science popularization; but others (popularizers without a definite academic position) are almost unknown.

As Johan Kärnfelt has clearly shown in his paper on popular astronomy in Sweden, professional scientists in the periphery often obtained extra income by providing a useful complement to their academic duties through their books and articles on popular science. Despite the positivistic rhetoric of progress, the economic factor was often more crucial than any kind of ideal to understanding the aims and motives of local experts. Generally speaking, in the periphery scientists used to work in very close liaison with amateurs, and the boundaries between professionalisation and amateurism were considerably blurred for a good part of the period under analysis.

Weak institutionalization probably meant a lack of capacity to have a voice in scientific centres, but on the other hand contributed to very fluid communication processes in which knowledge circulated easily in society. Popular science books were usually successful publishing enterprises and were widely read in public libraries; science and technology related articles were numerous and frequent in the daily press; scientific lectures were very well attended in cultural societies, athenaeums and working-class organizations; the boundaries between research papers, university textbooks, popular books, exhibitions and scientific theatres were not particularly well defined. There was no clear distinction between popular and educational science books. In countries with a low level of science professionalization, both played a very important role in the living and in the making of scientific careers. This can be considered a disadvantage in terms of the potential of these peripheral societies for making an impact on scientific centres, but it seems to be an advantage in terms of the dynamism of their local scientific culture as a whole.

As Josep Simon has demonstrated in his chapter, Adolphe Ganot's books travelled from France to Britain, for the distinction between centres and peripheries cannot be seen strictly in geographical terms – there are of course also centres and peripheries inside the supposed scientific centres. They were then read, reviewed and translated, in short, appropriated by different audiences around Europe. Drawing ideas from the history of the book and the history of reading practices, Simon's chapter helps us to understand how a relatively unknown French author had a major impact on the culture of physics in a broad sense in nineteenth-century Europe. He demonstrates convincingly how readers appropriated Ganot's books in different and contradictory ways, and how the meanings and uses of the text changed over time.

The historiographical debate on the existence of centres and peripheries is a highly sophisticated and controversial one. However, by giving voice to peripheral actors through their works on science popularization we can identify the importance attributed to international scientific achievements in different local contexts and in different historical periods. As Palmira Fontes da Costa has shown through the Marquise of Alorna's *Botanical Recreations*, popular science authors and books in the periphery reflected perceptions of backwardness and isolation very well. The book was written in exile, seeking to diffuse the new Linnaean botany from a cosmopolitan perspective, but at the same time being highly sensitive to the scientific isolation of Portugal in the eighteenth century. Geert Vanpaemel and Brigitte Van Tiggelen also discuss how Belgian popularizers hesitated between the diffusion of a 'standard' image of 'universal science' and the need to make local scientific actors known to the general public in order to serve the nationalistic purposes of a small country. Both popularizing endeavours – the *Bibliothèque Nationale* and the *Encyclopédie populaire* – are excellent examples of how the construction of the identity of a nation might rely on the appeal to international luminaries of science, but also on its local heroes and their stories that inspire patriotic feelings.

As Stefan Pohl-Valero's chapter stresses, it was not only a question of using a specific peripheral language in the process of science communication in the periphery, it was rather a matter of 'translating' theories and ideas from the centre to allow them to properly fit within the power relations and cultural constraints of the local context. In his case, local thermodynamics bypassed the concept of 'scientific materialism' in textbooks and popular books and articles to avoid a conflict with the powerful Spanish Catholic Church. Since, in that context, local actors discussing thermodynamics came not only from the university but also from religious institutions, it is obvious that the science–religion intersection took on a specific shape in the public arena of late-nineteenth-century Spain.

Enrique Perdiguero, José Pardo Tomás and Àlvar Martínez-Vidal's chapter shows how Catalan doctors used the popularization of medicine for the public legitimization of new medical specialities in their local context, and the maintenance of a certain degree of integration in the field of medicine. The rhetoric behind the presentation of the *Monografies Mèdiques* as a work for the general public aimed to boost the prestige of Catalan physicians. They contributed to the formation and reinforcement of a Catalan scientific culture and national identity, and used the Catalan language in their scientific articles. Within a context of weak professionalization, the use of minority languages was part of the strategy experts used to legitimise their activities, but at the same time to link their local interests to a general account of progress and modernization in reference to names and standards from the main scientific centres. Popularized medicine and technology were important contributions to the making of the modern 'peripheral' nation.

Centre–periphery, local–universal seem to be constant tensions in the way particular local actors build their accounts. As Matiana González-Silva's chapter convincingly analyses, a distant scientific subject from the centre which involved no local scientists from the periphery was often presented to the public as a controversial issue. However, once local scientists formed a significant part of the project, appropriation filters strongly acted into the peripheral public sphere. Here, the dominant view of popularization and the corporative interests of professional scientists in the local context were of major importance in deciding what kind of subjects and with what kind of emphasis should be presented to the general public in the daily press. Styles of scientific journalism differed in local communities also in terms of their relation with a specific international community of experts.

Many chapters included in this volume have also shown how the circulation of science and technology from centres to peripheries across Europe was not only related to theories, experiments and objects, but also to a variety of practices and spaces of popularization. Gábor Palló has shown how Urania, the scientific theatre of Budapest, was a particular appropriation of the Berliner Urania, a model that seems to have spread successfully in Eastern Europe. But Urania was more than a theatre; it was a local scientific society that aimed to spread the cause of science to the general public, which ranged from an elite learned public to non-urban populations and also working-class audiences, as the papers in the *Journal of the Urania Hungarian Scientific Theatre* clearly show. The involvement of politicians, businessmen, educators and scientists in Urania is indicative of different agendas and interests in the enterprise of science popularization.

Rikke Schmidt Kjærgaard has shown in her chapter how local exhibitions became a form of standard commodity for the popularization of science and technology, following the huge success of the London exhibition of 1851. Nevertheless, when we look in depth at each exhibition we can identify some particularities in each local context. After reading this case study, it is easy to conclude that there was no universal model for the holding of exhibitions. In the Danish case, journeys to international exhibitions in the 'centres' – London, Paris – acted as a driving force for local actors to invest in the design of their own local public events. Exhibition sites – pavilions, museums, zoos, botanical gardens – were clear adaptations from the London and Paris patterns, though appropriated differently in Copenhagen and Aarhus.

Could we perhaps conclude that, in countries that were scientific centres during a specific historical period, practices of popularization – including exhibitions – were more numerous and frequent than in peripheral countries? Maybe the latter were driven by different agendas and different local demands, and were more oriented towards utilitarianism? What was different in London and Paris in relation to Copenhagen and Aarhus? These are crucial questions, which will only be fully answered with further comparative research on the history of science and technology exhibitions both in the centres and the peripheries.

Although from the late-eighteenth century onwards popularizing science and technology in the European periphery had its similarities to other popularization practices in different European scientific centres – Paris, London, Berlin – the case studies in this volume have shown that some particularities should not be excluded. It formed a highly substantial part of the activities and duties of local experts. It reinforced the rhetoric against scientific backwardness and the need to modernise the nation through science and technology. It tended towards an uncritical hagiography of the great foreign luminaries of science, but at the same time reinforced local heroes and local culture. Local scientific controversies were submitted to the public arena in frequent contention with the supposed authority of scientific centres.

The wide circulation of popular books, articles, lectures and exhibitions across the nations, regions and cities of the European periphery enriched a scientific culture that enjoyed notable independence from the great names of the 'universal science' of the centres. In the same way, boundaries between expert and lay knowledge, between professional and amateur science, were loose and not well defined.

In the end, perhaps the history of science and technology in the European periphery is mainly the history of the communication practices – teaching and popularization – of local experts, who appropriated the great names and ideas from the centres, communicated their knowledge through local publishers in their local contexts and constituted our main primary sources for science and technology. This book is only a first attempt to recover this enormous historical heritage, which still deserves further attention throughout Europe as a whole.

Selected Bibliography

In the following pages we provide a selected bibliography of secondary sources for the benefit of those who wish to undertake research in the field. The bibliography is of course not exhaustive, but we have attempted to include a vast variety of sources – mainly in English, but also covering some other European languages. They deal with the various practices of science communication examined in this book, and cover different historiographical approaches.

Abir-Am, Pnina, G., *La mise en mémoire de la science. Pour une ethnographie des rites commémoratifs* (Paris: EAC, 1998).

Agassi, Joseph, *Science and Culture* (Dordrecht and Boston: Kluwer, 2003).

Alberti, Samuel, 'Objects and the Museum', *Isis*, 96 (2005): 559–71.

Alkon, Paul K., *Science Fiction before 1900: Imagination Discovers Technology* (New York and Toronto: Maxwell Macmillan, 1994).

Altholz, Joseph L., *The Religious Press in Britain, 1760–1900* (New York and London: Greenwood Press, 1989).

Altick, Richard D., *The English Common Reader: A Social History of the Mass Reading Public, 1800–1900* (Chicago: University of Chicago Press, 1957).

Altick, Richard D., 'English Publishing and the Mass Audience in 1852', in *Writers, Readers, and Occasions: Selected Essays on Victorian Literature and Life* (Columbus, OH: Ohio State University Press, 1989).

Altimore, Michael, 'The Social Construction of a Scientific Controversy: Comments on Press Coverage of the Recombinant DNA Debate', *Science, Technology and Human Values*, 7 (1982): 24–31.

Andersen, Casper and Hjermitslev, Hans Henrik, 'Directing Public Interest. Danish Newspaper Science, 1900-1903', *Centaurus*, 52 (2009) [forthcoming].

Anderson, Robert G.W., 'What Is Technology? Education through Museums in the Mid-Nineteenth Century', *British Journal for the History of Science*, 25 (1992): 169–84.

Andrews, James T., *Science for the Masses: The Bolshevik State and the Popular Imagination in Soviet Russia, 1917–1934* (College Station, TX: Texas A&M University Press, 2003).

Auerbach, Jeffrey A., *The Great Exhibition of 1851: A Nation on Display* (New Haven, CT: Yale University Press, 1999).

Barbier, Frédéric, 'Les marchés étrangers de la librairie française', in Martin, Henri-Jean (ed.), *Histoire de L'Edition Française* (Paris: Promodis, 1985).

Barton, Ruth, 'An Influential Set of Chaps: The X-Club and Royal Society Politics, 1864–85', *British Journal for the History of Science*, 23 (1990): 53–81.

Barton, Ruth, 'Just before *Nature*: The Purposes of Science and the Purposes of Popularization in Some English Popular Science Journals of the 1860s', *Annals of Science*, 55 (1998): 1–33.

Bauer, Martin (ed.), *Resistance to New Technology: Nuclear Power, Information Technology and Biotechnology* (Cambridge: Cambridge University Press, 1995).

Bauer, Martin W. and Bucchi, Massimiano (eds), *Journalism, Science and Society: Science Communication between News and Public Relations* (London: Routledge, 2006).

Bauer, Martin W., Petkova, Kristina and Boyadjieva, Pepka, 'Long-Term Trends in the Public Representation of Science across the "Iron Curtain": 1945–1995', *Social Studies of Science*, 36 (2006): 99–131.

Bauer, Martin W. and Schoon, Ingrid, 'Mapping Variety in Public Understanding of Science', *Public Understanding of Science*, 2 (1993): 141–55.

Bayertz, Kurt, 'Spreading the Spirit of Science: Social Determinants of the Popularization of Science in the 19th Century Germany', in Shinn, Terry and Whitley, Richard (eds), *Expository Science: Forms and Functions of Popularisation* (Dordrecht: Reidel, 1985).

Bedini, Silvio A., 'The Evolution of Science Museums', *Technology and Culture*, 6 (1965): 1–29.

Beer, Gillian, 'Science and Literature', in Olby, Robert, Cantor, Geoffrey, Christie, John and Hodge, Jonathan (eds), *Companion of the History of Modern Science* (London: Routledge, 1990).

Béguet, Bruno, Cantor, Marilyn and Le Men, Ségolène, *La science pour tous (Exposition, Paris, Musée d'Orsay)* (Paris: Réunion des musées nationaux, 1994).

Béguet, Bruno (ed.), *La science pour tous: sur la vulgarisation scientifique en France de 1850 à 1914* (Paris: Conservatoire national des arts et métiers, 1990).

Behari Lal, Gobind, 'Popularization of Science through News', *Philosophy of Science*, 12 (1945): 41–4.

Belhoste, Bruno, Balpe, Claudette and Laporte, Thierry (eds), *Les sciences dans l'enseignement secondaire français. Textes officiels* (Paris: INRP, 1995).

Ben-David, Joseph, *The Scientist's Role in Society: A Comparative Study* (Chicago and London: Prentice-Hall, 1971)

Benjamin, Marina (ed.), *A Question of Identity: Women, Science and Literature* (New Brunswick: Rutgers University Press, 1993).

Bennett, Jim, 'Museums and the History of Science: Practitioner's Postscript', *Isis*, 96 (2005): 602–8.

Bennett, Scott, 'Revolutions in Thought: Serial Publication and the Mass Market for Reading', in Shattock, Joanne and Wolff, Michael (eds), *The Victorian Periodical Press: Samplings and Soundings* (Leicester and Toronto: Leicester University Press, Toronto University Press, 1982).

Bennett, Tony, *The Birth of the Museum: History, Theory*, Politics (London: Routledge, 1995).

Bensaude-Vincent, Bernadette, 'La science populaire d'un philosophe', *Corpus*, 4 (1987): 143–67.

Bensaude-Vincent, Bernadette, 'Hélène Metzger's *La chimie*: A Popular Treatise', *History of Science*, 25 (1987): 71–84.

Bensaude-Vincent, Bernadette, 'La science populaire, ancêtre ou rivale de la vulgarisation?', *Protée, théorie et pratiques sémiotiques*, 16 (1988): 85–91.

Bensaude-Vincent, Bernadette, 'Camille Flammarion: prestige de la science populaire', *Romantisme*, 65 (1989): 93–104.

Bensaude-Vincent, Bernadette, 'L'Astronomie Populaire de Comte: Priorité Philosophique et Projet Scientifique', *Revue de Synthèse*, 112 (1991): 49–60.

Bensaude-Vincent, Bernadette, 'Un public pour la science: l' essor de la vulgarisation au XIXe siècle', *Réseaux*, 58 (1993): 47–66.

Bensaude-Vincent, Bernadette, 'Les savants et les autres', *Diogène*, 169 (1995): 136–55.

Bensaude-Vincent, Bernadette, 'In the Name of Science', in Krige, John and Pestre, Dominique, *Science in the Twentieth Century* (Amsterdam: Harwoor Academic Publishers, 1997).

Bensaude-Vincent, Bernadette, *L'opinion publique et la science: à chacun son ignorance* (Paris : Institut d'éd. Sanofi-Synthélabo , 2000).

Bensaude-Vincent, Bernadette, 'A Genealogy of the Increasing Gap between Science and the Public', *Public Understanding of Science*, 10 (2001): 99–113.

Bensaude-Vincent, Bernadette, *La science contre l'opinion. Histoire d'un divorce* (Paris: Les empêcheurs de penser en rond, 2003).

Bensaude-Vincent, Bernadette, Bertomeu, José Ramón and García Belmar, Antonio, *L' emergence d'une science des manuels. Les livres de chimie en France (1782–1852)* (Paris: Éd. des archives contemporaines, 2003).

Bensaude-Vincent, Bernadette, and Blondel, Christine (eds), *Vulgariser les sciences, 1919–1939: acteurs, projets, enjeux, Cahiers d'Histoire et de Philosophie des Sciences* [special issue], 24 (1988).

Bensaude-Vincent, Bernadette and Blondel, Christine, *Des savants face à l'occulte, 1870–1940* (Paris: La Découverte, 2002).

Bensaude-Vincent, Bernadette and Blondel, Christine (eds), *Science and Spectacle in the European Enlightenment* (Aldershot: Ashgate, 2008). Bensaude-Vincent, Bernadette and Rasmussen, Anne (eds), *La science populaire dans la presse et l'édition:XIXe et XXe siècles* (Paris: CNRS, 1997).

Bensaude-Vincent, Bernadette (ed.), *Sciences pour tous, Romantisme* [special issue], 65 (1989).

Birley, Graham, *Barbarians at Play: A Case-Study of Examinations in the Physical Sciences in Nineteenth Century Britain* (Walsall: University of Wolverhampton, 1996).

Blair, Ann, Topham, Jonathan R. and Daston, Lorraine, 'Scientific reading', *Isis*, 95 (2004): 420–448.

Blondel, Christine, 'Industrial Science as a "Show": A Case-Study of Georges Claude', in Shinn, Terry and Whitley, Richard (eds), *Expository Science: Forms and Functions of Popularisation* (Dordrecht: Reidel, 1985).

Bloom, Lisa, 'Constructing Whiteness: Popular Science and "National Geographic" in the Age of Multiculturalism', *Configurations: a Journal of Literature, Science and Technology*, 2 (1994): 15–32.

Bodmer, Walter F (ed.), *The Public Understanding of Science* (London: Royal Society, 1985).

Bostetter, Mary, 'The Journalism of Thomas Wakley', in Wiener, Joel H., *Innovators and Preachers: The Role of the Editor in Victorian Britain* (Westport, CT and London: Greenwood Press, 1985).

Bourdelais, Patrice (ed.), *Les hygiénistes. enjeux, modèles et pratiques* (Paris: Belin, 2001).

Bourdier, Pierre, 'The Uses of the "People"', in Bourdier, Pierre, *In Other Words: Essays Towards a Reflexive Sociology*, trans. Matthew Adamson (Stanford, CA: Stanford University Press, 1990).

Bourdier, Pierre, 'Did you say "popular"?', in Thompson, John B (ed.), *Language and Symbolic Power*, trans. Gino Raymond and Matthew Adamson (Cambridge, MA: Harvard University Press, 1991).

Bowler, Peter, 'Experts and Publishers: Writing Popular Science in Early Twentieth-Century Britain, Writing Popular History of Science Now', *British Journal for the History of Science*, 39 (2006): 159–87.

Bowler, Peter and Morus, Iwan R., 'Popular Science', in *Making Modern Science: A Historical Survey* (Chicago and London: University of Chicago Press, 2005).

Brain, Robert, *Going to the Fair: Readings in the Culture of Nineteenth-Century Exhibitions* (Cambridge: The Whipple Museum of the History of Science, 1993).

Brock, William H., 'The Development of Commercial Science Journals in Victorian Britain', in Meadows, Arthur J., *Development of Science Publishing in Europe* (Amsterdam: Elsevier Science Publishers, 1980).

Brock, William H., *Science for All: Studies in the History of Victorian Science and Education* (Aldershot: Ashgate, 1996).

Brock, William H. and Macleod, Roy M., *Natural Knowledge in Social Context: The Journals of Thomas Archer Hirst FRS* (London: Mansell, 1980).

Brock, William H. and Meadows, Arthur J., *The Lamp of Learning: Two Centuries of Publishing at Taylor & Francis* (Bristol: Taylor & Francis, 1998).

Broks, Peter, 'Science, Media and Culture: British Magazines, 1890–1914', *Public Understanding of Science*, 2 (1993): 123–39.

Broks, Peter, *Media Science before the Great War* (Basingstoke: MacMillan, 1996).

Broks, Peter, *Understanding Popular Science* (Buckingham: Open University Press, 2006).

Broman, Thomas, 'The Habermasian Public Sphere and Science in the Enlightenment', *History of Science*, 26 (1998): 123–49.

Browne, Janet, 'Squibs and Snobs; Science in Humorous British Undergraduate Magazines around 1830', *History of Science*, 30 (1992): 165–97.

Bucchi, Massimiano, 'La scienza e i mass media: la fusione fredda nei quotidiani Italiani', *Nuncius,* 11 (1996): 581–611.

Bucchi, Massimiano, *Science and the Media: Alternative Routes in Scientific Communication* (London and New York: Routledge, 2004).

Bucchi, Massimiano, 'Communicating Science', in *Science in Society: An Introduction to Social Studies of Science* (London and New York: Routledge, 2004).

Bucchi, Massimiano, 'Can Genetics Help Us Rethink Communication? Public Communication of Science as "A Double Helix"', *New Genetics and Society*, 23 (2004): 269–83.

Bucchi, Massimiano and Mazzolini, Renato G., 'Big Science, Little News: Science Coverage in the Italian Daily Press, 1964–1997', *Public Understanding of Science*, 12 (2003): 7–24.

Bucchi, Massimiano and Lewenstein, Bruce V. (eds) *Handbook of Public Communication of Science and Technology* (London: Routledge, 2008).

Bud, Robert and Cozzens, Susan., Invisible *Connections: Instruments, Institutions and Science* (Washington: SPIE, 1992).

Burbick, Joan, *Healing the Republic: The Language of Health and the Culture of Nationalism in Nineteenth-Century America* (Cambridge: Cambridge University Press, 1994).

Burke, Peter, *Popular Culture in Early Modern Europe* (London: Harper Torchbook, 1978).

Burnham, John C., *How Superstition Won and Science Lost: Popularizing Science and Health in the United States* (New Brunswick: Rutgers University Press, 1987).

Bynum, William F. and Porter, Roy (eds), *Medical Fringe and Medical Orthodoxy, 1750–1850* (London: Routledge, 1987).

Calhoun, Craig (ed.), *Habermas and the Public Sphere* (Cambridge, MA: The MIT Press, 1992).

Callon, Michel (ed.), *La science et ses réseaux. Genèse et circulation des faits scientifiques* (Paris: La Découverte, 1989).

Callon, Michel, 'The Role of Lay People in the Production and Dissemination of Scientific Knowledge', *Science Technology Society*, 4 (1999): 81-94.Cannon, Susan F., *Science in Culture: The Early Victorian Period* (New York: Science History Publications, 1978).

Cantor, Geoffrey, Dawson, Gowan, Gooday, Graeme, Noakes, Richard, Shuttleworth, Sally and Topham, Jonathan R., *Science in the Nineteenth*

Century Periodical: Reading the Magazine of Nature (Cambridge: Cambridge University Press, 2004).
Cantor, Geoffrey and Shuttleworth, Sally (eds), *Science Serialized: Representations of the Sciences in Nineteenth-Century Periodicals* (Cambridge and New York: The MIT Press, 2004).
Cardot, Fabienne, 'La vogue de l'électricité et les amateurs de science et technique dans la seconde moitié du XIXe siècle', *Bulletin d'Histoire de l'Électricité*, 10 (1987): 31–47.
Cardot, Fabienne, 'Le théâtre scientifique de Louis Figuier', *Romantisme*, 65 (1989): 59–68.
Caron, Jean-Claude, *Générations romantiques: Les étudiants de Paris et le Quartier Latin, 1814–1851* (Paris: Armand Colin, 1991).
Cavallo, Guglielmo and Chartier, Roger (eds), *A History of Reading in the West* (Amherst: University of Massachusetts Press, 2003).
Chaperon, Danielle, *Camille Flammarion: Entre astronomie et littérature* (Paris: Imago, 1997).
Chapple, John A.V., *Science and Literature in the Nineteenth Century* (London: Macmillan, 1986).
Chartier, Roger, 'Culture as Appropriation: Popular Cultural Uses in Early Modern France', in Kaplan, Steven L., *Understanding Popular Culture: Europe from the Middle Ages to the Nineteenth Century* (Berlin: Walter de Gruyter & Co, 1984).
Chartier, Roger, *Lectures et lecteurs dans la France d'Ancien Régime* (Paris: Seuil, 1987).
Chartier, Roger, *The Cultural Uses of Print in Early Modern Europe* (Princeton: Princeton University Press, 1987).
Chartier, Roger, 'Texts, Printings, Readings', in Hunt, Lynn (ed.), *The New Cultural History* (Berkeley and London: University of California Press, 1989).
Chartier, Roger, *The Order of Books: Authors and Libraries in Europe between the 14th and 18th Centuries* (Cambridge: Polity Press, 1994).
Chartier, Roger, *Forms and Meanings: Texts, Performances and Audiences from Codex to Computer* (Philadelphia: University of Pennsylvania Press, 1995).
Chartier, Roger, Martin, Henri-Jean and Vivet, Jean-Pierre (eds), *Histoire de l'édition française* (Paris: Promodis, 1983).
Chartier, Roger and Corsi, Pietro (eds), *Sciences et langues en Europe* (Paris: EHESS, 1996).
Chesneaux, Jean, *Jules Vernes. Une lecture politique* (Paris: François Maspero, 1982).
Christidou, Vasilia, Dimopoulos Kostas and Koulaidis Vasilis, 'Constructing Social Representations of Science and Technology: The Role of Metaphors in the Press and the Popular Scientific Magazines', *Public Understanding of Science*, 13 (2004): 347–62.
Clemens, Elisabeth S., 'Of Asteroids and Dinosaurs: The Role of the Press in the Shaping of Scientific Debate', *Social Studies of Science*, 16 (1989): 421–56.

Cohen, I. Bernard, 'The Fear and Distrust of Science in Historical Perspective: Some First Thoughts', in Markovits, Andrei A. and Deutsch, Karl W., *Fear of Science – Trust in Science: Conditions for Change in the Climate of Opinion* (Cambridge, MA: Oelgeschlager, Gunn & Hain, 1980).

Cohen, Yves and Drouin, Jean-Marc (eds), *Les amateurs de sciences et de techniques, Cahiers d'Histoire et de Philosophie des Sciences* [special issue], 27 (1989).

Collins, Harry, 'Certainty and the Public Understanding of Science: Science on Television', *Social Studies of Science*, 17 (1987): 689–713.

Cook, Guy, Robbins, Peter T. and Pieri, Elisa, '"Words of Mass Destruction": British Newspaper Coverage of the Genetically Modified Food Debate, Expert and Non-Expert Reactions', *Public Understanding of Science*, 15 (2006): 5–29.

Conner, Clifford D., *A People's History of Science. Miners, Midwives and 'Low Mechanics'* (New York: Nation Books, 2005).

Cooter, Roger, *The Cultural Meaning of Popular Science: Phrenology and the Organization of Consent in Nineteenth Century Britain* (Cambridge: Cambridge University Press, 1984).

Cooter, Roger, *Studies in the History of Alternative Medicine* (London: Palgrave Macmillan, 1988).

Cooter, Roger and Pumfrey, Stephen, 'Separate Spheres and Public Places: Reflections on the History of Science Popularization and Science in Popular Culture', *History of Science*, 32 (1994): 237–67.

Cooter, Roger, *Phrenology in Europe and America* (London: Routledge, 2001).

Cotardière (de la), Philippe and Fuentes, Patrick, *Camille Flammarion* (Paris: Flammarion, 1994).

Cotkin, George, 'The Socialist Popularization of Science in America, 1901 to the First World War', *History of Education Quarterly*, 24 (1984): 201–14.

Craveri, Benedetta, *La cultura de la conversación* (Madrid: Siruela, 2003).

Crease, Robert P., *The Play of Nature: Experimentation as Performance* (Bloomington and Indianapolis: Indiana University Press, 1993).

Crosland, Maurice, 'Popular Science and the Arts; Challenges to Cultural Authority in France under the Second Empire', *The British Journal for the History of Science*, 34 (2000): 301–22.

Crowe, Michael J., *The Extraterrestrial Life Debate, 1750–1900* (Cambridge: Dover Publications, 1986).

Cuny, Hillaire, *Camille Flammarion et l'astronomie populaire* (Paris: Seghers, 1964).

Curtis, Ron, 'Narrative Form and Normative Force: Baconian Story-Telling in Popular Science', *Social Studies of Science*, 24 (1994): 419–61.

Darnton, Robert, *La fin des Lumières. Le mesmérisme et la Révolution* (Paris: Odile Jacob, 1968).

Darnton, Robert, 'What Is the History of Books?', *Daedalus*, summer (1982): 65–83.

Darnton, Robert, *The Kiss of Lamourette: Reflections in Cultural History* (New York: W. W. Norton & Company, 1990).

Darnton, Robert, 'History of Reading', in Burke, Peter (ed.), *New Perspectives on Historical Writing* (Oxford: Pennsylvania State University Press, 1991).

Darnton, Robert, 'An Early Information Society: News and the Media in 18th Century Paris', *American Historical Review*, 105 (2000): 1–35.

Daston, Lorraine, 'Taking Note(s)', *Isis*, 95 (2004): 443–8.

Daston, Lorraine, *Things that Talk: Object Lessons from Art and Science* (New York: Zone Books, 2004).

Daum, Andreas W., *Wissenschaftspopularisierung im 19 Jahrhundert. Bürgerliche Kultur, naturwissenschaftliche Bildung und die deutsche Öffentlichkeit, 1848–1914* (2nd edn, Munich: R. Oldenburg Verlag, 2002).

Desmond, Adrian, 'Artisan Resistance and Evolution in Britain, 1819–1848', *Osiris*, 2nd series, 3 (1987): 77–110.

Desmond, Adrian, *The Politics of Evolution: Morphology, and Reform in Radical London* (Chicago: University of Chicago Press, 1989).

Despy-Meyer, Andrée (ed.), *Institutions and Societies for Teaching, Research and Popularisation* (Turnhout: Brepols, 2002).

Dickson, David, 'Science and Its Public: The Need for A "Third Way"', *Social Studies of Science*, 30 (2000): 917–23.

Dimopoulos, Kostas and Koulaidis, Vasilis, 'The Socio-Economic Constitution of Science and Technology in the Greek Press: An Analysis of Its Presentation', *Public Understanding of Science* (2002): 225–41.

Dornan, Christopher, 'Some Problems in Contextualizing the Issue of Science in the Media', in Scanlon, Eileen, Whitelegg, Elizabeth and Yates, Simeon (eds), *Communicating Science* (London: Routledge, 1999).

Douglas, Aileen, 'Popular Science and the Representation of Women: Fontenelle and After', *Eighteenth-Century Life*, 18 (1994): 1–14.

Dunwoody, Sharon and Peters, Hans Peter, 'Mass Media Coverage of Technological and Environmental Risks: A Survey of Research in the United States and Germany', *Public Understanding of Science*, 1 (1992): 199–230.

Durant, John, Bauer, Martin W. and Gaskell, George, *Biotechnology in the Public Sphere* (London: Science Museum, 1998).

Eamon, William, *Science and the Secrets of Nature* (Princeton: Princeton University Press, 1994).

Ehrhardt, George H., 'Psychical Research and the Press: Thoughts on Their Relationship, 1920–1934', *Newsletter for the History and Sociology of Marginality in Science*, 2 (1990): 1–3.

Einsiedel, Edna F., 'Framing Science and Technology', *Public Understanding of Science*, 11 (1992): 89–101.

Eisenstein, Elizabeth, *The Printing Press as Agent of Change* (Cambridge: Cambridge University Press, 1979).

Eliot, Simon, *Some Patterns and Trends in British Publishing, 1800–1919* (London: Bibliographical Society, 1994).

Eliot, Simon, 'Some Trends in British Book Production, 1800–1919', in Jordan, John O. and Patten, Robert L (eds), *Literature in the Marketplace: Nineteenth-Century British Publishing and Reading Practices* (Cambridge: Cambridge University Press, 1995).

Ellegård, Alvar, *Darwin and the General Reader: The Reception of Darwin's Theory of Evolution in the British Periodical Press, 1859–1872* (Chicago and London: Chicago University Press, 1990).

Elzinga, Aant, 'Einstein in the Land of Nobel: An Episode in the Interplay of Science, Politics, Epistemology and Popular Culture', in Gavroglu, Kostas et al (eds), *Physics, Philosophy and the Scientific Community* (Dordrecht: Kluwer Academic, 1995).

Felt, Ulrike, 'Fabricating Success Stories', *Public Understanding of Science*, 2 (1993): 375–90.

Felt, Ulrike, 'Sciences, Science Studies and their Publics: Speculating on Future Relations', in Joerges, Bernward and Nowotny, Helga (eds), *Social Studies of Science and Technology: Looking Back Ahead* (Dordrecht: Springer, 2003).

Ferguson, Christine, *Language, Science and Popular Fiction in the Victorian fin-de-siècle: The Brutal Tongue* (Aldershot: Ashgate, 2005).

Ferguson, Eugene S., 'Technical Museums and International Exhibitions', *Technology and Culture*, 6 (1965): 30–46.

Findling, John E. et al (ed.), *Historical Dictionary of World's Fairs and Expositions, 1851–1989* (London: Greenwood Press, 1990).

Fleck, Ludwick, *Genesis and Development of a Scientific Fact*, ed. Trenn, T.J. and Merton, R.K. (Chicago and London: University Of Chicago Press, 1979) [1st edn in German, Basle, 1935].

Fletcher, Ronald, *Science, Ideology and the Media: The Cyril Burt Scandal* (New Brunswick: Transaction Publishers, 1991).

Forgan, Sophie, 'Building the Museum: Knowledge, Conflict, and the Power of Place', *Isis*, 96 (2005): 572–85.

Fox, Robert, 'Les conférences mondaines sous le Second Empire', *Romantisme*, 65 (1989): 49–57

Fox, Robert and Gooday, Graeme (eds), *Physics in Oxford, 1839–1939: Laboratories, Learning and College Life* (Oxford: Oxford University Press, 2005).

Fox, Robert and Weisz, George (eds), *The Organization of Science and Technology in France, 1800–1914* (Cambridge: Cambridge University Press, 1980).

Frasca-Spada, Marina and Jardine, Nick, *Books and the Sciences in History* (Cambridge: Cambridge University Press, 2000).

Fuller, Gillian, 'Cultivating Science: Negotiating Discourse in the Popular Texts of Stephen Jay Gould', in Martin, James R. and Veel, Robert (eds), *Reading Science: Critical and Functional Perspectives on Discourses of Science* (London and New York: Routledge, 1998).

Fyfe, Aileen, 'Publishing and the Classics: Paley's Natural Theology and the Nineteenth-Century Scientific Canon', *Studies in History and Philosophy of Science*, 33 (2002): 729–51.

Fyfe, Aileen, *Science for Children* (Bristol: Toemmes, 2003).

Fyfe, Aileen, *Science and Salvation: Evangelical Popular Science Publishing in Victorian Britain* (Chicago and London: University of Chicago Press, 2004).

Fyfe, Aileen and Lightman, Bernard, *Science in the Marketplace: Nineteenth-Century Sites and Experiences* (Chicago: University of Chicago Press, 2007).

Garnham, Nicholas, 'The Media and the Public Sphere', in *Habermas and the Public Sphere: Studies in Contemporary German Social Thought* (Cambridge, MA: The MIT Press, 1993).

Gates, Barbara T. and Shteir, Ann B. (eds), *Natural Eloquence: Women Reinscribe Science* (Madison and London: University of Wisconsin Press, 1997).

Gavroglu, Kostas, 'The Centre from the Periphery: Appropriating Scientific Ideas during the Eighteenth Century', *Revue de la Maison Française*, 1 (2003): 11–32.

Gavroglu, Kostas (ed.), *The Sciences in the European Periphery during the Enlightenment* (Dordrecht: Springer, 1999).

Gavroglu, Kostas et al., 'Science and Technology in the European Periphery. Some Historiographical Reflections', *History of Science*, 46 (2008): 153–75.

George, Nadine. F., *Popular Science and Philosophy in France, 1850–1875* (New York: Cornell University, 1974).

Gieryn, Thomas F., 'Boundary-Work and the Demarcation of Science from Non-Science: Strains and Interests in Professional Ideologies of Scientists', *American Sociological Review*, 48 (1983): 781–95.

Gieryn, Thomas F., 'The (Cold) Fusion of Science, Mass Media and Politics', in *Cultural Boundaries of Science* (Chicago: University of Chicago Press, 1999).

Gillet, Michel, 'Machines des Romans Feuilletons', *Romantisme*, 41 (1983): 79–90.

Goldsmith, Maurice, *The Science Critic: A Critical Analysis of the Popular Presentation of Science* (London: Routledge & Kegan Paul Books Ltd, 1986).

Golinski, Jan, 'A Noble Spectacle: Phosphorus and the Public Culture of Science in the Early Royal Society', *Isis*, 80 (1989): 11–39.

Golinski, Jan, *Science as Public Culture: Chemistry and Enlightenment in Britain, 1760–1820* (Cambridge: Cambridge University Press, 1992).

Golinski, Jan, *Making Natural Knowledge: Constructivism and the History of Science* (Chicago: University of Chicago Press, 2005) [1st edn, Cambridge, 1998].

González-Silva, Matiana, 'Del factor sociológico al factor genético. Genes y enfermedad en las páginas de *El País* (1976–2002)', *Dynamis*, 25 (2005): 487–512.

González-Silva, Matiana, 'Funding through the Press: Genomics as a New Political Issue in the Pages of *El País*', in Herran, Néstor et al (eds), *Synergia: Jóvenes Investigadores en Historia de la Ciencia* (Madrid: CSIC, 2008).

González-Silva, Matiana and Herran, Néstor, 'Ideology, Elitism and Social Commitment: Alternative Images of Science in two fin-de-siècle Barcelona Newspapers', *Centaurus*. 52 (2009) [forthcoming].

Gooday, Graeme, *The Morals of Measurements: Accuracy, Irony and Trust in Late Victorian Electrical Practice* (Cambridge: Cambridge University Press, 2004).

Gooday, Graeme, 'Liars, Experts and Authorities', *History of Science*, 46 (2008): 431-56.

Gooding, David, Pinch, Trevor and Schaffer, Simon, *The Uses of Experiment: Studies in the Natural Sciences* (Cambridge: Cambridge University Press, 1989).

Goschler, Constantin (ed.), *Wissenschaft und Öffentlichkeit in Berlin, 1870–1930* (Stuttgart: Franz Steiner Verlag, 2000).

Govoni, Paola, 'La divulgazione scientifica: un genere marginale?', *Intersezioni*, 3 (1991): 553–64.

Govoni, Paola, '*Nature* à l'italienne. La presse de science populaire en Italie à la fin du XIXe siècle', in Bensaude-Vincent, Bernadette and Rasmussen, Anne (eds), *La science populaire dans la presse et l'édition* (Paris: CNRS, 1997).

Govoni, Paola, *Un pubblico per la scienza. La divulgazione scientifica nell'Italia in formazione* (Rome: Carocci, 2002).

Govoni, Paola, 'Historians of Science and the "Sobel Effect"', *Journal of Science Communication*, 4 (2005): 1–17.

Govoni, Paola (ed.), *Storia, scienza e società. Ricerche sulla scienza in Italia nell'età moderna e contemporanea* (Bologna: CIS - Dipartimento di Filosofia, 2006).

Greenberg, Mark L. and Schachterle, Lance (eds), *Literature and Technology* (Bethlehem: Lehigh University Press, 1992).

Greenhalgh, Paul, *Ephemeral Vistas: The Expositions Universelles, Great Expositions and World's Fairs, 1851–1939* (Manchester: Manchester University Press, 1988).

Gregory, Jane and Miller, Steve, *Science in Public: Communication, Culture and Credibility* (New York: Plenum Press, 1998).

Hagendijk, Rod and Meeus Jan, 'Blind Faith: Fact, Fiction and Fraud in Public Controversy over Science', *Public Understanding of Science*, 2 (1993): 391–415.

Hall, Stuart, 'The Rediscovery of "Ideology": Return of the Repressed in Media Studies', in Gurevitch, Michael, Bennett, Tony, Curran, James and Woollacott, Janet (eds), *Culture, Society and the Media* (London: Methuen, 1982).

Hamlin, Christopher, 'Games Editors Played or Knowledge Readers Made?', *Isis*, 96 (2005), 633–42.

Heesen, Anke Te, 'News, Paper, Scissors: Clippings in the Sciences and Arts around 1920', in *Things That Talk: Object Lessons from Art and Science* (Cambridge, MA: The MIT Press, 2004).
Henson, Louise, Cantor, Geoffrey et al (eds), *Culture and Science in the Nineteenth-Century Media* (Aldershot: Ashgate, 2004).
Hessenbruch, Arne, 'Science as Public Sphere: X-Rays between Spiritualism and Physics', in Goschler, Constantin (ed.), *Wissenschaft und Öffentlichkeit in Berlin 1870–1930* (Stuttgart: Franz Steiner Verlag, 2000).
Hilgartner, Stephen, 'The Dominant View of Popularization: Conceptual Problems, Political Uses', *Social Studies of Science*, 20 (1990): 519–39.
Hinkle, Gerald and Elliot, William R., 'Science Coverage in Three Newspapers and Three Supermarket Tabloids', *Journalism Quarterly*, 66 (1989): 353–8.
Hoage, Robert J. and Deiss, William A (eds), *New Worlds, New Animals: From Menagerie to Zoological Park in the Nineteenth Century* (Baltimore: The Johns Hopkins University Press, 1996).
Hochadel, Oliver, *Öffentliche Wissenschaft. Elektrizität in der deutschen Aufklärung* (Göttingen: Wallstein, 2003).
Hodge, Jonathan, 'The Universal Gestation of Nature: Chambers' "Vestiges" And "Explanations"', *Journal of the History of Biology*, 5 (1972): 127–51.
Holton, Gerald, 'How to Think About the "Anti-Science" Phenomenon', *Public Understanding of Science*, 1 (1992): 103–28.
Holton, Gerald and Blanpied, William A (eds), *Science and Its Public: The Changing Relationship* (Dordrecht: Reidel, 1976).
Homburg, Ernst, 'From Chemistry for the People to the Wonders of Technology: The Popularization of Chemistry in the Netherlands during the Nineteenth Century', *HYLE – International Journal for Philosophy of Chemistry*, 12 (2006): 163–91.
Howsam, Leslie, 'An Experiment with Science for the Nineteenth-Century Book Trade: The International Scientific Series', *British Journal of the History of Science*, 33 (2000): 187–207.
Howson, Albert Geoffrey and Kahane, Jean-Pierre (eds), *The Popularization of Mathematics* (Cambridge: Cambridge University Press, 2003).
Hudson, Kenneth, *A Social History of Museums: What the Visitors Thought* (London: Macmillan, 1975).
Hyden, Fiona and King, Chris, 'What the Papers Say: Science Coverage by UK National Newspapers', *School Science Review*, 88 (2006): 81–4.
Inge, Thomas (ed.), *American Popular Culture* (London: Greenwood Press, 1981).
Jacobi, Daniel, *La communication scientifique. Discours, figures, modèles* (Grenoble : Presses Universitaires de Grenoble, 1999).
Jacobi, Daniel and Shiele, Bernard (eds), *Vulgariser la science. Le procès de l'ignorance* (Seysell: Champ Vallon, 1988).

James, Frank A.J.L., 'Never Talk about Science, Show It to Them: The Lecture Theatre of the Royal Institution', *Interdisciplinary Science Reviews*, 27 (2002): 225–8.

James, Louis (ed.), *Print and the People, 1819–1851* (London: Allen Lane, 1976).

Jasanoff, Sheila (ed.), *States of Knowledge: The Co-Production of Science and Social Order* (London and New York: Routledge, 2004).

Jeanneret, Yves, *Écrire la science. Formes et enjeux de la vulgarisation* (Paris: Presses Universitaires de France, 1994).

Johns, Adrian, *The Nature of the Book: Print Knowledge in the Making* (Chicago: University of Chicago Press, 1998).

Jordanova, Ludmila (ed.), *Languages of Nature: Critical Essays on Science and Literature* (London: Free Association Books, 1986).

Jurdant, Baudouin, 'Popularization of Science as the Autobiography of Science', *Public Understanding of Science*, 2 (1993): 365–73.

Kaiser, David (ed.), *Pedagogy and the Practice of Science: Historical and Contemporary Perspectives* (Cambridge, MA: The MIT Press, 2005).

Kaplan, Steven L (ed.), *Understanding Popular Culture: Europe from Middle Ages to the Nineteenth Century* (Berlin and New York: Mouton Publications, 1984).

Kargon, Roger, *Science in Victorian Manchester: Enterprise and Expertise* (Manchester: Manchester University Press, 1977).

Kärnfelt, Johan, *Mellan Nytta Och Nöje: Ett Bidrag Till Populärvetenskapens Historia I Sverige* (Stockholm: Stehag, 2000).

Kern, Stephen, *The Culture of Time and Space, 1880–1918* (Cambridge, MA: Harvard University Press, 1983).

Kirby, David A., 'Science Consultants, Fictional Films, Scientific Practice', *Social Studies of Science*, 33 (2003): 231–68.

Kjærgaard, Peter C. and Papanelopoulou, Faidra, 'Making the Paper: Science and Technology in Spanish, Greek and Danish Newspapers around 1900', *Centaurus*, 52 (2009) [forthcoming].

Knight, David M., 'Scientific Lectures: A History of Performance', *Interdisciplinary Science Reviews*, 27 (2002): 217–24.

Knight, David M., 'Scientists and their Publics: Popularization of Science in the Nineteenth Century', in Nye, Mary Jo (ed.), *The Cambridge History of Science. Vol. 5: The Modern Physical and Mathematical Sciences* (Cambridge and New York: Cambridge University Press, 2002).

Knight, David M., *Public Understanding of Science: A History of Communicating Scientific Ideas* (London: Routledge, 2006).

Kohlstedt, Sally Gregory '"Thoughts in Things": Modernity, History, and North American Museums', *Isis*, 96 (2005): 602–8.

Kohring, Matthias and Matthes, Jörg, 'The Face(t)s of Biotech in the Nineties: How the German Press Framed Modern Biotechnology', *Public Understanding of Science*, 11 (2002): 143–54.

Kua, Eunice, Reder, Michael and Grossel, Martha J., 'Science in the News: A Study of Reporting Genomics', *Public Understanding of Science*, 13 (2004): 309–27.

Lafollette, Marcel C., *Making Science Our Own: Public Images of Science 1910–1955* (Chicago: University of Chicago Press, 1990).

Lafollette, Marcel C., *Science on the Air. Popularizers and Personalities on Radio and Early Television* (Chicago: University of Chicago Press, 2008)

Lafuente, Antonio, 'La cultura técnica como espectáculo. Las exposiciones universales', *Política Científica*, 31 (1992): 22–5.

Lafuente, Antonio and Elena, Alberto, 'Los científicos ante su imagen y su público', *Claves de Razón Práctica*, 67 (1996): 48–55.

Lafuente, Antonio and Pimentel, Juan, 'La construcción de un espacio público para la ciencia: escrituras y escenarios en la Ilustración española', in Peset, José Luis (ed.), *Historia de la Ciencia y de la Técnica en la Corona de Castilla. Siglo XVIII* (Salamanca: Junta de Castilla y León, 2002).

Lafuente, Antonio and Saraiva, Tiago (eds), *Imágenes de la ciencia en la España contemporánea* (Madrid: Fundación Telefónica, 2000).

Langlois, Claude and François Laplanche (dirs.), *La science catholique. L'Encyclopédie Théologique de Migne (1844–1873) entre apologétique et vulgarisation* (Paris : CERF, 1992).

Latour, Bruno, *Science in Action: How to Follow Scientists and Engineers through Society* (Milton Keynes: Open University Press, 1987).

Latour, Bruno, *The Pasteurization of France* (Cambridge, MA: Harvard University Press, 1988).

Laurent, John, 'Science, Society and Politics in Late-Nineteenth-Century England: A Further Look at Mechanics Institutes', *Social Studies of Science*, 14 (1984): 585–619.

Lee, Yaross Jurith, 'Fossil Feuds: Popular Science and the Rhetoric of Vernacular Humor', *Essays in Arts and Sciences*, 23 (1994): 1–19.

Levere, Trevor H. and Turner, Gerald L'E., *Discussing Chemistry and Steam: The Minutes of a Coffee House Philosophical Society* (Oxford: Oxford University Press, 2001).

Levi, Ragnar, *Medical Journalism: Exposing Fact, Fiction, Fraud* (Arnes, IA: Iowa State University Press, 2001).

Levine, George (ed.), *One Culture: Essays in Science and Literature* (Madison, WI: The University of Wisconsin Press, 1987).

Lewenstein, Bruce V., 'Was There Really a Popular Science "Boom"?' *Science, Technology and Human Values*, 12 (1987): 29–41.

Lewenstein, Bruce V., 'Magazine Publishing and Popular Science after World War II', *American Journalism*, 6 (1989): 218–34.

Lewenstein, Bruce V., 'Popularization', in Hessenbruch, Arne (ed.), *Reader's Guide to the History of Science* (London: Fitzroy, 2000).

Lewis, Elton, 'Einstein, General Relativity, and the German Press, 1919–1920', *Isis*, 77 (1986): 95–103.

Lightman, Bernard, 'The Visual Theology of Victorian Popularizers of Science: From Reverent Eye to Chemical Retina', *Isis*, 91 (2000): 651–80.

Lightman, Bernard (ed.), *Victorian Science in Context* (Chicago and London: University of Chicago Press, 1997).

Lightman, Bernard, *Victorian Popularizers of Science: Designing Nature for New Audiences* (Chicago: University of Chicago Press, 2007).

Lindee, Susan, M., 'The American Career of Jean Marcet's Conversations on Chemistry, 1806–1853', *Isis*, 82 (1991): 9–23.

Livingstone, David N., *Putting Science in Its Place: Geographies of Scientific Knowledge* (Chicago and London: University of Chicago Press, 2003).

Lundgren, Anders and Bensaude-Vincent, Bernadette (eds), *Communicating Chemistry: Textbooks and Their Audiences, 1789–1939* (Canton, MA: Science History Publications, 2000).

Luey, Beth, '"Leading the Public Gently": Popular Science Books in the 1950s', *Book History*, 2 (1999): 218–53.

Lusting, Harry, 'Science as Theatre', *American Scientist*, 90 (2002): 550–55.

Lynn, Michael R., 'Dividing the Enlightenment: Public Opinion and Popular Science in Old Regime France', *Isis*, 92 (2001): 34–54.

Macdonald, Sharon (ed.), *The Politics of Display: Museums, Science, Culture* (London: Routledge, 1998).

MacLeod, Roy M (ed.), *Days of Judgement: Science Examinations and the Organization of Knowledge in Late Victorian England* (Driffield: Studies in Education, 1982).

Macleod, Roy M., 'Evolutionism, Internationalism and Commercial Enterprise in Science: The International Scientific Series 1871–1910', in Meadows, Arthur J., *The Development of Science Publishing in Europe* (Amsterdam: Elsevier, 1980).

Madden, Lionel and Dixon, Diana, *The 19th Century Periodical Press in Britain: A Bibliography of Modern Studies, 1901–1971* (Toronto: Victorian Periodicals Newsletter, 1975).

Major, Ann M. and Atwood, L. Erwin, 'Environmental Risks in the News: Issues, Sources, Problems, and Values', *Public Understanding of Science*, 13 (2004): 295–308.

Markovits, Andrei A. and Deutsch, Karl W (eds), *Fear of Science – Trust in Science: Conditions for Change in the Climate of Opinion* (Cambridge, MA: Oelgeschlager, Gunn & Hain, 1980).

Martin, James R. and Veel, Robert (eds), *Reading Science: Critical and Functional Perspectives on Discourses of Science* (London and New York: Routledge, 1998).

Mazzolini, Renato G (ed.), *Non-Verbal Communication in Science Prior to 1900* (Florence: Olscki, 1993).

McCall, Robert B., 'Science and the Press: Like Oil and Water?' *American Psychologist*, 43 (1988): 87–94.

McLaughlin-Jenkins, Erin, 'Common Knowledge: Science and the Late Victorian Working-Class Press', *History of Science*, 39 (2001): 445–65.

McRae, Murdo W. (ed.), *The Literature of Science: Perspectives on Popular Scientific Writing* (Athens: University of Georgia Press, 1993).

Meadows, Arthur J., *Communication in Science* (London: Buteerworths, 1962).

Meadows, Arthur J (ed.), *Development of Science Publishing in Europe* (Amsterdam: Elsevier, 1980).

Meadows, Arthur J., 'Histoire Succinte de la Vulgarisation Scientifique', in Spurgeon, David (ed.), *La vulgarisation scientifique: son histoire, ses succès, ses echecs*, *Impact science et société* [special issue], 36 (1986).

Meadows, Arthur J., *The Victorian Scientist: The Growth of a Profession* (London: British Library, 2004).

Mergoupi-Savaidou, Eirini, Papanelopoulou Faidra and Tzokas, Spyros, 'The Public Image(s) of Science and Technology in the Greek Daily Press, 1908-1910', *Centaurus*, 52 (2009) [forthcoming].

Mertens, Joost, 'Technology as the Science of Industrial Arts: Louis-Sébastien Lenormand (1757–1837) and the Popularization of Technology', *History and Technology*, 18 (2000): 203–31.

Midgley, Mary, Science as Salvation: A Modern Myth and Its Meaning (London: Routledge, 1992).

Miller, David P., 'The "Sobel Effect": The Amazing Tale of How Multitudes of Popular Writers Pinched All the Best Stories in the History of Science and Became Rich and Famous While Historians Languished in Accustomed Poverty and Obscurity, and How This Transformed the World. A Reflection on a Publishing Phenomenon', *Metascience*, 11 (2002): 185–200.

Morrell, Jack, 'Brains of Britain', *Social Studies of Science*, 16 (1986): 735–45.

Morrell, Jack, 'Wissenschaft in Worstedpolis: Public Science in Bradford, 1800–1850', *The British Journal for the History of Science*, 18 (1995): 1–23.

Morrell, Jack and Thackray, Arnold, *Gentlemen of Science: Early Years of the British Association for the Advancement of Science* (Oxford: Oxford University Press, 1981).

Morton, Alan Q (ed.), *Science Lecturing in the Eighteenth Century*, *British Journal for the History of Science* [special issue], 28 (1995).

Morus, Iwan Rhys, *Frankenstein's Children: Electricity, Exhibition and Experiment in Early-Nineteenth-Century London* (Princeton, NJ: Princenton University Press, 1998).

Morus, Iwan R., 'Seeing and Believing Science', *Isis*, 97 (2006): 101-110.

Mussell, James, Science, *Time and Space in the Late Nineteenth-Century Periodical Press: Movable Types* (Aldershot: Ashgate, 2007).

Myers, Greg, 'Nineteenth-Century Popularizations of Thermodynamics and the Rhetoric of Social Prophecy', in Brantlinger, Patrick (ed.), *Energy and Entropy: Science and Culture in Victorian Britain* (Bloomington, IN: Indiana University Press, 1989).

Myers, Greg, 'Science for Women and Children: The Dialogue of Popular Science in the Nineteenth Century', in Christie, John, and Shuttleworth, Sally (eds), *Nature Transfigured: Science and Literature, 1700–1900* (Manchester: Manchester University Press, 1989).

Myers, Greg, 'Fictionality, Demonstration, and a Forum for Popular Science: Jane Marcet's "Conversations on Chemistry"', in Gates, Barbara T. and Shteir Ann B (eds), *Natural Eloquence: Women Reinscribe Science* (Madison, WI: University of Wisconsin Press, 1997).

Myers, Greg, 'Discourse Studies of Scientific Popularization: Questioning the Boundaries', *Discourse Studies*, 5 (2003): 265–79.

Nani, Michele, 'Editoria e culture scientifiche nell'Italia postunitaria. Appunti sulle edizioni Dumolard', *Ricerche storiche*, 2 (1999): 257–97.

Naumann, Barbara (ed.), 'Science and Literature' *Science in Context* [special issue], 18 (2005).

Neidhardt, Friedhelm, 'The Public as a Communication System', *Public Understanding of Science*, 2 (1993): 339–50.

Nelkin, Dorothy, *Selling Science: How the Press Covers Science and Technology* (New York: W.H. Freeman, 1995).

Nerlich, Brigitte, Johnson, Susan and Clarke, David D., 'The First "Designer Baby": The Role of Narratives, Clichés and Metaphors in the Year 2000 Media Debate', *Science as Culture*, 12 (2003): 471–98.

Nicolson, Majorie H., *Science and Imagination* (Ithaca, NY: Cornell University Press, 1956).

Nicolson, Majorie H., *Voyages to the Moon* (New York: Macmillan, 1960).

Niderst, Alain (ed.), 'La diffusion des sciences au XVIIIe siècle'., *Revue d'Histoire des Sciences* [special issue], 44 (1991).

Nies, Betsy L., *Eugenic Fantasies: Racial Ideology in the Literature and Popular Culture of the 1920s* (New York: Routledge, 2002).

Nieto-Galan, Agustí, 'The Images of Science in Modern Spain: Rethinking the "Polémica"', in Gavroglu, Kostas (ed.), *The Sciences in the European Periphery during the Enlightenment* (Dordrecht: Springer, 1999).

Nieto-Galan, Agustí and Papanelopoulou, Faidra 'Science, Technology and the Public in the European Periphery: A Report of the 5th Step Meeting', *Journal of Science Communication*, 5 (2006): 2–6.

Nieto-Galan, Agustí and Roca-Rosell, Antoni (eds), *La Reial Acadèmia de Ciències i Arts de Barcelona als segles XVIII i XIX. Història, ciència i societat* (Barcelona: IEC-RACAB, 2000).

Nikolow, Sybilla and Schirrmacher, Arne (eds) *Wissenschaft und Öffentlichkeit als Ressourcen füreinander* (Frankfurt a.M.: Campus, 2007).

Noakes, Richard, 'Science in Mid-Victorian Punch', *Endeavour*, 26 (2002): 92–6.

Nowotny, Helga, 'Socially Distributed Knowledge: Five Spaces for Science to Meet the Public', *Public Understanding of Science*, 2 (1993): 307–19.

Nowotny, Helga, Scott, Peter and Gibbons, Michael, *Re-thinking Science: Knowledge and the Public in an Age of Uncertainty* (Cambridge: Polity Press, 2001).

O'Mahony, Patrick and Schäfer, Mike Steffen, 'The "Book of Life" in the Press: Comparing German and Irish Media Discourse on Human Genome Research', *Social Studies of Science*, 35 (2005): 99–130.

Olesko, Kathryn, *Physics as a Calling: Discipline and Practice in the Königsberg Seminar for Physics* (Ithaca, NY: Cornell University Press, 1991).

Olesko, Kathryn, 'Tacit Knowledge and School Formation', *Osiris*, 8 (1993): 21–48.

Olesko, Kathryn, 'Science Pedagogy as a Category of Historical Analysis: Past, Present, & Future', *Science & Education,* 15 (2006): 863-880.

Ophir, Adi, Shapin, Steven and Schaffer, Simon, 'The Place of Knowledge: The Spatial Setting and Its Relation to the Production of Knowledge', *Science in Context* [special issue], 4 (1991).

Ordóñez, Javier and Elena, Alberto (eds), *La ciencia y su público: perspectivas históricas* (Madrid: CSIC, 1990).

Oudshoorn, Nelly and Pinch, Trevor (eds), *How Users Matter: The Co-Construction of Users and Technologies* (Cambridge, MA: The MIT Press, 2003).

Paradis, James G., 'Satire and Science in Victorian Culture', in Lightman, Bernard, *Victorian Science in Context* (Chicago and London: University of Chicago Press, 1997).

Paradis, James G. and Postlewait, Thomas (eds), *Victorian Science and Victorian Values: Literary Perspectives* (New York: Rutgers University Press, 1981).

Pardo Tomás, José, *Un lugar para la ciencia. Escenarios de práctica científica en la sociedad hispana del siglo XVI* (La Orotava: Fundación de Historia de la Ciencia, 2006).

Paul, Harry, 'The Debate over the Bankruptcy of Science', *French Historical Studies*, 5 (1968): 299–327.

Paul, Harry, *The Edge of Contingency: French Catholic Reaction to Scientific Change from Darwin to Duhem* (Gainesville, FL: University Presses of Florida, 1979).

Penley, Constance, *NASA/Trek: Popular Science and Sex in America* (New York: Verso, 1997).

Perdiguero, Enrique, 'The Popularization of Medicine during the Spanish Enlightenment', in Porter, Roy (ed.), *The Popularization of Medicine, 1650–1850* (London: Routledge, 1992).

Pestre, Dominique, 'Les revues de vulgarisation scientifique en France, 1918–1940: un panorama', *Cahiers d'Histoire et de Philosophie des Sciences*, 24 (1987): 71–81.

Pestre, Dominique, *Science, argent et politique. Essai d'intersection* (Paris: INRA, 2003).

Petit, Annie, 'La diffusion des savoirs comme devoir positiviste', *Romantisme*, 65 (1989): 7–26.

Petkova, Kristina and Boyadjieva, Pepka, 'The Image of the Scientist and Its Functions', *Public Understanding of Science*, 3 (1994): 215–24.

Pickstone, John V., *Ways of Knowing: A New History of Science, Technology and Medicine* (Manchester: Manchester University Press, 2000).

Poirier, Jacques and Langlois, Claude (eds), *Raspail et la vulgarisation médicale* (Paris: Vrin, 1988).

Porter, Roy (ed.), *Patients and Practitioners: Lay Perceptions of Medicine in Pre-Industrial Society* (Cambridge: Cambridge University Press, 1985).

Porter, Roy (ed.), *The Popularization of Medicine, 1650–1850* (London: Routledge, 1992).

Purbrick, Louise (ed), *The Great Exhibition of 1851. New interdisciplinary Essays* (Manchester: Manchester University Press, 2001).

Pyenson, Lewis and Sheets-Pyenson, Susan, *Servants of Nature: A History of Scientific Institutions, Enterprises and Sensibilities* (London: Norton, 1999).

Raichvarg, Daniel, *Science et spectacle. Figures d'une rencontre* (Nice: Z éditions, 1993).

Raichvarg, Daniel, *Sciences pour tous?* (Paris: Gallimard, 2005).

Raichvarg, Daniel and Jacques Jean, *Savants et ignorants: une histoire de la vulgarisation des sciences* (Paris: Le Seuil, 1991).

Ramsey, Matthew, *Professional and Popular Medicine in France, 1770–1830: The Social World of Medical Practice* (Cambridge: Cambridge University Press, 1988).

Ravetz, Jerôme R., *Scientific Knowledge and its Social Problems* (Oxford: Clarendon Press, 1971).

Rebérioux, Madeleine (ed.), *Mise en scène et vulgarisation: l'Exposition Universelle de 1889, Le Mouvement Social* [special issue], 149 (1989).

Rey, Roseline, 'La vulgarisation médicale au XVIIIe siècle: Le cas des dictionnaires portatifs de santé', in Niderst, Alain (ed.), *La diffusion des sciences au XVIIIe siècle, Revue d'Histoire des Sciences* [special issue], 44 (1991): 413–33.Ritvo, Harriet, *The Animal Estate: The English and other Creatures in the Victorian Age* (Cambridge, MA: Harvard University Press, 1987).

Rivers, Isabel (ed.), *Books and their Readers in Eighteenth-Century England* (Leicester: Leceister University Press, 1982).

Roberts, Marie Mulrey and Porter, Roy (eds), *Literature and Medicine During the Eighteenth Century* (London: Routledge, 1993).

Roldán Vera, Eugenia, *The British Book Trade and Spanish American Independence: Education and Knowledge Transmission in Transcontinental Perspective* (Aldershot: Ashgate, 2003).

Roll-Hansen, Nills, 'Science, Politics, and the Mass Media: On Biased Communication of Environmental Issues', *Science, Technology and Human Values*, 19 (1994): 324–41.

Roqué, Xavier, 'Premsa i Cultura de la Ciència a Catalunya', in *Actes de les III Trobades d'Història de la Ciència í de la Tècnica* (SCHCT: Barcelona, 1995).

Roqueplo, Philippe, *Le partage du savoir: science, culture, vulgarisation* (Paris: Le Seuil, 1974).
Rothfels, Nigel, *Savages and Beasts: The Birth of the Modern Zoo* (Baltimore: The Johns Hopkins University Press, 2002).
Rothfield, Lawrence, *Vital Signs: Medical Realism in Nineteenth-Century Fiction* (Princeton: Princeton University Press, 1992).
Rousseau, G.S., 'Science Books and their Readers in the Eighteenth Century', in Rivers, Isabel (ed.), *Books and their Readers in Eighteenth-Century England* (Leicester: Leicester University Press, 1982).
Rudwick, Martin, 'Charles Darwin in London: The Integration of Public and Private Science', *Isis*, 73 (1982): 182–206.
Russell, Miles, *Digging Holes in Popular Culture: Archaeology and Science Fiction* (Oxford: Oxbow, 2002).
Rydell, Robert W., *All the World's a Fair: Visions of Empire at American International Expositions, 1876–1916* (Chicago: University Of Chicago Press, 1984).
Rydell, Robert W., 'Museums and Cultural History. A Review Article', *Comparative Studies in Society and History*, 34 (1992): 242–7.
Rydell, Robert W., *World of Fairs: The Century-of-Progress Expositions* (Chicago: University Of Chicago Press, 1993).
Sadoul, Jacques, *Histoire de la science-fiction moderne: 1911–1971* (Paris: A. Michel, 1973).
Schaffer, Simon, 'Natural Philosophy and Public Spectacle in the Eighteenth Century', *History of Science*, 21 (1983): 1–43.
Schaffer, Simon, 'Comets and the World's End', in Howe, L. and Wain, A (eds), *Predicting the Future* (Cambridge: Cambridge University Press, 1993).
Schroeder-Gudehus, Brigitte, Rasmussen, Anne and Bolenz, Eckhard (eds), *La société industrielle et ses musées: demande sociale et choix politiques* (Paris: EAC, 1992).
Schroeder-Gudehus, Brigitte and Rasmussen, Anne, *Les fastes du progrès.Le guide des expositions universelles 1851–1992* (Paris: Flammarion, 1992).
Schuster, John A. and Yeo, Richard (eds), *The Politics and Rhetoric of Scientific Method: Historical Studies* (Dordrecht and Boston: Reidel, 1986).
Schwartz, Joel S., 'Robert Chambers and Thomas Henry Huxley, Science Correspondents: The Popularization and Dissemination of Nineteenth Century Natural Science', *Journal of the History of Biology*, 32 (1999): 343–83.
Schwarz, Frederic D., 'We should All be Friends', *Technology and Culture*, 48 (2007): 407–10.
Secord, Anne, 'Science in the Pub: Artisan Botanists in Early-Nineteenth Lancashire', *History of Science*, 32 (1994): 269–315.
Secord, Anne, 'Botany on a Plate: Pleasure and the Power of Pictures in Promoting Early-Nineteenth-Century Scientific Knowledge', *Isis*, 93 (2002): 28–57.
Secord, James, 'Newton in the Nursery', *History of Science*, 23 (1985): 127–51.

Secord, James, *Victorian Sensation: The Extraordinary Publication, Reception, and Secret Authorship of 'Vestiges of the Natural History of Creation'* (Chicago: University Of Chicago Press, 2000).

Secord, James, 'Knowledge in Transit', *Isis*, 95 (2004): 654–72.

Seed, David, *A Companion to Science Fiction* (Oxford: Blackwell, 2005).

Sehgal, Narender K., Sangwan, Satpal and Mahanti, Subodh (eds), *Uncharted Terrains: Essays on Science Popularisation in Pre-Independence India* (New Delhi: Vigyan Prasar, 2000).

Seppänen, Jane and Väliverronen, Esa, 'Visualizing Biodiversity: The Role of Photographs in Environmental Discourse', *Science as Culture*, 12 (2003): 59–85.

Shapin, Steven, 'The House of Experiment in Seventeenth-Century England', *Isis*, 79 (1988): 373–404.

Shapin, Steven, 'Science and the Public', in Olby, Robert, Cantor, Geoffrey, Christie, John and Hodge, Jonathan (eds), *Companion to the History of Modern Science* (London: Routledge, 1990).

Shapin, Steven and Barnes, Barry, 'Science, Nature and Control: Interpreting Mechanics Institutes', *Social Studies of Science*, 7 (1977): 31–74.

Shea, William R., 'The History of Science and the Image of Science', *Notes and Records of the Royal Society of London*, 55 (2001): 3–7.

Sheets-Pyenson, Susan, *Low Scientific Culture in London and Paris, 1820–1875* (University of Pennsylvania, PhD dissertation, 1976).

Sheets-Pyenson, Susan, 'Popular Science Periodicals in Paris and London: The Emergence of a Low Scientific Culture, 1820–1875', *Annals of Science*, 42 (1985): 549–72.

Sheets-Pyenson, Susan, 'Cathedrals of Science: The Development of Colonial Natural History Museums during the Late-Nineteenth-Century', *History of Science*, 25 (1987): 279–300.

Shermer, Michael B., 'This View of Science: Stephen Jay Gould as Historian of Science and Scientific Historian, Popular Scientist and Scientific Popularizer', *Social Studies of Science*, 32 (2002): 498–524.

Shiach, Morag, *Discourse on Popular Culture: Class, Gender and History in Cultural Analysis, 1730 to the Present* (Cambridge: Polity Press, 1989).

Shibley, Ivan, 'Using Newspapers to Examine the Nature of Science', *Science and Education*, 12 (2003): 691–702.

Shils, Edward, *Center and Periphery: Essays in Macrosociology* (Chicago and London: University of Chicago Press, 1975).

Shils, Edward and Roszak, Theodore, 'Science and Its Public: The Changing Relationship', *Daedalus: Journal of the American Academy of Arts and Sciences*, 103 (1974): 1–22 [?].

Shinn, Terry and Whitley, Richard (eds), *Expository Science: Forms and Functions of Popularisation* (Dordrecht, Boston and London: Reidel, 1985).

Shteir, Ann B., *Cultivating Women, Cultivating Science: Flora's Daughters and Botany in England, 1760 to 1860* (Baltimore: The Johns Hopkins University Press, 1996).

Silverstone, Roger, 'Communicating Science to the Public', *Science, Technology and Human Values*, 16 (1991): 106–10.

Simon, Josep, 'The Baillières: The Franco-British Book Trade and the Transit of Knowledge', in Fox, Robert and Joly, Bernard (eds.), *Franco-British Interactions in Science since the Seventeenth Century* (London: College Publishers, 2009) [forthcoming].

Sismondo, Sergio, 'Expertise and the Public Understanding of Science', in Sismondo, Sergio (ed.), *An Introduction to Science and Technology Studies* (Oxford: Blackwell, 2004).

Spurgeon, David (ed.), *La vulgarisation scientifique: son histoire, ses succès, ses échecs, Impact, Science et Société* [special issue], 36 (1986).

Stahl, William A., 'Venerating the Black Box: Magic in Media Discourse on Technology', *Science, Technology and Human Values*, 20 (1995): 234–58.

Staley, Richard, 'Interdisciplinary Atoms? Exploring Twentieth-Century Culture through Einstein', *British Journal for the History of Science*, 36 (2003): 221–30.

Stevens, Rosemary, *American Medicine and the Public Interest* (Berkeley and London: University of California Press, 1998).

Stewart, Larry, *The Rise of Public Science: Rhetoric, Technology and Natural Philosophy in Newtonian Britain, 1660–1750* (Cambridge: Cambridge University Press, 1992).

Stichweh, Rudolf, *Zur Entstehung des modernen Systems wissenschaftliche Disziplinen: Physik in Deutschland* (Frankfurt: Suhrkamp , 1992).

Taton, René (ed.), *Enseignement et diffusion des sciences en France au XVIII Siècle* (Paris : Hermann, 1964).

Tesnière, Valérie, 'Le livre scientifique en France au XIXe siecle', *Romantisme*, 80 (1993): 67–77.

Tesnière, Valérie, 'Diffuser la science', in Barbier, Frédéric, Parent-Charon, Annie et al (eds), *Le livre et l'historien. Études offertes en l'honneur du professeur Henri-Jean Martin* (Geneva: Droz, 1997).

Thoman, Elizabeth, 'Media, Technology and Culture: Re-Imagining the American Dream', *Bulletin of Science, Technology and Society*, 13 (1993): 20–27.

Thompson, John B., *The Media and Modernity: A Social History of the Media* (Cambridge: Polity Press, 1995).

Topham, Jonathan R., 'Science and Popular Education in the 1930s: The Role of the Bridgewater Treatises', *The British Journal for the History of Science*, 25 (1992): 387–430.

Topham, Jonathan R., 'Beyond The "Common Context": The Production and Reading of the Bridgewater Treatises', *Isis*, 89 (1998): 233–62.

Topham, Jonathan R., 'BJHS Special Section: Book History and the Sciences. Introduction', *The British Journal for the History of Science*, 33 (2000): 155–8.

Topham, Jonathan R., 'Scientific Publishing and the Reading of Science in Nineteenth-Century Britain: A Historiographical Survey and Guide to Sources', *Studies in History and Philosophy of Science*, 31 (2000): 559–612.

Topham, Jonathan R., 'A View from the Industrial Age', *Isis*, 95 (2004): 431–42.

Topham, Jonathan R., 'Thomas Byerley, John Limbird, and the Production of Cheap Periodicals in Regency Britain', *Book History*, 8 (2005): 75–106.

Turner, Frank Miller, *Between Science and Religion: The Reaction to Scientific Naturalism in Late Victorian Britain* (New Haven: Yale University Press, 1974).

Turner, Frank Miller, 'Public Science in Britain, 1880–1919', *Isis*, 71 (1980): 589–607.

Turner, Stephen, 'What Is the Problem with Experts?', *Social Studies of Science*, 31 (2001): 123–49.

Turney, Jon, *Frankenstein's Footsteps: Science, Genetics and Popular Culture* (New Haven, CT and London: Yale University Press, 1998).

Väliverronen, Esa, 'Stories of the "Medicine Cow": Representations of Future Promises in Media Discourse', *Public Understanding of Science*, 13 (2004): 363–77.

Van Dijck, José, *Imagenation: Popular Images of Genetics* (New York: New York University Press, 1998).

Van Riper, Bowdoin A., *Science in Popular Culture: A Reference Guide* (Westport, CT: Greenwood Press, 2002).

Van Wyhe, John, *Phrenology and the Origins of Victorian Naturalism* (Aldershot: Ashgate, 2004).

Versins, Pierre, *Encyclopédie de l'utopie des voyages extraordinaires et de la science fiction* (Lausanne: L'age de l'homme, 1984).

Warwick, Andrew, *Masters of Theory; Cambridge and the Rise of Mathematical Physics* (Chicago: University of Chicago Press, 2003).

Weingart, Peter, *Die Wissenschaft der Öffentlichkeit. Essays zum Verhältnis von Wissenschaft, Medien und Öffentlichkeit* (Weilerwist: Velbrück, 2001).

Winter, Alison, 'Mesmerism and Popular Culture in Early Victorian England', *History of Science*, 32 (1994): 317–43.

Wise, Norton M., 'Mediating Machines', *Science in Context*, 2 (1988): 77–113.

Withers, Charles W.J., 'Towards a History of Geography in the Public Sphere', *History of Science*, 36 (1998): 45–78.

Woodward, John and Richards, David (eds), *Health Care and Popular Medicine in Nineteenth-Century England: Essays in the Social History of Medicine* (London: Croom Helm, 1977).

Wynne, Brian, 'Public Understanding of Science', in Jasanoff, Sheila et al (eds), *Handbook of Science and Technology Studies* (Thousand Oaks, CA: Sage Publications, 1995).

Yeo, Eileen and Yeo, Stephen (eds.), *Popular Culture and Class Conflict, 1590–1914: Explorations in the History of Labour and Leisure* (Brighton: The Harvest Press, 1981).

Yeo, Richard, 'Reading Encyclopedias: Science and the Organization of Knowledge in British Dictionaries of Arts and Sciences', *Isis*, 82 (1991): 24–49.
Yeo, Richard, *Encyclopaedic Visions: Scientific Dictionaries and the Enlightenment Culture* (Cambridge: Cambridge University Press, 2001).

Index